MW01613814

Computer-assisted Reasoning in Cluster Analysis

Computer-assisted Reasoning in Cluster Analysis

E. Backer

Department of Electrical Engineering
Delft University of Technology
Delft, The Netherlands

with a foreword by Abraham Kandel

Prentice Hall
New York   London   Toronto   Sydney   Tokyo   Singapore
in association with Vereniging voor Studie en Studentenbelangen te Delft

First published 1995 by
Prentice Hall International (UK) Limited
Campus 400, Maylands Avenue
Hemel Hempstead
Hertfordshire, HP2 7EZ
A division of
Simon & Schuster International Group

© Prentice Hall International 1995

All rights reserved. No part of this publication may be reproduced,
stored in a retrieval system, or transmitted, in any form, or by any
means, electronic, mechanical, photocopying, recording or otherwise,
without prior permission, in writing, from the publisher.
For permission within the United States of America
contact Prentice Hall Inc., Englewood Cliffs, NJ 07632

Printed and bound in Great Britain by Bookcraft

Library of Congress Cataloging-in-Publication Data

Backer, E.
   Computer-assisted reasoning in cluster analysis / E. Backer ; with
foreword by Abraham Kandel.
   Includes bibliographical references and index.
   ISBN 0-13-341884-7 $80.00 (approx.)
   1. Cluster analysis–Data processing.   2. Expert systems (Computer
science)   I. Title.
QA278.2.B33   1995
519.5'35–dc20   94-33560
                                                                  CIP

British Library Cataloguing in Publication Data

A catalogue record for this book is available from
the British Library

ISBN 0-13-341884-7

---

1   2   3   4   5   99   98   97   96   95

To Anil K. Jain
Michigan State University
for a lifetime of friendship and lessons taught

To the memory of Richard C. Dubes

The validation of clustering structures is the most difficult and frustrating part of cluster analysis. Without a strong effort in this direction, cluster analysis will remain a black art accessible only to those true believers who have experience and great courage.
Jain and Dubes, 1988

Contents

PREFACE   XIII

FOREWORD BY ABRAHAM KANDEL   XVII

PART 1: RECONNAISSANCE OF THE DOMAIN

1   CLUSTER ANALYSIS: WHY, WHAT, HOW, WHEN   3
    1.1   Introduction   3
           1.1.1 Uses of cluster analysis   4
           1.1.2 Basic concepts in clustering methodology   4
           1.1.3 General reading in cluster analysis   8
           1.1.4 From numbers to structure: an example   8
           1.1.5 EDAPLUS: a first glance   10
           1.1.6 Concluding remarks   17
    1.2   Organization of the book   17

2   METHODS IN CLUSTERING DATA   21
    2.1   Data representation   21
           2.1.1 Pattern matrix   21
           2.1.2 Proximity matrix   22
           2.1.3 Missing data   25
           2.1.4 Mixed data   29
    2.2   Linear projection   29
           2.2.1 Eigenvector projection   30
           2.2.2 Discriminant analysis   32
    2.3   Nonlinear projection   34
           2.3.1 Multi-dimensional scaling   35
           2.3.2 Sammon's algorithm   35
           2.3.3 Frame method   37
           2.3.4 Triangulation method   37
           2.3.5 Projection using simulated annealing   38
    2.4   Hierarchical clustering   39
           2.4.1 Single-link clustering   40
           2.4.2 Complete-link clustering   40
           2.4.3 Average-link clustering   41

2.5   Partitional clustering   43
        2.5.1 Square-error clustering   43
        2.5.2 *K*-means clustering   44
        2.5.3 Crucial parameters   45
2.6   Minimum spanning tree   47
2.7   Cluster validity   49
        2.7.1 Hierarchical structures   51
        2.7.2 Partitional structures   53
        2.7.3 Individual structures   54
        2.7.4 Monte Carlo analysis   54
2.8   Clustering tendency   54
2.9   Concluding remarks   56
Citations to be remembered   57

3   EXAMPLES AND DISCUSSION   58
3.1   The transitivity property   58
3.2   Normalization of data   63
3.3   Discussion of problems in cluster analysis   65
3.4   Practical guidelines   69
3.5   Concluding remarks   70

4   FRAMING OF CLUSTER ANALYSIS STUDIES   71
4.1   Research framing   71
4.2   Data framing   73
4.3   Belief framing   74
4.4   Concluding remarks   75

5   APPROXIMATE REASONING IN CLUSTER ANALYSIS:
    A BRANCH OF FUZZY LOGIC   76
5.1   The impact of this chapter   76
5.2   Fuzzy clustering: a historical review   78
5.3   Basic definitions   81
        5.3.1 Fuzzy relations   82
        5.3.2 Fuzzy partitions   86
        5.3.3 Fuzzy relations in fuzzy partitions   96
5.4   More on fuzzy relations   97
        5.4.1 Transitive closure   99
        5.4.2 Hierarchical structures   100
5.5   Generalized approximation   101
        5.5.1 Fuzzy graphs   101
        5.5.2 The approximation approach   105

5.6	Strong and weak patterns in clustering: a fuzzy relation	109
5.7	Goal-directed methods of comparison	113
5.8	Fuzzy classification inference	115
	5.8.1 Classification of patterns	115
5.9	Concluding remarks	120

PART 2: RECONSIDERATION OF THE TASK

6	INDETERMINACY AND UNCERTAINTY IN CLUSTER ANALYSIS	123
6.1	Preliminary remarks	123
6.2	Indeterminacy and uncertainty	130
6.3	Disclosing cluster structures: some experiments	135
	6.3.1 Repeated experimentation	135
	6.3.2 Lessons from the experiments	143
	6.3.3 Formalization from observations	146
6.4	Concluding remarks	149

7	THE NEED FOR AND RELEVANCE OF COMPUTER-ASSISTED REASONING IN CLUSTER ANALYSIS	150
7.1	The processing paradigm	150
	7.1.1 Knowledge-based systems (expert systems)	152
7.2	The data–information–knowledge paradigm or modelling paradigm	154
	7.2.1 Knowledge systems versus conventional systems	155
7.3	Concluding remarks	156

PART 3: COMPUTER-ASSISTED REASONING IN CLUSTER ANALYSIS

8	KNOWLEDGE-BASED EXPERT SYSTEMS: AN INTRODUCTION	161
8.1	General concepts of expert systems	164
8.2	Elements of an expert system	164
8.3	Representation of knowledge	167
	8.3.1 Taxonomy of knowledge (an example)	167
	8.3.2 Semantic nets (Stillings, 1987)	170
	8.3.3 Object–attribute–value triples, (Barr, 1981)	171
	8.3.4 Frames (Minsky, 1975)	171
	8.3.5 Predicate logic	172
8.4	Methods of inference	173
	8.4.1 Decision trees	173
	8.4.2 Rules of inference	174
	8.4.3 Forward and backward chaining	177
8.5	Reasoning under uncertainty	178
	8.5.1 About uncertainty	178

		8.5.2 Uncertainty in inference	180
		8.5.3 Degrees of belief	182
	8.6	Inexact reasoning	183
		8.6.1 Certainty factors	184
		8.6.2 Dempster–Shafer theory (Shafer, 1976)	185
	8.7	Approximate reasoning	188
		8.7.1 Fuzzy sets and natural language	188
		8.7.2 Fuzzy logic and fuzzy rules	193
	8.8	Concluding remarks	196

9   RULE BASE DEVELOPMENTS FOR CLUSTER ANALYSIS   197

	9.1	Historical review	198
		9.1.1 Knowledge-based clustering: the development of conceptual clustering	198
		9.1.2 Knowledge-based clustering: the development of computer-assisted reasoning in cluster analysis	199
	9.2	A language for rule bases	200
	9.3	Knowledge-based cluster analysis: a simple example	207
		9.3.1 Summary of underlying ideas	207
		9.3.2 The knowledge base	209
	9.4	Cluster-oriented reasoning	213
		9.4.1 Constructing rules from observations	214
		9.4.2 An example	216
		9.4.3 K-independent support rules	222
		9.4.4 Diagnostic rules	228
	9.5	A design example	229
		9.5.1 Planning	229
		9.5.2 Knowledge definition	229
		9.5.3 Knowledge design	230
		9.5.4 The program code	231
		9.5.5 Verification and evaluation	231
		9.5.6 An operational example	232
	9.6	Learning	241
		9.6.1 Rule induction by concept learning	243
		9.6.2 Learning (fuzzy) rules from examples	245
		9.6.3 Artificial neural network (ANN) learning	250
	9.7	Intelligent hybrid systems (IHS)	254
	9.8	Concluding remarks	255

10   CASE STUDY: ANALYSIS OF DELPHINID SONAR SOUND SIGNALS   256

| | 10.1 | Introduction | 257 |

	Evolution	257
	Adaptation	257
	Two-component sonar	257
10.2	The research problem	258
	10.2.1 Framing and validating Qr	260
	10.2.2 Framing and validating Hf	260
	10.2.3 Framing and validating Hr	262
	10.2.4 Implementing primary and secondary validation in the knowledge base	263
10.3	The analysis	266
	10.3.1 Analysis of Qr	268
	10.3.2 Analysis of Hf	269
	10.3.3 Analysis of Hr	270
10.4	Computer-assisted analysis	272
	10.4.1 Conceptual associations	272
	10.4.2 Capturing expert intuition	273
	10.4.3 A knowledge-based test for confirmation and falsification	274
10.5	Concluding remarks	274

GLOSSARY OF KEY TERMS	276

REFERENCES	280

APPENDIX

A: INCREMENTAL HIERARCHICAL KNOWLEDGE ORGANIZATION:	
A PROPOSAL AND AN EXAMPLE	293
A1	Introduction	293
A2	Statement of the problem	294
A3	Learning and incremental learning	296
A4	Hierarchical knowledge organization and a model for inexact reasoning	303
A5	The upper and lower bound formalism compared and contrasted with the Dempster–Shafer scheme	313
A6	An example	323

B: EDAPLUS FUNCTIONALITY	334
B1	Commands and functions	334
B2	Useful actions and logging	349
B3	The rule base editor	356
B4	Stand-alone programs	360
B5	Data	361

INDEX	363

Preface

Widely applicable in research, methods of cluster analysis are used to determine clusters of similar objects. The results of a cluster analysis are generally found to be determined by a complex reasoning process which may differ from one analysis to another. The daily practice of cluster analysis includes choosing the appropriate attributes, coefficients and techniques, interpretation of intermediate results, and validating the ultimate findings. These issues are part of such a complex reasoning process.

Today it is felt to be natural to associate such a reasoning process with knowledge-based systems, expert systems, or generally with computer-assisted decision support systems. As knowledge-based systems have experienced tremendous growth and popularity since the early 1980s, and the need and relevance of computer-assisted reasoning in cluster analysis is so significantly present, the time is right to bridge both methodologies.

Purpose of the book

The objective of this book is to provide a comprehensive introduction to the methodology of cluster analysis followed by a goal-oriented introduction to knowledge-based expert systems and reasoning under uncertainty. In between, we identify indeterminacy and uncertainty while performing a cluster analysis, explain the concept of fuzzy sets and fuzzy logic in cluster analysis, and associate the processing paradigm and the modelling paradigm with cluster-oriented reasoning.

The book is designed to inform students about applied computer-assisted reasoning and cluster analysis. The material is written at graduate level suitable for majors in computer science, engineering and other fields who are interested in applying cluster analysis in their research.

The book includes hands-on experience through the EDAPLUS software package that accompanies the book. Nearly all the algorithms covered in the text, as well as rule-based interpretation, are offered by EDAPLUS. The package may serve to reinforce, clarify and exemplify the theoretical concepts developed throughout the book. EDAPLUS was developed at the Department of Electrical Engineering, Delft University of Technology, by students taking part in courses on cluster analysis. As such, it is primarily meant for educational support. For real research applications it has a number of severe limitations.

Courses which have a term project associated with them, may use EDAPLUS as an excellent vehicle to further the understanding of cluster analysis by developing knowledge bases of the student's own choice.

Organization of the material

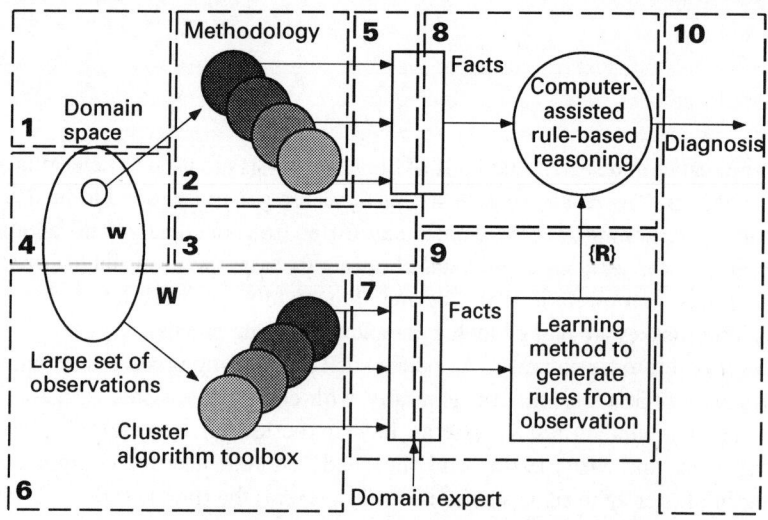

Figure 0.1: Schematic overview of the book's chapters:
Part 1: Reconnaissance of the domain:
1. cluster analysis: why, what, how, when; 2. methods in clustering data; 3. examples and discussion; 4. framing of cluster analysis studies; 5. fuzzy clustering: a branch of fuzzy logic.
Part 2: Reconsideration of the task:
6. indeterminacy and uncertainty in cluster analysis; 7. the need for and relevance of computer-assisted analysis.
Part 3: Computer-assisted reasoning in cluster analysis:
8. knowledge-based expert systems: an introduction; 9. rule base developments for cluster analysis; 10. case study: analysis of delphinid sonar sounds signals

The material is divided into three parts and an appendix. The organization of the book is depicted in Figure 0.1. The numbers 1–10 correspond to the Chapters 1–10, respectively. The figure reveals a message: a single set of collected data, w, can only be diagnosed by means of computer-assisted rule-based reasoning if we are able to generate rules from a large number of varied observed data sets, W, provided that learning the rules is controlled by cluster analysis expertise. While Part 1 introduces cluster analysis (1–5), Part 2 proceeds to develop a feeling for diversity, imprecision and complex decision-making in cluster analysis (6–7). Finally, Part 3, adresses the potentiality of computer-assisted rule-based reasoning (8–10).

It is important to appreciate that the work reported here is not intended to be a complete account either of cluster analysis or of computer knowledge-based reasoning. In fact it attempts to show how to approach computer-assisted reasoning in cluster analysis.

Part 1: *Reconnaissance of the domain.*

This part (Chapters 1–5) presents the basic methodology of cluster analysis, discusses the variety of methods and techniques, and developes the methodology of fuzzy clustering in greater depth. It is important to note that the book by A.K. Jain and R.C. Dubes, *Algorithms for clustering data* covers the methods and techniques more broadly, and therefore should be considered as the most important reference today within the scope of this book. Likewise, the book by J.C. Bezdek, *Pattern recognition with fuzzy objective function algorithms,* is still regarded as a key book on fuzzy clustering. A selected reprint volume on *Fuzzy models for Pattern Recognition,* edited by J.C. Bezdek and S.K. Pal, contains most of the prominent early contributions in the field of fuzzy cluster analysis.

Finally, in Part 1, we begin to tackle the framing of the cluster analysis itself in the context of the problem to be analyzed in the domain of application. The depth of Chapter 4 is relatively modest, mainly because it is at this stage just preluding its use in Chapter 10, at the end of this book. However, major reference is made to the book by H.C. Romesburg, *Cluster analysis for researchers,* which is much more domain-research-oriented than this book intends to be.

Part 2: *Reconsideration of the task.*

Two chapters discuss the problem of what to formalize in cluster analysis (Chapter 6) and how to formalize cluster-oriented reasoning (Chapter 7). Through a series of experiments, we try to explain how clustering structure can be classified and what lessons can be drawn from experiments. The impact of repeated experimentation and simulation with random data, the contrast between numerical precision as offered by algorithmic processing, and approximate reasoning in the interpretation of the results, leads to 'full' understanding of the difficulties of a cluster analysis. In view of that, Part 2 provides an answer to the question 'Why a knowledge-based expert system approach?'.

Part 3: *Computer-assisted reasoning in cluster analysis.*

Part 3 offers an introduction to knowledge-based expert systems (Chapter 8) and presents a comprehensive account of the principles of the development of rule bases for cluster analysis (Chapter 9). The topics discussed include knowledge representation, methods of inference, and inexact reasoning. The design of rule bases and how to learn the rules benefits from functional and conceptual associations that enable the transition from data to information, and from data to knowledge, as presented in Chapter 7.

The book by J. Giarratano and G. Riley, *Expert systems principles and programming,* is a most valuable reference book for this part.

Part 3 ends with a case study (Chapter 10), in which we describe a research application which illustrates how to frame and validate the analysis and belief, how to choose the options, and how to make the decisions in practice. We selected the research of Kamminga on dolphin sonar sound production for navigation in relation to the mass-stranding phenomenon. In the discussion, we emphasize primary validation of the immediate facts and secondary validation with respect to the research goal.

Appendix A: *Incremental hierarchical knowledge organization: a proposal and an example.*
Appendix A reports on a proposal for an incremental learning scheme with the goal of inducing fuzzy rules from examples of data, which are organized hierarchically. The hierarchical knowledge organization supports a model for inexact reasoning even if the fuzzy rules are still subject to modification. Because of the fact that the upper and lower bound formalism resembles the Dempster–Shafer scheme, a comparison with it is presented. The appendix ends with an illustrative example (based on examples that also appear in Chapter 9).

Appendix B: *EDAPLUS command and function summary.*
In Appendix B, a variety of functions for performing useful actions in EDAPLUS are outlined and illustrated. All commands are described and demonstrated. Moreover, the EDAPLUS functionality with respect to editing and manipulating rule bases is dealt with. Appendix B ends with the description of some stand-alone programs which are useful when carrying out exercises and experiments with EDAPLUS.

There are data to experiment with on the accompanying diskette.

Acknowledgements

Without the stimulating environment of the Information Theory Group at Delft University of Technology, this book would never have appeared. I thank all the students who have taken part in the course on cluster analysis and who have contributed so substantially to the development of EDAPLUS over the past four years.

I also thank Cees Kamminga for guiding me through the fascinating world of dolphins and his research associated with them.

Eric Backer
Delft University of Technology
September 1994

Foreword

Over the past decade, computer-assisted reasoning in cluster analysis has evolved into a substantial body of methods and techniques providing us with a variety of tools for the management of uncertainty in pattern classification. The author of the present volume, Professor Eric Backer, has played a major role in the development of both the theory and its applications, especially through his elegant and highly skilful use of the dual concepts of knowledge-based reasoning under uncertainty and cluster analysis.

As pointed out by Professor Backer, the results of a cluster analysis are generally found to be determined by a complex reasoning process which may differ from one analysis to another. In the evolution of cluster analysis, concern with automated reasoning under uncertainty is almost invariably a concomitant of maturation. Since the early days of pattern recognition the development of theories relating to the management of uncertainty as well as to computer-assisted reasoning have captured a major portion of our research interests.

As in many other fields of science, in those early days it was traditional to deal with uncertainty through the use of probability theory. In recent years, however, it has become increasingly clear that there are some important facets of uncertainty which do not lend themselves to analysis by classical probability-based techniques. Early attempts to equip machines with computer-assisted reasoning as well as humanlike intelligence have revealed the need for heuristic knowledge incorporating tools for the management of uncertainty in non-deterministic environments.

In examining the contents of Professor Backer's volume I was struck by the vastness of the progress made in the realm of computer-assisted reasoning and uncertainty management in cluster analysis in the past decade.

Extensive research in this area has led to many investigations for new theoretical principles, paradigms and techniques which can be exploited to conceive, design and build intelligent systems applied in the field of cluster analysis and having high MIQ (Machine Intelligence Quotient).

The paradigm shift from traditional to computer-assisted reasoning is reflected in the orientation and contents of Professor Backer's text. Viewed in this perspective, Professor Backer's authoritative exposition of the topic provides the reader with an extensive field of knowledge. One cannot but be greatly impressed by the methodical presentation of the material and his success in making the principles and the ideas easy to understand and simple to implement. This text is likely to have a very important impact on the way in which computer-assisted reasoning methods are dealt with in pattern recognition and cluster analysis.

Professor Backer deserves much credit for producing this important contribution; his text is a must for anyone who is interested in a comprehensive study of this field. The material herein presented provides an invaluable reference for those interested in acquainting themselves with the field of computer-assisted reasoning in cluster analysis and we owe Professor Backer our thanks for enthusiastically going through the quest of developing and extending this important area of science.

June 1994

Abe Kandel
Computer Science and Engineering
University of South Florida
Tampa, Florida 33620

Part 1
Reconnaissance of the domain

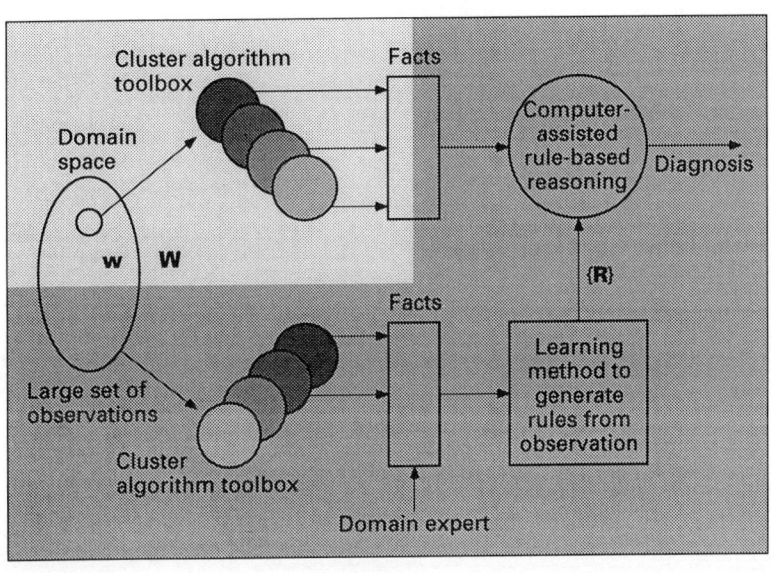

1 | Cluster Analysis: why, what, how, when

Cluster analysis is a tool of exploratory data analysis. It tries to find the intrinsic structure of data by organizing patterns into groups or clusters. The general structure of a clustering technique may be described as follows. One has a sample set of objects or patterns that are described by a number of parameters. ... A measure of distance (or dissimilarity) between objects is introduced. Using the measure, one should determine a criterion that expresses the idea of clustering, i.e., one of partitioning the set of objects into separated clusters. Minimization of the criterion should lead to optimal clusters; (Brailovsky, 1991).

1.1 Introduction *(Jain and Dubes, 1988)*

'The act of sorting similar things into catagories is one of the most primitive and common pursuits of men. Our languages consist of words which help us to recognize and discuss the different types of events, objects and people we encounter. Each noun is a label used to describe a class of things which have certain striking features in common. We need to be able to classify our activities and surroundings simply in order to make life manageable, since it would be impossible to treat everything we encounter as unique.'

The task of classifying things is not complete – far from it. In this age of intensive research and the explosion of information the need to categorize and name new phenomena is as important as ever. Nor are classification problems confined to particular disciplines. They arise in all fields of science and their solution becomes increasingly more difficult as the volume of relevant data expands', Jain and Dubes (1988).

Cluster analysis is an exploratory method for helping to solve classification problems. Its use is appropriate when little or nothing is known about the category structure in a body of data. The object of a cluster analysis is to sort a sample of cases under consideration into groups such that the degree of association is high between members of the same group and low between members of different groups.

Cluster analysis is a tool of discovery. It can be used to reveal associations and structures in data which, though not previously conceived, are nevertheless sensible and useful when found. The results can contribute to the development of a classification scheme; they may suggest general models to describe other samples and ultimately the parent population; or they may simply provide definitions of size and measures of change in what previously were just notional categories.

1.1.1 Uses of cluster analysis (Hartigan, 1975; Romesburg, 1984)

Cluster analysis is a modern statistical method of partitioning an observed sample population into disjoint or overlapping homogeneous classes, to produce an operational classification. This classification may help to:
– formulate hypotheses concerning the origin of the population (e.g. evolution studies);
– describe the sample in terms of a typology (e.g. market analysis);
– predict the future behaviour of population types (e.g. modelling economic prospects);
– optimize a functional process (e.g. information retrieval);
– assist identification (e.g. diagnosing);
– measure the differential effects of treatments on classes within the population (e.g. analysis of variance).

1.1.2 Basic concepts in clustering methodology (Everitt, 1974)

A cluster is comprised of a number of *similar* objects collected or grouped together. The following definitions may be functional:
– A cluster is a set of entities which are alike, while entities from different clusters are not alike.
– A cluster is an aggregation of points in the test space such that the *distance* between any two points in the cluster is less than the distance between any point in the cluster and any point not in it;
– Clusters may be described as connected regions of a *multi-dimensional space* containing a relatively high *density* of points, separated from other such regions by a region containing a relatively low density of points.

The last two definitions assume that the objects to be clustered are represented in the measurement space. While it is easy to give a functional definition of a cluster, it is very difficult to give an operational definition of a cluster.

Clustering (intrinsic classification) is called *unsupervised learning* in pattern recognition because no category labels denoting an a priori partition of the objects are used. An intrinsic classification has only an unlabelled *pattern matrix* or a *proximity matrix* as an input. An intrinsic classification can be subdivided into *hierarchical* and *partitional* classifications by the type of structure imposed on the data. A hierarchical classification is a nested sequence of partitions, whereas a partition classification is a single partition. Thus a hierarchical classification is a special sequence of partitional classifications (Sneath and Sokal, 1973).

Several algorithms can be proposed to express the same classification. One frequently uses an algorithm to express a clustering method, then examines various computer implementations of the method.

A hierarchical clustering method is a procedure for transforming a *proximity matrix* into a sequence of nested partitions. A hierarchical clustering algorithm is the specification of steps for performing a hierarchical clustering. It is often convenient to characterize a hierarchical clustering method by writing down an algorithm, but the algorithm should be separated from the method itself. An important characteristic of hierarchical clustering methods is the visual impact of the *dendrogram*, which enables a data analyst to see how objects are being merged into clusters or split at successive levels of proximity. The data analyst can then try to decide whether the entire dendrogram describes the data or can select a clustering, at some fixed level of proximity, which makes sense for the application in hand.

We refer to non-hierarchical clustering methods as partitional clustering methods. They generate a single partition of the data in an attempt to recover natural groups present in the data.

Hierarchical methods generally require only the *proximity matrix* among the objects, whereas partitional techniques expect the data in the form of a *pattern matrix*.

Partitional techniques are used frequently in engineering applications where single partitions are important. In image processing, segmenting the image in meaningful segments or regions is in fact a partitional grouping of individual pixels. Therefore, image segmentation is intrinsically a (constrained) partitional clustering problem. Partitional clustering methods are especially appropriate for the efficient representation and compression of large data bases (Anderberg, 1973). Dendrograms with more than a few hundred patterns are impractical.

The problem of partitional clustering can be formally stated as follows. Given *n* patterns in a *d*-dimensional metric space, determine a partition of the patterns into *K* groups, or clusters, such that the patterns in a cluster are more similar to each other than to patterns in different clusters. The value of *K* may or may not be specified. A clustering criterion, such as the *square-error criterion*, must be adopted. Evaluation of all possible partitions containing *K* clusters, and picking that partition that optimizes the criterion, will only be feasible for small *n* and *K*. A variety of partitional algorithms are designed to avoid the combinatorial explosion, but may converge to a *local minimum* of the criterion function.

There are general guidelines for deciding which technique is best for what data. However, each single technique comes with a single result. You have to know the structure of the data in order to justify the proper choice of the technique used. But the true structure is unknown, so this appears to be an endless problem. Very often *multiple testing* is the only way out. Then, we face the problem of combining and *interpretation* of (intermediate) results. Internal statistics (indices) and additional testing form the basis for what has been identified to be one of the most subtle problems: *validation*.

From the foregoing it is clear that using a particular method for clustering data, together with appropriate validation of its result, is just part of the problem.

Figure 1.1 shows the major steps in *exploratory data analysis* (EDA) (Tukey, 1977), in which cluster analysis appears to be a major tool. The process of data collection, data representation, data grouping, and data interpretation is a sort of endless loop. Each time, new insights are obtained and new ideas generated. A substantial amount of a priori information (*problem knowledge, domain knowledge*) is needed in order to guide such a process.

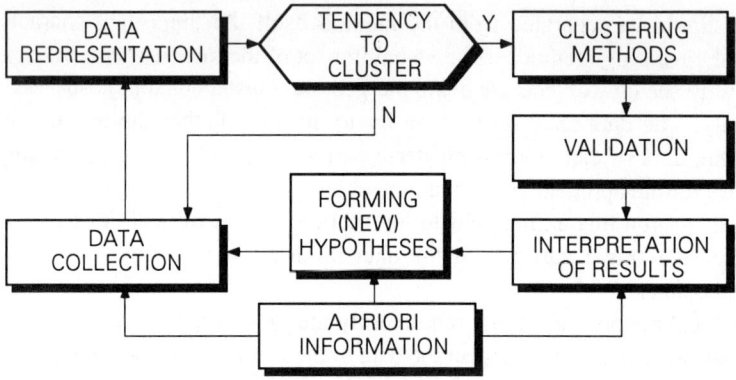

Figure 1.1: Major steps in exploratory data analysis (EDA) (Tukey, 1977; Jain and Dubes, 1988).

Testing the *tendency* to cluster (are the data random or does some justification exist for grouping) is again a subtle issue which arises before a particular method is used and its result validated.

Here it is important to note that user-specified parameters introduce *subjectivity*. Very often a prior knowledge and some *intuition* are needed to specify appropriately the free parameters. Moreover, it is also important to note that each algorithm has its own pecularities depending on the type of data. If those artefacts occur we know that the result may be misleading.

Evidently, the foregoing process incorporates *heuristic* and informal (*approximate*) *reasoning*. The difference between formal and informal analysis is interesting, as is the difference between mathematical (purely statistical) and heuristic and approximate reasoning. Figure 1.2 shows a comparison of formal and informal analysis, and their interrelation.

Multivariate data analysis is typically model-driven. The inferences are true as long as the formal model is true (Johnson and Wichern, 1982). So, statistical reasoning is *model-driven*. In exploratory data analysis we may have just a vague idea about the underlying data model. Inferences are very much determined by diagnosing the various representations, both visually and numerically. So, approximate reasoning is *data-driven*.

In summary, Figure 1.3 relates the major steps to be considered when undertaking an exploratory data analysis whose central component is a cluster analysis. As mentioned before, the process is an endless loop in which new insights are obtained and new ideas generated each time through the loop. The end result could be the design of an experiment that uses standard statistical tools to come to decisions about the phenomenon being studied. One might derive enough information about the phenomenon from an exploratory data analysis itself to draw *informal conclusions*.

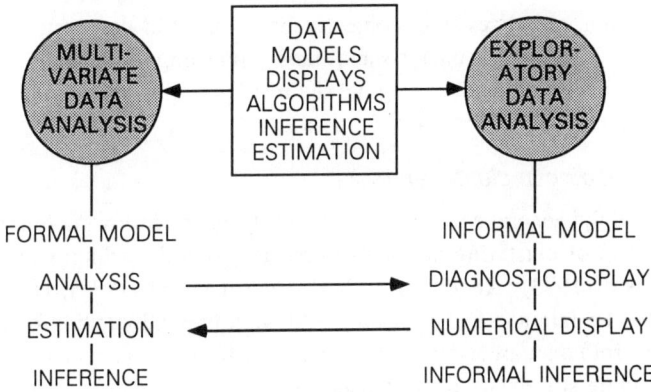

Figure 1.2: Formal versus informal analysis; mathematical versus heuristic and approximate reasoning.

The major steps can be identified as follows:

- *Data collection.* The careful recording of data in accordance with the standards of the application area.
- *Initial screening.* Raw data usually need some massaging (normalization); quick feature visualization.
- *Representation.* Choosing proximity index; perform linear and nonlinear projection.
- *Clustering tendency.* Are the data random or does some justification exist for clustering?
- *Clustering strategy.* The major question is the choice between hierarchical and partitional procedures.

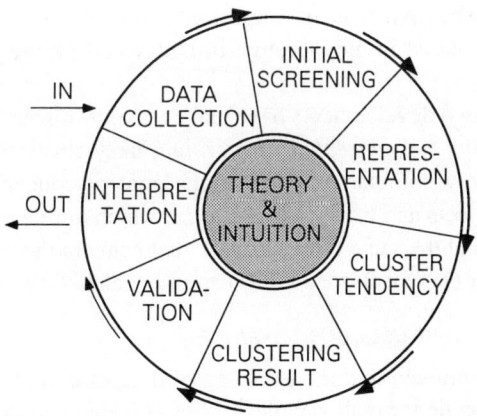

Figure 1.3: Clustering methodology; the expert is in the middle.

- *Validation.* Internal indices assess objectively the merit of the clustering. Validation often involves Monte Carlo analysis and statistical testing.
- *Interpretation.* How does one integrate the results of cluster analysis with previous studies and draw conclusions about the data?

1.1.3 General reading in cluster analysis

Having discussed the general concept of cluster analysis, the following references pertain to various aspects of clustering and are offered as general reading: Cormack (1971); Sneath and Sokal (1973); Anderberg (1973); Everitt (1974, 1979); Hartigan (1975); Romesburg (1984); Jain and Dubes (1988). Although they all attempt to review cluster analysis, either from an application point of view, or from a conceptual or algorithmical point of view, there still exists a strong need for guidance in making intelligent choices among the bewildering array of alternatives. Hundreds of specific references will be mentioned as we consider some of these alternatives in the chapters that follow.

Even though cluster analysis has come to development in various disciplines of science, we still consider clustering as a unique topic in statistical pattern recognition, often refered to as unsupervised classification or *learning*. This in contrast to supervised classification or *recognition*. Therefore, the traditional literature on pattern recognition (Duda and Hart, 1973; Devijver and Kittler, 1982; Fu, 1982) offers well-written treatments of the subject matter. There, cluster analysis – as a means to define pattern classes algorithmically – precedes the design of a pattern classification system. Moreover, cluster analysis is often used for pattern display in two dimensions in order to visualize class separability while maintaining useful physical insights into the measured quantities involved. As an example, the interactive pattern analysis and recognition system developed by Sammon (1969) aiming at preserving the data structure in displaying the data in lower dimensions, is still relevant to applied pattern recognition today.

As *image analysis* is the joint effort of image processing and pattern recognition, cluster analysis has shown to be of importance for image segmentation (*spatial clustering*) and has appeared in the literature in Diday and Simon (1976); Jain (1986); Gerbrands et al. (1986).

In the last decade, two developments have make a major impact in the field of cluster analysis: *fuzzy clustering* and *conceptual clustering*. Fuzzy clustering allows the classes of objects to have fuzzy boundaries, whereas the goal of conceptual clustering is to construct classes of objects and to discover the *rules* which define relationships between objects in these classes at the same time. General treatments in those areas are offered by Ruspini (1969); Backer (1978); Bezdek (1981) and Michalski (1980), respectively.

1.1.4 From numbers to structure: an example

It has already been mentioned that the grouping of data essentially implies making the structure of the data explicit. Small samples of low-dimensional data can be structured (or grouped) by visual means, as indicated by Figure 1.4. Visual parameters are sized by data entries. However, for much larger quantities of data, this visualization becomes impractical.

As an example, if we represent the data as they have been recorded in the domain of application just as a list of numbers, structure will hardly be visible. Figure 1.5 shows such list of collected data. It is the very aim of cluster analysis to represent the structure in such a way that the investigator may gain a clear insight into the structure which may be present in the data by inspecting the statistics of a partition and the resulting dendrogram.

5.0	3.6	1.4	0.2
4.9	3.1	1.5	0.1
6.5	2.8	4.6	1.5
5.2	2.7	3.9	1.4
6.5	3.0	6.6	2.1
7.2	3.6	6.1	2.5

Figure 1.4: Visualization of small samples of low–dimensional data.

17			
4			
−133	−83	76	34
−164	−84	67	31
−154	−85	68	28
−173	−95	64	23
−123	−50	81	38
−172	−69	81	52
−133	−71	73	48
−121	−57	67	25
−118	−53	67	21
−94	−24	60	14
−94	−38	64	25
−71	−45	57	32
−94	−56	61	8
−89	−35	64	21
−72	−36	43	14
−94	−30	90	47
−94	−67	56	19

Figure 1.5: Recorded data in four dimensions.

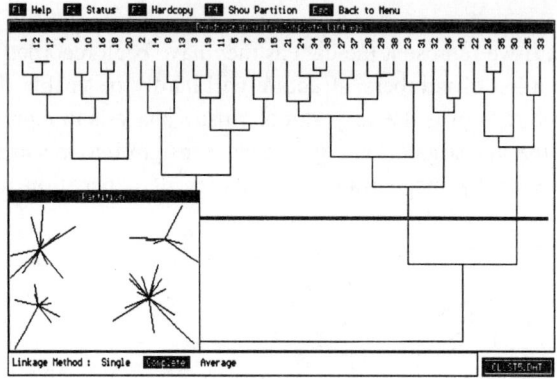

Figure 1.6: Visualization by hierarchical and partitional structure.

Cutting the dendogram at any level, we observe the corresponding partition. The dendogram itself provides 'local' and 'global' structure at the same time. Through the displayed pattern numbers, the user can keep track on how the recorded input data fit into the structure. For example Figure 1.6, a resulting tree and partitioning, provide a better insight into the structure of the data in figure 1.5.

1.1.5 EDAPLUS: a first glance

Let us consider the clustering processing tree as presented in Figure 1.7 in somewhat greater detail.

The actual measurements or recorded data are commonly considered as raw data. For reasons that will become clearer at a later stage, some *standardization* of the data is necessary, though at the same time it should be kept in mind that specific domain knowledge is always needed to justify the specific choices made here. In other words, *scaling* or *normalization* are very subtle issues.

To employ clustering techniques, the data have to be represented in the form of a *pattern matrix* (each row of the matrix contains all variables (recorded standardized measurements, features) of one single pattern (object, entity)) or in the form of a *proximity matrix* (a relation matrix which contains the values of the degree of association between pairs of objects). A variety of techniques are available to convert a pattern matrix into a proximity matrix and the reverse.

The next step includes some sort of inspection of the data: *visualization* by mapping the data onto a two-dimensional space (*projection*). Again, a large variety of projection techniques are available to display the data in the form of a two-dimensional scatter plot. One of the purposes is to provide the investigator with insight into what can be expected of point compactness, density, cluster shape and so on, and what cluster technique could be applied best. As mentioned before, two kinds of methodology are at the investigator's disposal: partitional and hierarchical clustering techniques.

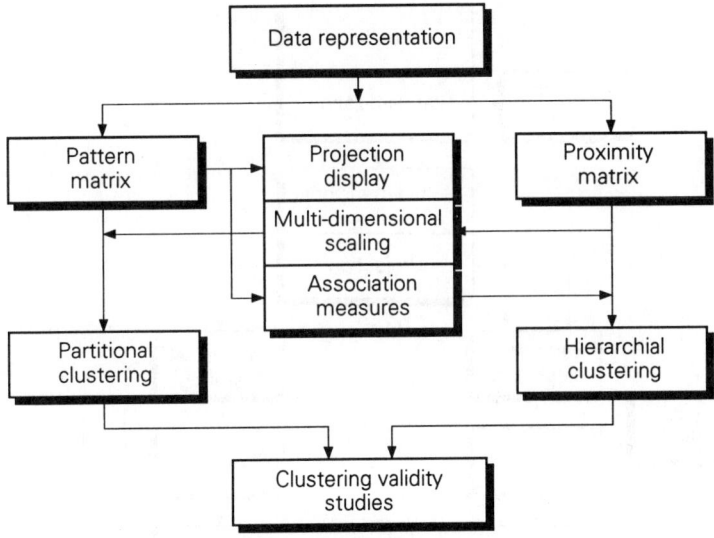

Figure 1.7: The clustering processing tree.

Computational complexity (the number of objects and the dimensionality) may influence the ultimate choice but modern computer facilities often allow the investigator to apply both methodologies. Interpreting the results and combining them leads to the final conclusion about the nature of the structure in view of the specific domain of application which the analysis was meant for.

In order to justify the ultimate findings, a large number of statistics, additional testing, and procedures are available to validate these findings. Again, *validation* is known to be one of the most subtle problems in cluster analysis.

The above clustering processing tree leads automatically to the backbone of a software package such as EDAPLUS. As can be seen, the functional scheme of EDAPLUS, as presented in Figure 1.8, follows the above methodology. In the following pages, some of its features are briefly outlined. For reasons that will become clear at a later stage, raw data are standardized and transformed (projected), enabling two-dimensional visual inspection. Based on the standardization, *random data* are generated – as many data points in as many dimensions as the data themselves – in order to perform validation studies. Clearly, all pieces of evidence (statistics and visual observations) are then combined in rule-generated reasoning to yield a final judgement about the structure as well as its validity.

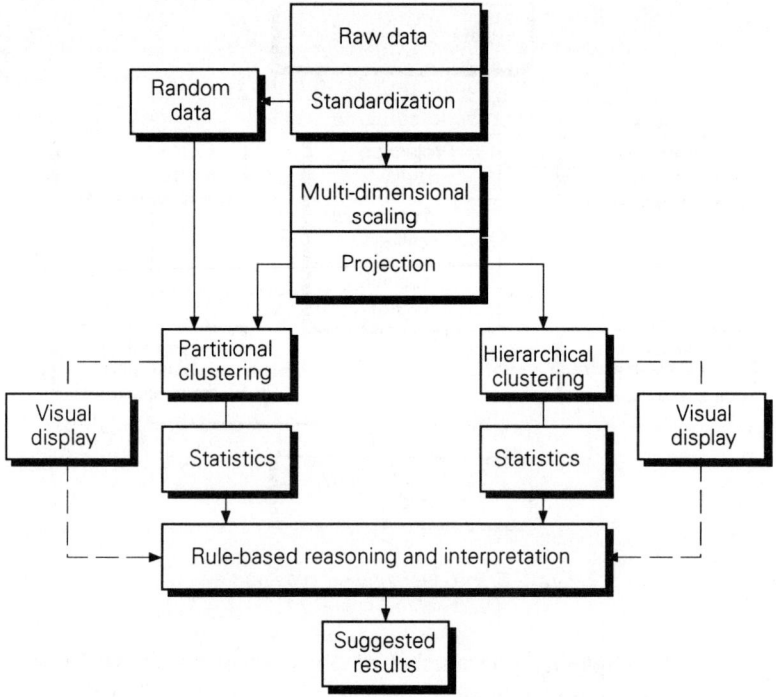

Figure 1.8: The functional scheme of EDAPLUS.

Simple and intuitive

In contrast to one of the most powerful *software packages* for cluster analysis, Clustan 3, (Wishart, 1986), developed at the University of St Andrews, EDAPLUS is a public domain program package developed at the Delft University of Technology, offering most of the basic cluster analysis tools and providing automatic *rule-based reasoning* about the validity of the results obtained. Due to its help facilities and menu options, it is easy to learn to use the main program because it is both simple and intuitive. The intention of the package is to support the material covered in this book.

How to run EDAPLUS

EDAPLUS may be run on a standard IBM PC or compatible with at least 1024 Kb of memory. The system should run from version DOS 2.0 and up. EDAPLUS is distributed on a single 1.44 Mb diskette. Insert the diskette in drive A: and type the command EDAPLUS.

How to use EDAPLUS

The program appears in two forms: as EDA (an earlier version) including projection capability though not suporting rule-based reasoning, and as EDAPLUS including rule-based interpretation though not suporting projection.

The use of EDAPLUS only requires a properly formatted data file (*name*.DAT). For reasons of limited computer requirements imposed on the system of the potential user, the number of patterns is assumed to be less than or equal to 100, and the dimensionality less than or equal to 10.

See Appendix B for detailed commands, functions, and useful actions.

Overview of major system commands (see Appendix B for details)
Figure 1.9 shows the simple nesting of the major menus, both for user interaction and selection of algorithmic options. Note that rule-generated interpretation is behind the status function key.

The following keys are of immediate importance:

projection screen	*partitional screen*	*dendrogram screen*
F1 help	**F1** help	**F1** help
F2 status	**F2** status	**F2** status
F3 hardcopy	**F3** hardcopy	**F3** hardcopy
F4 points/labels	**F4** show cluster	**F4** show partition
F5 projection	**F5** expand	**F5** statistics
F6 options	**F6** options	

Any analysis will start with selection of the data to be considered, as illustrated in Figure 1.10. Data (*name*.DAT) and corresponding dissimilarity matrices (*name*.DIS) may be selected from a file directory using into cursor keys.

The help facility (F1)
Throughout the program, at any level of operation, on-line HELP is available. Appropriate reference is made to the page number of the book. WHAT and HOW information is provided.

Consulting the system
Once a data base has been loaded the analysis can be started.
–   Selecting an analysis mode or run '*number of clusters*' which responds with the most likely number of clusters based on the *Davies–Bouldin statistic* performed on the complete link hierarchy.
–   All significant output will be written to a special file as well as to the data status to be inspected on the screen. This file is called *name*.STA and may be examined later by using the DOS-command TYPE.
–   All graphic results are kept in different screens and windows and remain available throughout the analysis. Figure 1.11 shows the layout of the upper and lower partitional screen, and the dendrogram screen.

Main menu:
<LOAD>
command;
directory mask

hierarchical
methods;

partitional
methods;

projections
(EDA only);

statistics

partitional
function keys;
F6: partitional
select options

F6: partitional
select options

F2: status
information;
rule-based
reasoning;
hierarchical
function keys;
hierarchical
select options

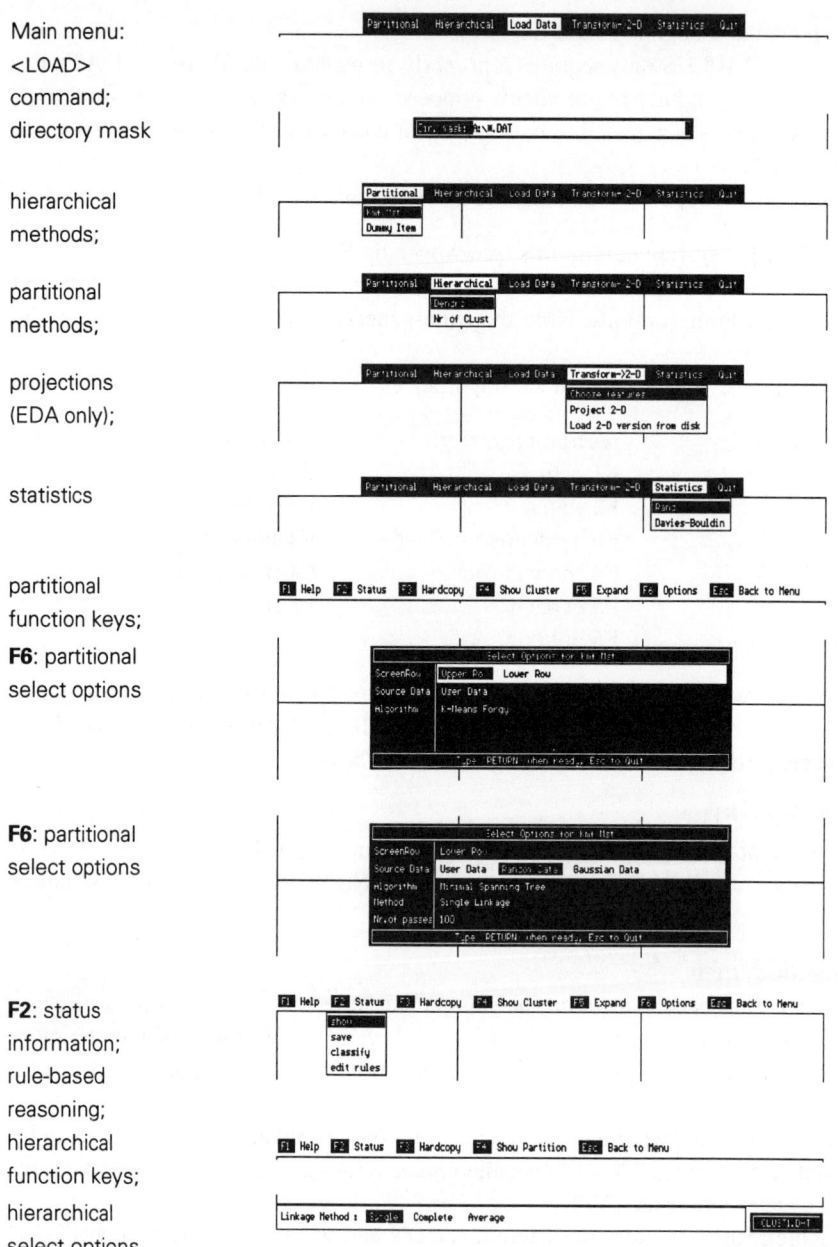

Figure 1.9: Overview of the major system commands of EDAPLUS (version 1.0).
See Appendix B for details.

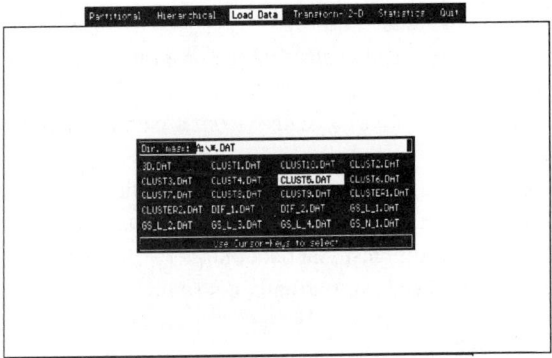

Figure 1.10: Selection of data from the file directory.

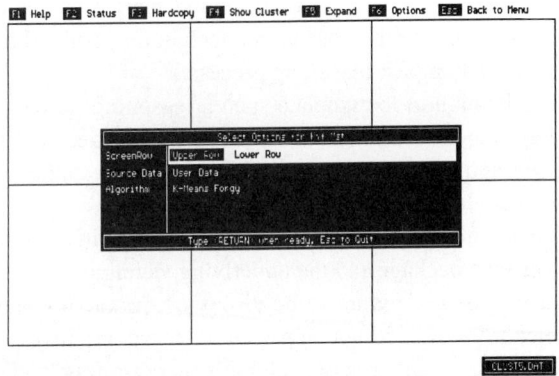

a.

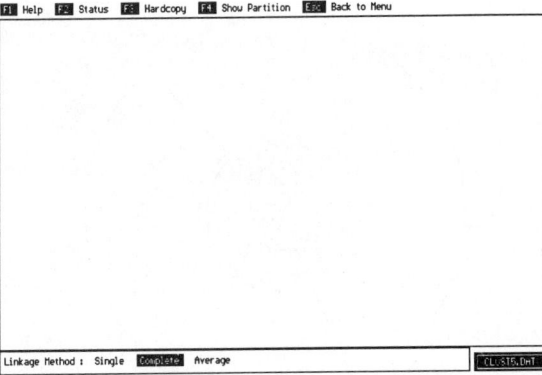

b.

Figure 1.11: General screen layout of EDAPLUS; see Appendix B for details.
(a): Partitional/projection/minimum spanning tree screen; (b): dendrogram screen.

Available stand-alone programs

In support of EDA or EDAPLUS, five stand-alone programs are available. These are:

- DATA: converts any input raw pattern matrix into an EDAPLUS proximity matrix;
- MISDAT: computes a dissimilarity matrix from a pattern matrix containing missing values;
- FEATSEL: selects feature subsets out of a given feature set by sequential forward, sequential backward, or branch and bound feature evaluation;
- PROJECT: projects multi-dimensional data onto a two-dimensional space;
- DATGEN: generates uniformly or normally distributed data.

Intelligence in EDAPLUS

The expert user of a statistical package for exploratory data analysis generally has a keen feeling of determining the order in which procedures and validation have to take place. Therefore, the results of any kind of data analysis are very much determined by a complex reasoning process which may differ from one analysis to another. Interpretation of intermediate results, choosing the appropriate techniques, and validating the ultimate findings are part of such a complex reasoning process.

Today it is felt to be natural to associate such a reasoning process with knowledge-based systems. Indeed, the need for more intelligence in statistical software packages has long been recognized and is a topic of interest among both statisticians and artificial intelligence researchers (Gale, 1986).

EDAPLUS offers an informal knowledge-based reasoning in order to make it feasible for 'naive' users to use the package and the underlying techniques.

Notice that in using the techniques to be discussed, problem or domain knowledge will be indispensable; however, will not be part of the reasoning here.

Figure 1.12 shows the limited scope of built-in reasoning compared to the full endless-loop reasoning as presented in Figure 1.3.

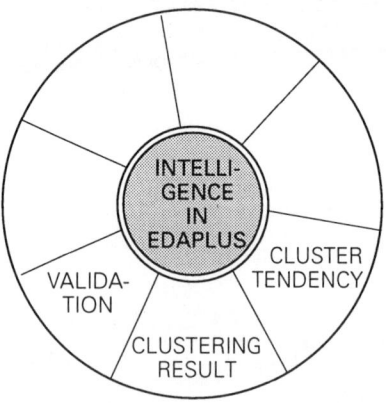

Figure 1.12: Intelligence in EDAPLUS.

All reasoning is limited to proper use of the methods involved and the impact of their results to clustering tendency, and cluster validation. As such, initiation rules on the basis of internal statistics, support rules based on different tests, and diagnostic rules (combining all suggested evidence) are incorporated in a multi-layer decision network.

1.1.6 Concluding remarks

In this introduction, we have identified exploratory data analysis as a scientific discipline which unravels structure in data by multiple statistical testing, validation and complex reasoning. Cluster analysis is one of the major tools of this discipline. Within the area, many techniques are available; however, significant problems do arise:

– choosing the appropriate techniques and testing;
– interpreting the intermediate results;
– knowing how to proceed in the analysis;
– combining different results; and
– validating the ultimate findings.

We do have a coherent body of knowledge about the global methodologies; however, subjectivism and intuition still play an important role. Moreover, due to algorithmic artefacts, interpretation of the empirical evidences appears quite difficult. As has been pointed out, the goals:

– discovery of natural structure of collected data;
– generating ideas about the underlying data distributions; and
– testing and forming of hypotheses

are self-evident.

However, discovery and generation of ideas belong to intuitive processes. Even if internal statistics are used to test a hypothesis, the degree of acceptance is highly user-dependent. Apart from being a domain expert, the user needs to have a keen feeling for the use of the statistical tools in order to use a data analysis toolbox appropriately. The traditional toolbox generally fails in arranging the order of steps, in guiding the choice of details, and in untangling relations among data. This is particularly true for the 'naive' user. The combination of both, being an expert in data analysis and being a domain expert seldom occurs. This is precisely the reason why automatic informal inference by heuristic and approximate reasoning within the context of algorithmic data analysis is very much desired: to prevent inappropriate use of algorithms and to avoid misleading conclusions based on single algorithmic results.

1.2 Organization of the book

Applying the methodology of cluster analysis in the context of exploratory data analysis can be divided into three principal sub-problems: (1) gaining understanding of available *methods* and techniques, knowing the algorithmic pecularities in a wide range of data shapes, and specifying and interpreting the algorithmic output; (2) *framing* the analysis itself in the context of the problem to be analyzed in the domain of application; and (3) applying *rules* and *strategy* to combine and to support the intermediate results of the methods and techniques involved, aiming at drawing informal *conclusions* about the structure of collected data in the context of the problem posed in the domain of application.

The material in the book is organized in essentially the same order. The book is subdivided into three parts:

Part 1   introduces cluster analysis (Chapters 1–5);
Part 2   proceeds to supply a feeling for diversity, imprecision and complex decision making in cluster analysis (Chapters 6 and 7);
Part 3   adresses the potentiality of computer-assisted rule-based reasoning (Chapters 8–10).

As discussed in Section 1.1.2, methods can be devided into *hierarchical* and *partitional* methods. Only the major techniques are presented in Chapter 2, as well as their impact on related problems like *clustering tendency* and *validation*.

It is important to note that (Jain and Dubes, 1988) covers methods and techniques more broadly, and therefore should be considered as the major and most important reference today within the scope of the topical field of this book.

However, here the philosophy of the EDAPLUS program becomes clear. The presentation of any algorithm, even if accompanied with many examples, still remains static. Through experimentation with the implemented algorithm under consideration, a better understanding of its capabilities is achieved. The user may inspect the performance of the algorithms using supplied test data, or using self-collected data. EDAPLUS offers the appropriate environment to do so. As such, EDAPLUS serves as an aid to reinforce, clarify and exemplify the concepts developed throughout the book.

A discussion of the methods and related questions, as well as their effect on the data analysis itself, is given in Chapter 3. The chapter ends with some intuitively reasonable procedures for evaluating the stability and usefulness of the solutions found.

In Chapter 4, we try to identify the purpose of cluster analysis: that is, what the researcher intends to achieve. Making decisions about what has to be achieved, how goals are to be achieved and choosing the options available is called *framing*. Framing and planning are discussed in this chapter.

Even if methods and techniques are stated precisely, cluster analysis has been developed starting from a highly informal and imprecise working definition of a cluster. When dealing with a partition in the form of a set of compact, disjoint, well-separated clusters, there will be no *ambiguity* and *uncertainty* in the assignment of patterns to any of the clusters. However, if the clusters are touching and do have no sharp boundaries, the assignment of patterns to any of the clusters turns out to be non-unique, ambiguous, uncertain, or vague. Such partitions are better described by replacing ordinary clusters (pattern subsets) by fuzzy pattern subsets. The theory of *fuzzy sets* provides the mathematical tools to reformulate the process of clustering data. This is the subject of Chapter 5. Fuzzy sets are likely to find increasing use in applications involving imprecise and incomplete information, common-sense reasoning, and complex concepts. Without doubt, cluster analysis is such an application. At the same time, utilizing the theory of fuzzy sets, approximate reasoning is to be considered as a branch of *fuzzy logic* and can be viewed as a more expressive mathematical language for representation of uncertain or vague structures.

In Chapter 6, we discuss the problem of indetermination and uncertainty in cluster analysis originating from the fact that there exists no single algorithm that may uncover all possible structures. Solving this problem is demonstrated to be a solution in which the strenghts of different methods, techniques and algorithms are used, aiming at carefully combining weighted pieces of evidence provided by them. Repeated experimentation, *simulation* with random data, and approximate reasoning are offered as the keystones of a combined approach. Experiments show that numerical precision is useless and overdone.

Next, in Chapter 7, we identify the ingredients for, and the underlying paradigms of the development of computer-assisted reasoning in cluster analysis. The previous chapters have shown to us the following:

1. Cluster analysis is a very complex process in which intuition and past experience play a dominant role.
2. The intent of cluster analysis is not to produce numerical precision but rather global approximate qualification of detected structures.
3. Cluster analysis is not known by a single general strategy but by a variety of possible (feasible) *data-dependent strategies* each of which may sometimes lead to true insight, and sometimes to misleading interpretation which can only be detected and unravelled by using one's feeling, intuition and past experience, Jain and Dubes (1988).
4. One might learn to recognize the characteristics in the data views, the plots of diferent indices and statistics by eye, but it is difficult to recognize automatically without any further support in the decision procedure; the naive users from other disciplines should be assisted by experts in finding valid solutions, Jain and Dubes (1988).

In view of these facts, it is natural to associate such a reasoning process with expert systems: *intelligent computer programs* that use expert knowledge and inference procedures to solve problems (here the problem of carrying out a validated cluster analysis) that are difficult enough to require significant human expertise for their solution (Barr and Feigenbaum, 1981, 1982).

So, Chapter 8 proceeds with discussing the general concepts of knowledge-based expert systems, representation of knowledge and methods of inference and reasoning under uncertainty.

Next, Chapter 9 presents the general guidelines for developing rule bases to assist the cluster analyst in making valid diagnoses about the data under consideration. To facilitate exemplification and understanding, a simple language is described in which rule bases can be developed and can be used directly in EDAPLUS.

As discussed before (Chapters 6 and 7), probably one of the most important questions to be answered is how rules can be learned from observations and what to do with exceptions. Learning rules from observations is covered by the remainder of Chapter 9.

Finally, in Chapter 10, we describe a research application which illustrates how to frame and validate the usage of cluster analysis. More specifically, we illustrate how hierarchical trees and partitions produced by cluster analysis may become information. That is, we show the steps that link the gathering of data and the generation of information needed to draw conclusions.

The research of Dudok van Heel and Kamminga on dolphin sonar sound production for navigation in relation to the mass stranding phenomenon includes a series of different intermediate research goals for which also different 'data to thought' paths apply (Kamminga, 1994).

The book ends with an Appendix summarizing the commands and functions for performing useful actions in EDAPLUS. Examples are given in order to demonstrate the main functionality of the package.

When using the EDAPLUS diskette for learning and excercises, read the README.TXT file first and consult the Appendix for commands and functionality.

It is important to appreciate that EDAPLUS is not intended to be a research tool. For real research applications it has a number of severe limitations. Its purpose is to serve as an easy-to-use vehicle to reinforce, clarify and exemplify the methodology, concepts and algorithms through simple experimentation.

Througout the book, EDAPLUS comments and messages will appear in this format.

Format of the EDAPLUS comments and messages throughout the book.

2 | Methods in clustering data

The reasons for this variety of methods are probably twofold. To begin with, automatic classification is a very young scientific discipline in vigorous development, as can be seen from the thousands of articles scattered over many periodicals (mostly journals of statistics, biology, psychometrics, computer science, and marketing). Nowadays, automatic classification is establishing itself as an independent discipline. ... The second main reason for the diversity of algorithms is that there exists no general definition of a cluster, and in fact there are several kinds of them: spherical clusters, drawn-out clusters, linear clusters, and so on. Moreover, different applications make use of different data types, such as continuous variables, discrete variables, similarities and dissimilarities. Therefore one needs different clustering methods in order to adopt to the kind of application and the type of clusters sought (Kaufman and Rousseeuw, 1990).

In Chapter 1, we identified cluster analysis as one of the major tools in explorative data analysis. Within that area, many techniques and methods are available. Most of them are well documented (see Anderberg, 1973; Sneath and Sokal, 1973; Everitt, 1974; Hartigan, 1975; Romesburg, 1984; Jain and Dubes, 1988). Hundreds of specific implementations are brought forward in the many scientific disciplines from which cluster analysis methods emerged.

In this chapter, we restrict ourselves to some of the basic methods and algorithms. Our main objective will be the development of a basic toolbox such that the algorithmic output can be used for computer-assisted reasoning. Most of the methods covered are offered in EDAPLUS. The methods to be discussed include data representation, linear and nonlinear projection, hierarchical and partitional algorithms, and validity measures based on internal and external indices, as well as on Monte Carlo experimentation.

2.1 Data representation

Clustering algorithms group objects based on indices of proximity between pairs of objects. A set of objects comprises the raw data for a cluster analysis and can be described by two standard formats:
- a pattern matrix (Figure 2.1);
- a proximity matrix (Figure 2.2).

2.1.1 Pattern matrix

If each object in a set of n objects is represented by a set of d measurements (or attributes or scores), each object is represented by a pattern, or d-place vector. The set itself is

n

d

x_{11}	x_{12}	x_{13}	x_{14}	·	·	·	x_{1d}	L
x_{21}	x_{22}	x_{23}	x_{24}	·	·	·	x_{2d}	L
x_{31}	x_{32}	x_{33}	x_{34}	·	·	·	x_{3d}	L
·								
·								
·								
x_{n1}	x_{n2}	x_{n3}	x_{n4}	·	·	·	x_{nd}	L

n: number of patterns
d: number of measurements (dimensionality)
L: labels (optional)

Figure 2.1: Formatting of a pattern matrix.

viewed as an $n \times d$ *pattern matrix*. Each row of this matrix defines a pattern and each column denotes a feature or measurement. Figure 2.1 illustrates the formatting of such a pattern matrix. Normally, labels (L) are not known a priori. However, if they are known a priori (or obtained by any other clustering method) or even if they are partially known, they can be added to each measurement row (of length d). Labels are required for discriminant analysis, and measurement (attribute, feature) selection.

The d features are usually pictured as a set of orthogonal axes. The n patterns are then points embedded in a d-dimensional space called a *pattern space*. A cluster can be visualized as a collection of patterns which are close to one another or which satisfy some spatial relationship. The task of a clustering algorithm is to identify such natural grouping in spaces of many dimensions.

2.1.2 Proximity matrix

Clustering methods require that an index of proximity (likeness, affinity, association) be established between pairs of objects. The index can be computed from a pattern matrix or can be formed from raw data.

A proximity matrix [$d(i,j)$] gathers the pairwise indices of proximity in a matrix in which each row and column represents a pattern. A proximity index can represent either a *similarity* or a *dissimilarity*. The more the ith and jth objects resemble one another, the larger a similarity index and the smaller a dissimilarity index. For example, Euclidean distance between two patterns in a pattern space is a dissimilarity index, whereas the correlation is a similarity index.

The formatting of a proximity matrix is shown in Figure 2.2. Apart from the number of patterns n, only the upper triangle of the proximity matrix is required as the proximity matrix is assumed to be symmetric. If the true dimensionality d of the pattern space is known, its value may be input optionally.

A thorough review of *measures of association* and their interrelationships is provided by (Anderberg, 1973).

n
d

$d(1,1)$	$d(1,2)$	·	·	·	$d(1,n)$
	$d(2,2)$	$d(2,3)$	·	·	$d(2,n)$
·			·		
·				·	
·					$d(n,n)$

n: number of patterns
d: number of measurements (dimensionality)

Figure 2.2: Formatting of a proximity matrix.

A proximity index between the ith and kth pattern is denoted $d(i,k)$ and must satisfy the following properties:

1. (a) For a dissimilarity $d(i,i) = 0$, for all i.
   (b) For a similarity:$d(i,i) \geq$ max $d(i,k)$, for all i.
2. $d(i,k) = d(k,i)$, for all (i,k).
3. $d(i,k) > 0$, for all (i,k).

Ratio types (distances)
Let $[x_{ij}]$ be pattern matrix, where x is the jth feature for the ith pattern. The ith pattern, which is the ith row of the pattern matrix, is denoted by the column vector $\mathbf{x}_i$.

$$\mathbf{x}_i = (x_{i1}, x_{i2}, \ldots, x_{id})^{\mathrm{T}}, \ i = 1, 2, \ldots, n$$

The most common metric is the *Euclidean distance*:

$$d(i,k) = \left[\sum_{j=1}^{d} (x_{ij} - x_{kj})^2 \right]^{\frac{1}{2}}$$

satisfying the additional metric properties:

4. $d(i,k) = 0$ only if $\mathbf{x}_i = \mathbf{x}_k$.
5. $d(i,k) \geq d(i,m) + d(m,k)$, for all (i,k,m).

EDAPLUS also incorporates the squared *Mahalanobis distance*. The expression for the squared Mahalanobis distance between patterns $\mathbf{x}_i$ and $\mathbf{x}_k$ is

$$d(i,k) = (\mathbf{x}_i - \mathbf{x}_k)^{\mathrm{T}} R^{-1} (\mathbf{x}_i - \mathbf{x}_k)$$

where R is the pooled sample covariance matrix (Duda and Hart, 1973).

Nominal types (matching coefficients)
Many actual measurements, especially data collected from human subjects, are *binary*

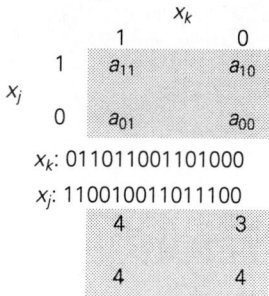

Figure 2.3: Counting correspondence between two binary measurement vectors.

and nominal. Matching coefficients are proximity indices for such data. If x_i and x_k are two binary vectors to be related, then the number of corresponding 1's is denoted as a_{11}. Likewise, a_{00} expresses the number of corresponding 0's. Figure 2.3 illustrates counting the corresponding or differing binary measurements.

Let $d = a_{11} + a_{10} + a_{01} + a_{00}$. Then three common *matching coefficients* for x_i and x_k are:

- Simple matching coefficient: $d(i,k) = (a_{00} + a_{11})/d$
- Jaccard coefficient: $d(i,k) = a_{11}/(d - a_{00})$
- Russell–Rao coefficient: $d(i,k) = a_{11}/d$

Distance measures and matching coefficients are instrumental in producing dissimilarity matrices from existing pattern matrices. In Section 2.3 the reverse operation will be explained. Figure 2.4 shows the conversion of different data formats schematically using EDAPLUS or the stand-alone program, DATA.

A stand-alone program, DATA, is available on diskette.

The module DATA converts all kinds of raw data into EDAPLUS format.

All distance measures and matching coefficients documented above are available to produce the desired pattern matrix and dissimilarity matrix.

The program is started up by typing **DATA** <RETURN>.

The stand-alone program, DATA.

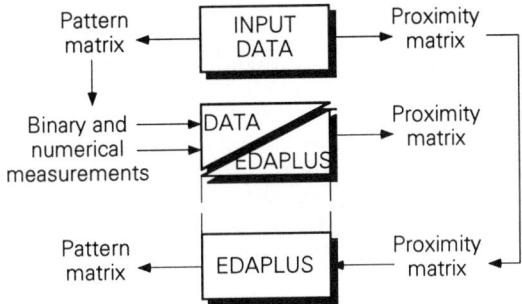

Figure 2.4: Schematic relation between pattern matrices and proximity matrices (dissimilarity matrices).

   The module DATA is required if a binary pattern matrix is considered to be the input. If the pattern matrix consists of numerical measurements, EDAPLUS generates automatically a dissimilarity matrix based on the Euclidean distance.

   In Figure 2.5, an example is given of a binary pattern matrix. Corresponding similarity matrices are obtained using the simple matching coefficient, the Jaccard matching coefficient and the Russell–Rao coefficient. Clearly, the resulting matrices will differ. Basically, it is the choice of the domain expert which association measure is to be preferred. Only within the domain of application can the meaning of the coefficients a_{11} and a_{00} and their impact on the desired pattern matching be justified.

   Two particular situations may occur in practical applications: missing data and mixed data. In these cases, additional processing is required in order to enable later clustering analysis.

2.1.3 Missing data

In practice, measurement pattern vectors could be incomplete because of errors, failures, or unavailability of information. Then the question arises whether incomplete pattern vectors should be discarded or missing entries should be estimated. Dixon (1979) describes several simple and general techniques for handling such *missing values*. Jain and Dubes (1988) summarize these techniques as follows:

1. Simply delete the pattern vectors or features that contain missing values. This technique does not lead to the most efficient utilization of the data and should be used only in situations where the number of missing values is very small.
2. Suppose that the jth feature value in the ith pattern vector is missing. Find the K nearest neighbours of $\mathbf{x}_i$ and replace the missing value x_{ij} by the average of the j-th feature of the K nearest neighbours. The value of K should be a function of the size of the pattern matrix.

Pattern matrix

```
10
6
0  1  0  0  1  1
0  1  1  1  1  0
0  0  1  0  1  0
0  0  1  1  1  1
0  0  0  0  1  1
0  0  1  1  0  1
0  0  0  1  0  1
0  0  0  1  1  1
0  0  1  1  0  0
0  1  1  1  0  1
```

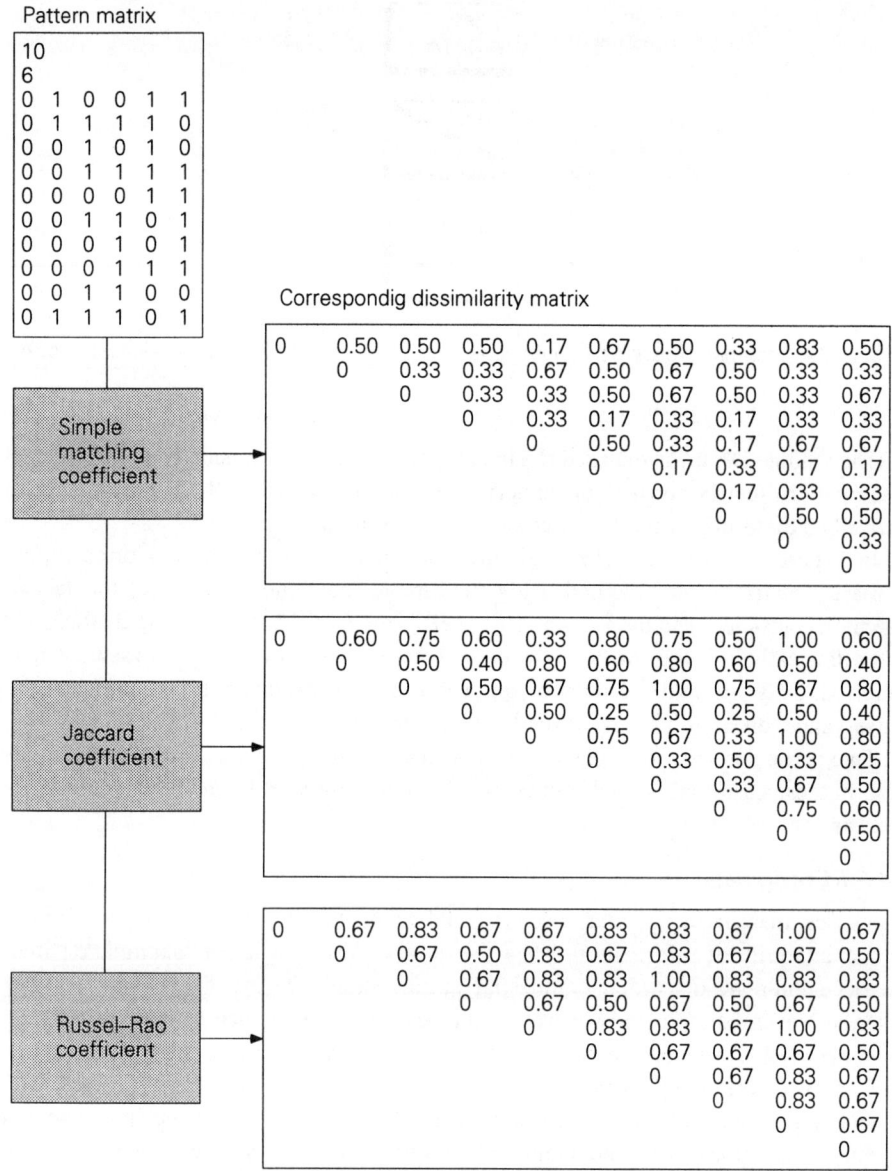

Correspondig dissimilarity matrix

Simple matching coefficient

0	0.50	0.50	0.50	0.17	0.67	0.50	0.33	0.83	0.50
	0	0.33	0.33	0.67	0.50	0.67	0.50	0.33	0.33
		0	0.33	0.33	0.50	0.67	0.50	0.33	0.67
			0	0.33	0.17	0.33	0.17	0.33	0.33
				0	0.50	0.33	0.17	0.67	0.67
					0	0.17	0.33	0.17	0.17
						0	0.17	0.33	0.33
							0	0.50	0.33
								0	0.33
									0

Jaccard coefficient

0	0.60	0.75	0.60	0.33	0.80	0.75	0.50	1.00	0.60
	0	0.50	0.40	0.80	0.60	0.80	0.60	0.50	0.40
		0	0.50	0.67	0.75	1.00	0.75	0.67	0.80
			0	0.50	0.25	0.50	0.25	0.50	0.40
				0	0.75	0.67	0.33	1.00	0.80
					0	0.33	0.50	0.33	0.25
						0	0.33	0.67	0.50
							0	0.75	0.60
								0	0.50
									0

Russel–Rao coefficient

0	0.67	0.83	0.67	0.67	0.83	0.83	0.67	1.00	0.67
	0	0.67	0.50	0.83	0.67	0.83	0.67	0.67	0.50
		0	0.67	0.83	0.83	1.00	0.83	0.83	0.83
			0	0.67	0.50	0.67	0.50	0.67	0.50
				0	0.83	0.83	0.67	1.00	0.83
					0	0.67	0.67	0.67	0.50
						0	0.67	0.83	0.67
							0	0.83	0.67
								0	0.67
									0

Figure 2.5: An example of the use of the major association measures in the case of a binary input measurement vector.

Original pattern matrix

10			
4			
−133	−83	76	34
−164	−84	67	31
−154	−85	68	28
−183	−103	49	24
−173	−95	64	23
−94	−24	60	14
−94	−38	64	25
−71	−45	57	32
−94	−56	61	8
−89	−35	64	21

Corresponding dissimilarity matrix

0	32.4	23.3	61.1	44.8	75.2	61.4	75.2	56.1	67.5
	0	10.5	33.1	16.6	94.0	84.0	101.3	79.1	90.2
		0	39.3	22.4	87.1	76.4	92.9	69.9	82.4
			0	19.7	119.9	111.2	126.6	102.7	117.0
				0	106.7	97.4	114.2	89.4	103.2
					0	18.2	36.1	32.6	14.5
						0	26.0	24.9	7.1
							0	35.2	24.4
								0	25.4
									0

Pattern matrix with missing data

10			
4			
−133	−83	76	34
−164	−84	67	31
−154	*	68	28
−183	−103	49	24
−173	−95	*	23
−94	−24	60	14
*	−38	64	25
−71	−45	57	32
−94	−56	*	8
−89	−35	64	21

Estimated dissimilarity matrix using Dixon method 3

0	32.4	26.9	61.1	49.9	75.2	54.8	75.2	62.5	67.5
	0	12.1	33.1	18.9	94.0	53.7	101.3	91.0	90.2
		0	40.3	27.8	71.7	7.1	96.8	89.4	75.6
			0	14.8	119.9	77.0	126.6	117.7	117.0
				0	123.1	80.7	131.6	103.2	119.3
					0	21.1	36.1	37.6	14.5
						0	14.0	35.0	5.8
							0	40.4	24.4
								0	29.1
									0

Estimated dissimilarity matrix using Dixon method 4

0	32.4	41.8	61.1	44.2	75.2	68.9	75.2	54.9	67.5
	0	36.3	33.1	18.9	94.0	68.3	101.3	79.4	90.2
		0	49.3	41.0	71.2	61.1	90.7	72.8	74.2
			0	16.0	119.9	83.4	126.6	102.3	117.0
				0	107.0	76.4	114.3	89.9	103.6
					0	53.2	36.1	33.9	14.5
						0	51.4	56.6	50.2
							0	36.3	24.4
								0	26.9
									0

Figure 2.6: Numerical example of Dixon methods 3 and 4 for handling missing data using MISDAT.

3. The distance between two vectors $\mathbf{x}_i$ and $\mathbf{x}_k$ containing missing values is computed as follows. First define the distance d_j between the two patterns along the jth feature:

$$d_j = \begin{cases} 0 & \text{if } x_{ij} \text{ or } x_{kj} \text{ is missing} \\ x_{ij} - x_{kj} & \text{otherwise} \end{cases}$$

Then the distance between $\mathbf{x}_i$ and $\mathbf{x}_k$ is written as

$$d(i,k) = \frac{d}{d - d_0} \sum_j d_j^2$$

where d_0 is the number of features missing in $\mathbf{x}_i$ or $\mathbf{x}_k$ or both. Note that if there are no missing values, then $d(i,k)$ is the squared Euclidean distance.

4. Let $\overline{d_j}$ denote the average distance between all pairs of patterns along the jth feature defined as follows:

$$\overline{d_j} = \frac{2}{n(n-1)} \sum_{i=2}^{n} \sum_{k=1}^{i-1} |x_{ij} - x_{kj}|$$

where n is the number of patterns. Now define the distance between two patterns along the jth feature as

$$d_j = \begin{cases} \overline{d_j} & \text{if } x_{ij} \text{ or } x_{kj} \text{ is missing} \\ x_{ij} - x_{kj} & \text{otherwise} \end{cases}$$

Finally, the distance between patterns $\mathbf{x}_i$ and $\mathbf{x}_k$ is written as

$$d(i,k) = \sum_j d_j^2$$

Based on experimental results (Dixon, 1979), method 3 is recommended as the best overall method. Methods 3 and 4 are available in the stand-alone program, MISDAT, on the accompanying diskette.

> A stand-alone program, MISDAT, is available on diskette.
>
> The module MISDAT computes a dissimilarity matrix from pattern matrices containing missing values based on Dixon methods 3 and 4.
>
> The program is started up by typing **MISDAT** <RETURN>.

The stand-alone program, MISDAT

In Figure 2.6, an example of a pattern matrix with missing data is presented. The results of the estimated dissimilarity matrices (using Dixon methods 3 and 4, respectively) are given and are to be compared with the true dissimilarity matrix corresponding with the original pattern matrix. From this experiment, the result of method 3 seems to be slightly preferable over the result from method 4.

2.1.4 Mixed data

As mentioned before, mixed data often occur in practical applications. Basically, two options are available for handling mixed data.

1. Convert the numerical data into interval classes coded with 1's and 0's. Consequently, treat them as binary measurements along with the original binary measurements. Converting numerical data into a lower representation implies loss of information. Moreover, the length of coding with 1's and 0's implies a weighting effect between the numerical part and the binary part in the mixed data representation.

2. Compute both the proximity matrix from the numerical measurements and the proximity matrix from the binary measurements. Let $p_n(i,j)$ be the proximity index obtained from the numerical part, and $p_b(i,j)$ the proximity index from the binary part. Then an overall proximity relation can be achieved by using a convex combination as follows:

$$p(i,j) = \alpha\, p_n(i,j) + (1 - \alpha)\, p_b(i,j)$$

Again, α implies a weighting between the numerical part and the binary part in the mixed data. Only within the context of the domain of application can the proper choice of such weighting be made.

In Figure 2.7, an example of mixed data is presented. The problem is treated by method 2. As mentioned before, no attempt is made to combine the two matrices because no domain information is available regarding proper weighting between the numerical and the binary part in the mixed input data and scaling of the dissimilarity indices from the numerical measurements on the basis of which rescaling on [0,1] could take place. The resulting Jaccard coefficients of the binary measurements can easily be converted into a dissimilarity matrix.

2.2 Linear projection

Projection algorithms map a set of n d-dimensional patterns onto an m-dimensional space, where $m < d$. The main motivation for studying projection algorithms in the context of cluster analysis is to permit *visual examination* of multivariate data, so $m = 2$ in our discussion, Andrews (1972).

A linear projection expresses the m new features as linear combinations of the original d features:

$$\mathbf{y}_i = H\,\mathbf{x}_i, \text{ for } i = 1,2,\ldots,n$$

Pattern matrix with mixed data

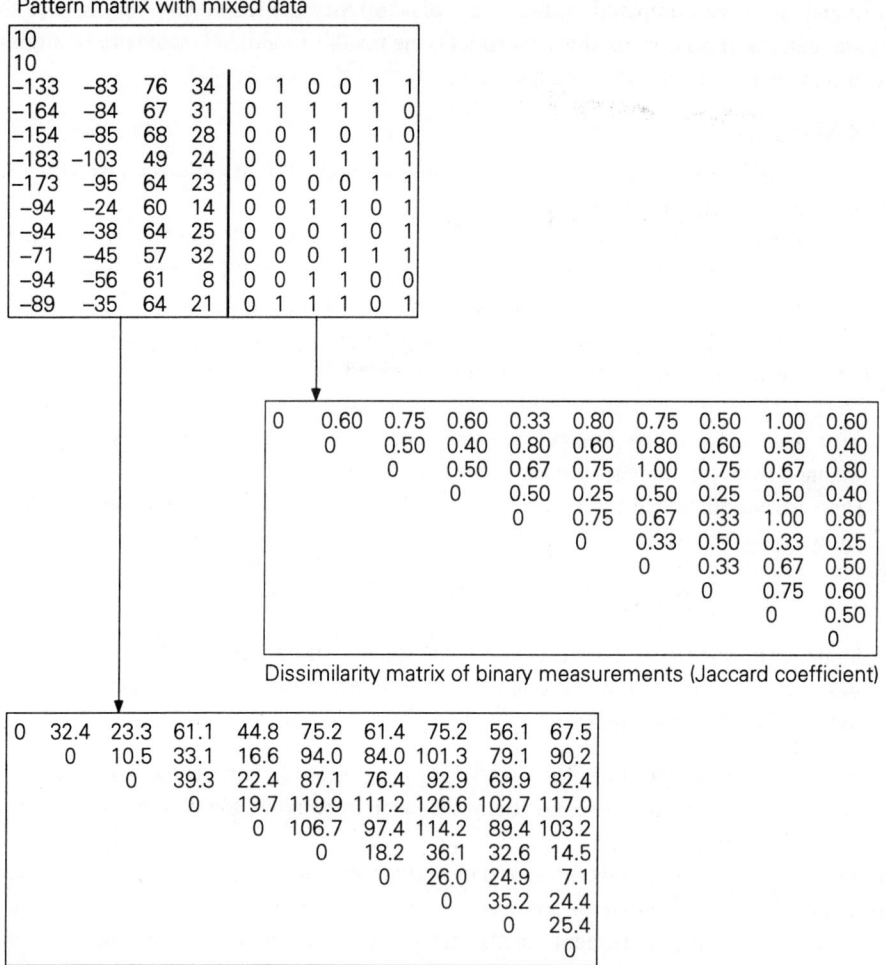

Dissimilarity matrix of binary measurements (Jaccard coefficient)

Dissimilarity matrix of numerical measurements (Euclidean distance)

Figure 2.7: An example of handling mixed data.

Here, $\mathbf{y}$ is an m-place column vector, $\mathbf{x}$ is a d-place column vector, and H is an $m \times d$ matrix.

If *no category information* is available, the eigenvector projection is commonly used. Discriminant analysis is a popular linear mapping when *category labels* are available.

2.2.1 Eigenvector projection (Duda and Hart, 1973)

The eigenvectors of the covariance matrix R define a linear projection that replaces the features in the raw data with *uncorrelated* features. Since R is a $d \times d$ positive definite matrix, its eigenvalues are real and can be labelled so that

$$\lambda_1 \geq \lambda_2 \geq \ldots \geq \lambda_d \geq 0$$

A set of corresponding eigenvectors $c_1, c_2, \ldots, c_d$, are labelled accordingly. The $m \times d$ matrix of transformation H is defined from the eigenvectors of the covariance matrix R as follows:

$$H_m = \begin{bmatrix} c_1^T \\ c_2^T \\ . \\ . \\ . \\ c_m^T \end{bmatrix}$$

yielding the linear projection $\mathbf{y}_i = H\,\mathbf{x}_i$ for $i = 1,\ldots,n$.

The linear eigenvector projection on a two-dimensional plane is a projection on the plane formed by the two eigenvectors corresponding with the two largest eigenvalues. EDAPLUS offers an environment to study the properties of the above projection; see Figure 2.8.

The linear projection on the eigenvectors corresponding to the two largest eigenvalues (λ_1 and λ_2) preserves the highest amount of variance present in the pattern space. Since pattern numbers are not included in the input data format, they are internally generated in order to identify individual data points.

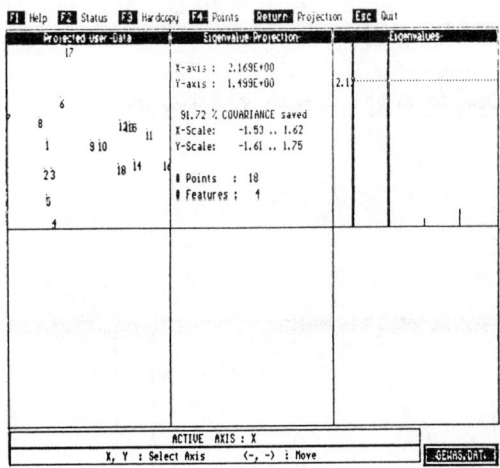

Figure 2.8: EDAPLUS example of eigenvector projection (upper screen).

Eigenvector projection (Figure 2.8) and discriminant projection (Figure 2.9) on standard printer output (hardcopy **F3**).

The pattern numbering is internally generated. The projection on the eigenvectors corresponding to the two largest eigenvalues aims to preserve a maximum of variance.

If pattern labels are available, then the discriminant projection aimins to preserve maximim separability between the groups corresponding to the labels.

Linear projections in EDAPLUS.

2.2.2 Discriminant analysis (Duda and Hart, 1973)

The discriminant analysis projection maximizes the *between-group scatter*, S_B, while holding the *within-group scatter*, S_W, constant. An important result of discriminant analysis demonstrates the existence of a $(K-1) \times d$ matrix, H_0, with a very interesting property. For each class, H_0 projects each pattern into a $(K-1)$-dimensional subspace by

$$\mathbf{y}_j = H_0\, \mathbf{x}_j \,, j = 1,2,\ldots,n_i \text{ and } i = 1,\ldots,K$$

where K is the number of categories.

Writing $S = S_W + S_B$, then the ratio of scatters $|S_W|/|S|$ remains constant. This ratio is called the *Wilks lambda statistic*. The rows of H_0 are the eigenvectors corresponding to the $K-1$ non-zero eigenvalues of $S_W^{-1}S_B$.

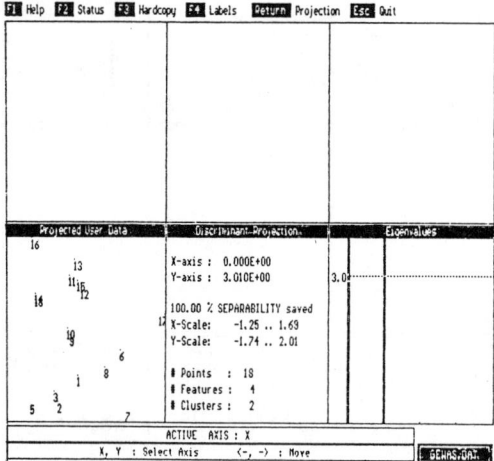

Figure 2.9: EDAPLUS example of discriminant projection (lower screen).

The linear discriminant projection on a two-dimensional plane is a projection on the plane formed by the two eigenvectors corresponding with the two largest eigenvalues of $S_W^{-1}S_B$.

EDAPLUS offers an environment to study the properties of the above projection; see Figure 2.9. Note that in the two-class case, only one non-zero eigenvalue exists.

From the above, it is clear that pattern labels (L) are required in order to compute S_W and S_B. From the definition of scatter, preserving maximum separability is in the squared error sense.

Projection on pairs of features (Devijver and Kittler, 1982)
If the dimensionality d is (much) larger than 2, one can always project the data on the plane formed by any pair of features (measurements). Sometimes, a priori information from the application domain may make such projection useful. However, in general, the loss of information may be dramatic and visual inspection may have no value. Generally, we have no a priori knowledge which pair of features is best to use as a projection plane.

If labels (L) are provided (partly), we are able to select a 'best' pair of features to project on. Fundamentals of and solutions for feature selection aiming at 'best' separability in terms of pattern classification are dealt with in depth by Devijver and Kittler (1982). The *branch and bound algorithm* (Fukunaga and Koontz, 1970) is one of the most popular solutions because it reaches the optimal feature subset in the sense of a given monotonic criterion function without complete enumeration of all possibilities.

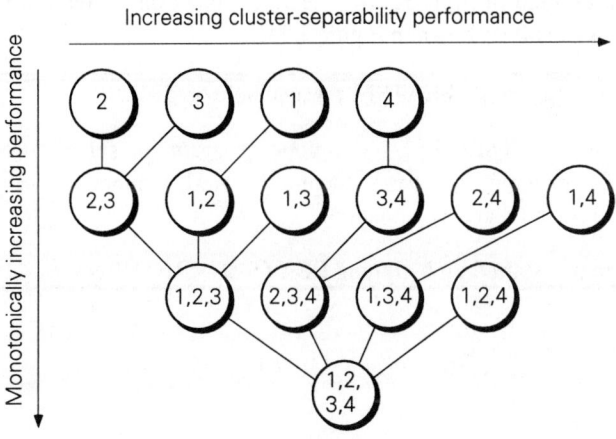

Figure 2.10: An example of an ordered complete search tree for feature subsets (GEWAS.DAT).

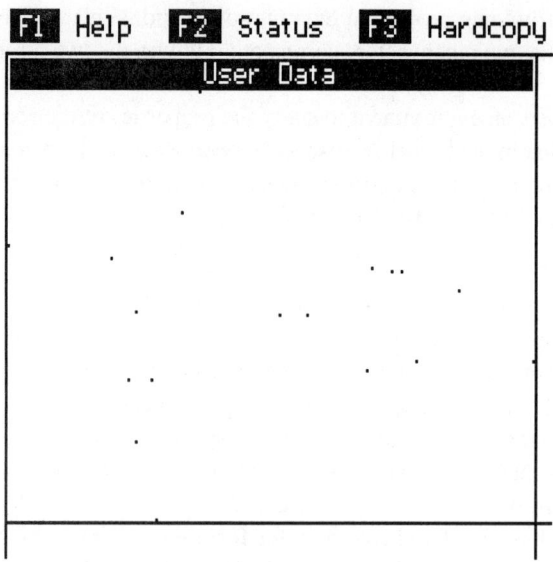

Figure 2.11: Projection of GEWAS.DAT on feature pair (1,4).

For the given data example (GEWAS.DAT) with dimension 4, the ordered complete search tree is presented in Figure 2.10. Clearly, projection on feature pair (1,4) is the 'best' solution. Its result is shown in Figure 2.11.

A stand-alone program, FEATSEL, is available on diskette.

The module FEATSEL is a stand-alone program for selection of feature subsets. The program includes sequential forward, sequential backward, and branch and bound feature evaluation.

The program is started up by typing **FEATSEL** <RETURN>.

The stand-alone program, FEATSEL.

2.3 Nonlinear projection

Most nonlinear projection algorithms are based on maximizing or minimizing a function of a large number of variables. This optimization problem is data-dependent and does not involve an explicit mapping function. Nonlinear projection algorithms are expensive to use, so several heuristics are employed to reduce the search time for the optimal solution.

Eigenvector projection preserves variance, while discriminant projection preserves class separability. Nonlinear projection may preserve any structural property implicitly defined by its optimization function. Classical projection methods are due to Sammon (1969), Fukunaga and Koontz (1970) and Lee et al. (1977). Here, it is assumed that the data are presented as a pattern matrix. If the analysis starts with a proximity matrix, multidimensional scaling is applicable.

2.3.1 Multi-dimensional scaling (Kruskal, 1964)

Multi-dimensional scaling is a generic name for a body of procedures and algorithms that start with an ordinal proximity matrix and generate configurations of points in one, two or three dimensions. The basic Kruskal algorithm reads as follows (Kruskal, 1964).

We begin with the $n \times n$ proximity matrix of dissimilarities $[d(i,j)]$ and search for an m-dimensional configuration of points $(\mathbf{x}_1, \mathbf{x}_2,...,\mathbf{x}_n)^T$, where

$$\mathbf{x}_i = [x_{i1}, x_{i2},...,x_{im}]^T$$

for which the rank order of the distances $[D(i,j)]$ in the configuration space 'matches' the proximities.

Let the measure of distance in the configuration space be the Euclidean distance, thus

$$D(i,j) = \left[\sum |x_{ik} - x_{jk}|^2 \right]^{\frac{1}{2}}$$

A perfect match occurs when the rank orders of the entries in the matrix $[D(i,j)]$ match those in the matrix $[d(i,j)]$. The degree to which the two sets of rank orders agree is measured by Kruskal's *stress*. The match between the dissimilarity matrix and the distance matrix is perfect if the $M = n(n-1)/2$ interpoint distances correspond monotonically with the M dissimilarity coeffecients: $f(d) = D$. The stress F is minimized in the least-squares sense, and is defined as

$$F = \min \left[\sum_i \sum_j [f(d(i,j)) - D(i,j)]^2 \right]^{\frac{1}{2}}$$

where s is a scale factor. The minimization of F is with respect to the best monotonic function $f(\bullet)$.

Figure 2.12 shows the result of multi-dimensional scaling. The method is known to be computationally complex and thus time-consuming. However, it produces results is very often quite reliable.

2.3.2 Sammon's algorithm (Sammon, 1969)

Sammon (1969) proposed a nonlinear technique that tries to create a two-dimensional configuration of points in which interpattern distances are preserved. Let $d(i,j)$ denote the distance between patterns $\mathbf{x}_i$ and $\mathbf{x}_j$ in the d-dimensional space. Let $D(i,j)$ be the distance between the points in the lower-dimensional space.

Similar to the stress criterion, Sammon defined the mean square error E as follows:

$$E = \frac{1}{\sum_{i<j} \sum d(i,j)} \sum_{i<j} \sum \frac{[d(i,j) - D(i,j)]^2}{d(i,j)} .$$

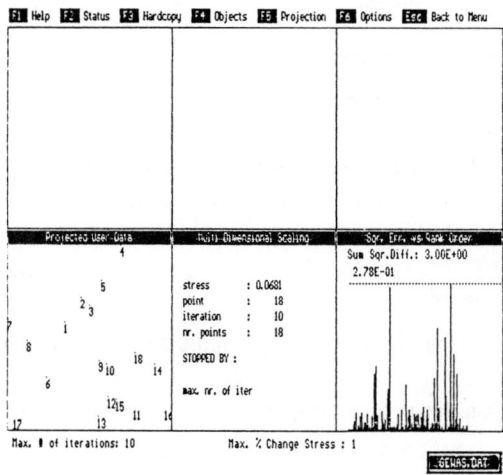

Figure 2.12: EDAPLUS example of multi-dimensional scaling (lower screen).

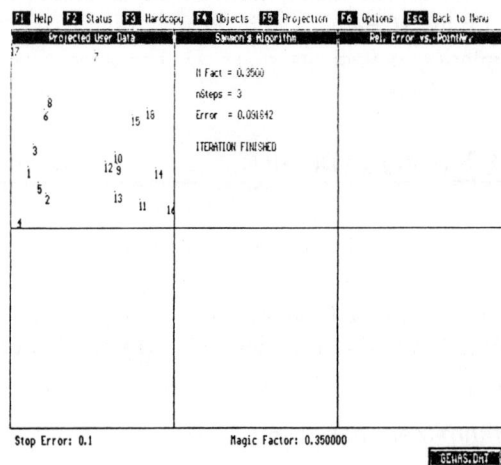

Figure 2.13: EDAPLUS example of Sammon nonlinear mapping (upper screen).

Multi-dimensional scaling (Figure 2.12) and Sammon's nonlinear mapping (Figure 2.13) on standard printer output (hardcopy **F3**).

The pattern numbering is internally generated. Together with Kruskal's stress, graphical output is generated to visualize the degree to which the rank orders of the entries $D(i,j)$ match those of $d(i,j)$. Both projections aim to preserve the interpattern distances.

Nonlinear projections in EDAPLUS.

Sammon's algorithm starts with a random configuration of n patterns in m dimensions and uses the method of steepest descent to reconfigure the patterns so as to minimize E in an iterative fashion.

The result of Sammon's algorithm for our experimental data is presented in Figure 2.13. The algorithm is less time-consuming. However, note that nonlinear iterative methods are time-consuming by definition for larger amounts of data.

2.3.3 *Frame method* (Chang and Lee, 1973)

The frame method, proposed by Chang and Lee (1973), defines a frame from a representative number n' of patterns and creates a two-dimensional configuration of the frame. The remaining $(n - n')$ patterns are then projected one by one, adjusting their distances only with respect to this fixed frame. The frame method achieves considerable savings in computation time and memory requirements. While Sammon's algorithm tries to preserve all $n(n - 1)/2$ interpattern distances, the frame method attempts to preserve only $[n'(n' - 1)/2 + n'(n - n')]$ interpattern distances.

2.3.4 *Triangulation method* (Lee et al., 1977)

While the above algorithms try to preserve the global structure of the d-dimensional patterns, the triangulation method (Lee et al., 1977), exactly preserves only local structure in the data. The patterns are represented in the plane sequentially in such a way that exactly $(2n - 3)$ of the $n(n - 1)/2$ possible interpattern distances are preserved. Whenever a point representing a pattern is placed in the plane, its distance to two patterns represented previously can be exactly preserved. The triangulation method preserves the $(n - 1)$ distances in the minimum spanning tree (MST) of the n d-dimensional patterns. That is, the MST of the patterns in the d-dimensional space is the same as the MST of the projected patterns in two dimensions. The triangulation method is much faster than Sammon's algorithm and the frame method.

> Neither the frame method nor the triangulation method are implemented in EDAPLUS. For research applications, this is unfortunate because the frame method is less time-consuming than Sammon's algorithm and the triangulation method preserves the MST distances which is often a favourable property.

Frame method and triangulation method.

Comparing the above projections (Figures 2.8, 2.9, 2.12 and 2.13) we observe common global data characteristics. However, local properties differ significantly.

Data inspection by using linear and nonlinear 2D mappings may guide us in gaining insight into the intrinsic dimensionality, the cluster shapes and the number of significant clusters, but any justification can hardly be given without further analysis.

2.3.5 Projection using simulated annealing (Klein and Dubes, 1989)

In their paper, Klein and Dubes (1989) examine whether *simulated annealing* provides a practical solution to the problem of mapping data onto a two-dimensional space to allow visual inspection of the projected patterns for structure.

Simulated annealing is a method of function optimization that tries to avoid the pitfalls inherent in the methods mentioned above; i.e. it seeks the global or near-global minimum of a function without becoming trapped in a local minimum.

Simulated annealing is one algorithm in the class of *stochastic relaxation* algorithms designed to optimize functions of several hundred variables or more, and is especially attractive when the functions are not smooth.

Simulated annealing is based on an analogous process in metallurgy in which a material is heated and cooled in sequence to create a strong, stable configuration.

The two main parts of simulated annealing are the *cooling schedule* and the definition of perturbation, or *move*. The parameters of the cooling schedule are starting temperature, the rule of decreasing temperature, the run length at each temperature, and stopping temperature. A move should select a next state that is close to the current state in the state space. The annealing algorithm models the minimization of a function of many variables as a *Markov chain* of many states. The Markov chain is simulated and allowed to run until it reaches steady state.

The *optimization problem* begins with a cost function of many variables, such as stress or square error, which has known analytical form but is otherwise unrestricted. The minimum corresponds to the most stable state of the underlying system of variables. Only state changes corresponding to decreases in cost are accepted using a standard algorithm, such as steepest descent. Simulated annealing also accepts such state changes but, in addition, accepts states which increase cost with a probability determined by a new parameter called temperature.

As stated before, the *mapping problem* creates a configuration of N points in two dimensions whose interpoint distances try to match the distances among the given patterns in a high dimensional space.

Klein and Dubes compared their simulated annealing mapping to the standard gradient descent mapping (Sammon) and to a non-iterative linear mapping based on eigenvalues of a covariance matrix (eigenvalue projection). The annealing algorithm produced good results, better in some cases than the other algorithms. However, the results – though sometimes more reliable – were produced at an almost prohibitive computational *cost*. At present, much research on speeding up cooling schedules is going on (parallellism) aiming at a simulated annealing approach that is feasible in practice.

2.4 Hierarchical clustering

A hierarchical clustering is a sequence of partitions in which each partition is nested into the next partition in the sequence (see Figure 2.14). An *agglomerative* algorithm for hierarchical clustering starts with the disjoint clustering, which places each of the n objects in an individual cluster. The cluster algorithm dictates how the proximity matrix should be interpreted to merge two or more of these trivial clusters. A divisive algorithm performs the task in the reverse order.

A picture of a hierarchical clustering is much easier for a human being to comprehend than is a list of abstract symbols. A *dendrogram* is a special type of tree structure that provides a convenient picture of a hierarchical clustering.

Here, we discuss algorithms for hierarchical clustering in terms of a scheme for updating the proximity matrix. This approach was suggested by King (1967) and popularized by Johnson (1967), who formalized the procedure.

The $n \times n$ proximity matrix is $D = [d(i,j)]$. The clusterings are assigned sequence numbers $0, 1,\ldots, (n-1)$ and $L(k)$ is the level of the kth clustering. A cluster with

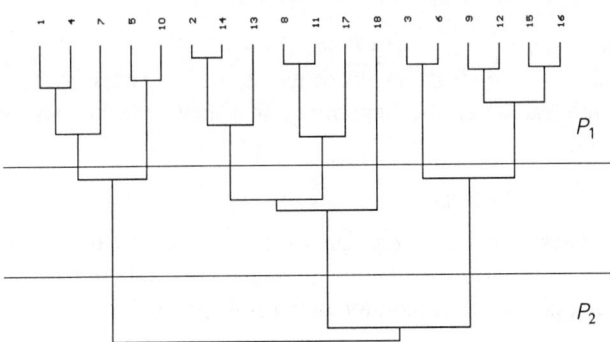

Figure 2.14: Hierarchical nesting of partitions: partition P_1 is nested in partition P_2.

sequence number m is denoted as (m) and the proximity between clusters (r) and (s) is denoted $d[(r),(s)]$.

Step 1. Begin with the disjoint clustering having level $L(0) = 0$ and sequence number $m = 0$.

Step 2. Find the least dissimilar pair of clusters in the current clustering, say pair $\{(r),(s)\}$, according to

$$d[(r),(s)] = \min \{d[(i),(j)]\}$$

where the minimum is over all pairs of clusters in the current clustering.

Step 3. Increment the sequence number: $m \leftarrow m + 1$. Merge clusters (r) and (s) into a single cluster to form the next clustering m. Set the level of this clustering to

$$L(m) = d[(r),(s)]$$

Step 4. Update the proximity matrix, D, by deleting the rows and columns corresponding to clusters (r) and (s) and adding a row and column corresponding to the newly formed cluster. The proximity between the new cluster, denoted (r,s), and the old cluster (k) is defined as follows (Lance and Williams, 1967).

$$d[(k),(r,s)] = \alpha_r d[(k),(r)] + \alpha_s d[(k),(s)] + \beta \, d[(r),(s)] + \gamma |d[(k),(r) - d[(k),(s)]|$$

Step 5. If all objects are in one cluster, stop, else go to step 2.

2.4.1 Single-link clustering

Single-link clusters are based on connectedness and are characterized by *minimum path length* among all pairs of objects in the cluster. The update of the proximity matrix is given by

$$d[(k),(r,s)] = \min \{d[(k),(r)],d[(k),(s)]\}$$

which corresponds to $\alpha_r = \frac{1}{2}$, $\alpha_s = \frac{1}{2}$, $\beta = 0$ and $\gamma = -\frac{1}{2}$.

Figure 2.15 shows a dendrogram result using single linkage. Note that automatic scaling may be misleading in the beginning. In Chapter 3, we shall encounter this problem in greater detail.

2.4.2 Complete-link clusering

Complete-link clusters are based on *complete subgraphs* where the diameter of a complete subgraph is the largest proximity among all proximities for pairs of objects in the subgraph. The update of the proximity matrix is given by

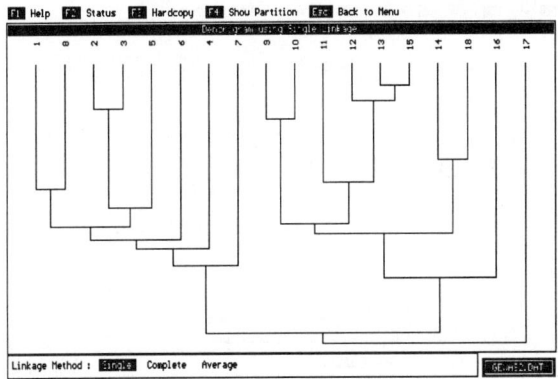

Figure 2.15: Dendrogram using single-linkage.

$$d[(k),(r,s)] = \max \{d[(k),(r)],d[(k),(s)]\}$$

which corresponds to $\alpha_r = \frac{1}{2}$, $\alpha_s = \frac{1}{2}$, $\beta = 0$ and $\gamma = \frac{1}{2}$. Figure 2.16 shows the result of complete linkage.

2.4.3 Average-link clustering

When measuring the dissimilarity between an existing cluster and a prospective cluster, the single-link method finds the closest pair of objects in the two clusters, the complete-link method finds the most distant pair, and the average-link method uses arithmetic averages of the dissimilarities. The update of the proximity matrix is given by

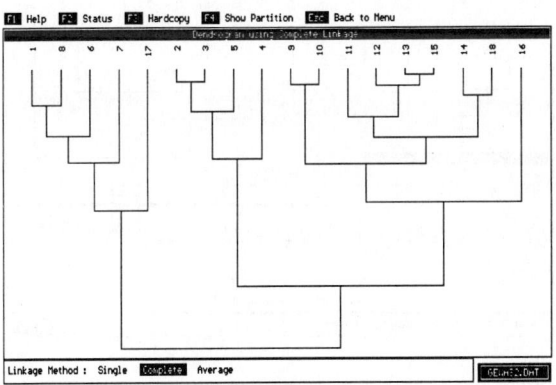

Figure 2.16: Dendrogram using complete linkage.

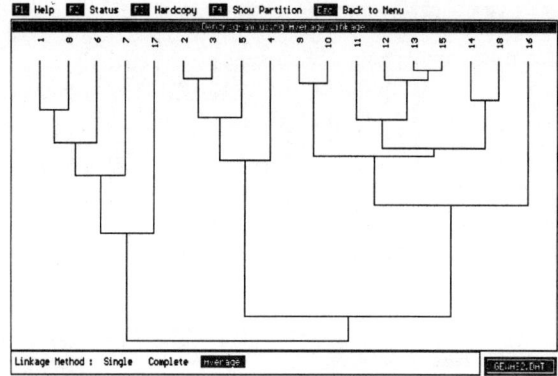

Figure 2.17: Dendrogram using average linkage.

$$\alpha_r = \frac{n_r}{n_r + n_s} \quad \alpha_s = \frac{n_s}{n_r + n_s} \quad \beta = 0 \quad \gamma = 0$$

where n_r is the number of objects in cluster (r) and n_s is the number of objects in cluster (s). The result of average linkage is shown in Figure 2.17.

If the data are two-dimensional or there exists a two-dimensional projection of the data under consideration, all dendrograms can be interpreted at any level of the dendrogram by inspecting the corresponding partition in the form cluster plot of the data. This is shown in Figure 2.18.

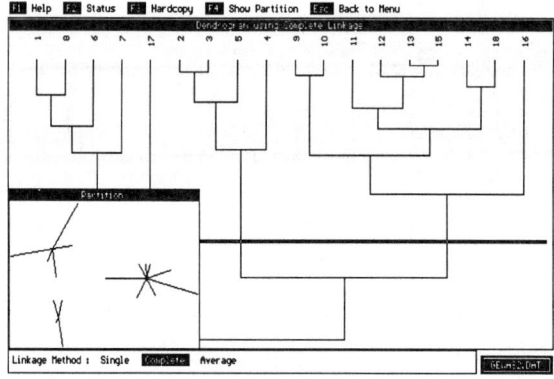

Figure 2.18: Dendrogram and generated partition corresponding to the cutting level as indicated.

2.5 Partitional clustering

The problem of partitional clustering can be formally stated as follows. Given n patterns in a d-dimensional metric space, determine a partition of the patterns into K groups, or clusters, such that the patterns in a cluster are more similar to each other than to patterns in other clusters. The value of K may or may not specified. A *clustering criterion*, such as *square error*, must be adopted.

2.5.1 Square-error clustering (Anderberg, 1973)

The most commonly used partitional clustering strategy is based on the square error criterion. The general objective is to obtain that partition which, for a fixed number of clusters, minimizes the square error.

Suppose that the given set of n patterns in d dimensions has somehow been partitioned into K clusters $\{C_1, C_2,..., C_K\}$ such that cluster C_K has n_K patterns and each pattern is in exactly one cluster, so that

$$\sum_{i=1}^{K} n_i = n$$

The mean vector, or centre, of cluster C_K is defined as the centroid of the cluster, or

$$\mathbf{m}^{(K)} = \frac{1}{n_K} \sum_{i=1}^{n_K} \mathbf{x}_i^{(K)}$$

where $\mathbf{x}_i^{(K)}$ is the ith pattern belonging to cluster C_K. The square error for cluster C_K is the sum of the squared Euclidean distances between each pattern in C_K and its cluster centre $\mathbf{m}^{(K)}$:

$$e_K^2 = \sum_{i=1}^{n_K} (\mathbf{x}_i^{(K)} - \mathbf{m}^{(K)})^{\mathrm{T}} (\mathbf{x}_j^{(K)} - \mathbf{m}^{(K)})$$

This square error is also called the *within-cluster variation*. The Mahalanobis distance can also be used to define square error.

The square error for the entire clustering containing K clusters is the sum of the within-cluster variation:

$$E_K^2 = \sum_{i=1}^{K} e_i^2$$

2.5.2 K-means clustering

The basic idea of an iterative clustering algorithm is to start with an initial partition and assign patterns to clusters so as to reduce the square error. The square error tends to decrease as the number of clusters increases and can be minimized only for a fixed number of clusters.

The process of the above iterative data grouping can be understood on the basis of the visualization of the data as given before. This visualization is shown in Figure 2.19. Note that the reallocations are controlled numerically by the above criterion function.

A general algorithm for iterative partitional clustering method is given below (Anderberg, 1973).

Step 1. Select an *initial partition* with *K* clusters. Repeat steps 2–5 until the cluster membership stabilizes.

Step 2. Generate a new partition by assigning each pattern to its closest cluster centre.

Step 3. Compute new cluster centres as the centroids of the clusters.

Step 4. Repeat steps 2 and 3 until an optimum value of the criterion function is found.

Step 5. Adjust the number of clusters by merging and splitting existing clusters or by removing small clusters, or outliers.

In the above algorithm some crucial parameters are involved. Some of them are briefly discussed.

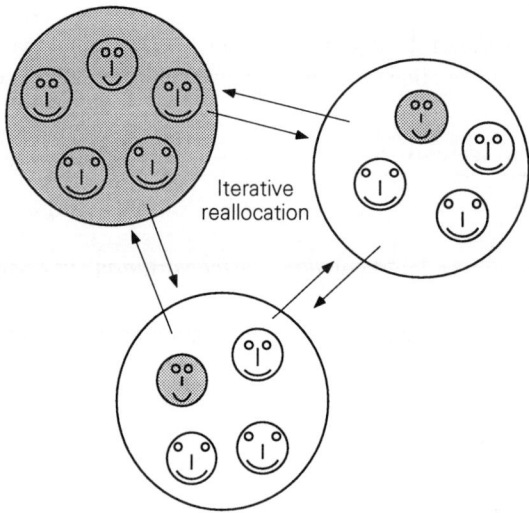

Figure 2.19: Visualization of iterative data grouping.

2.5.3 Crucial parameters

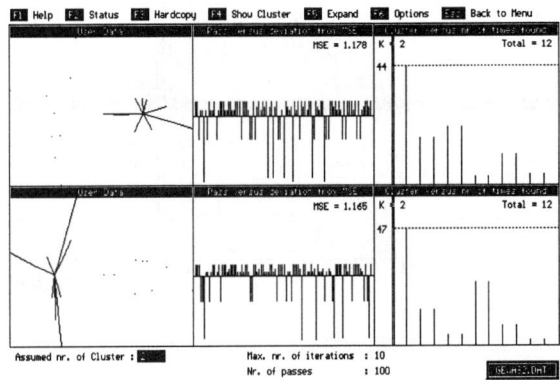

Figure 2.20: EDAPLUS example of partitional clustering for K = 2.

Initial partition

An initial partition can be formed by first specifying a set of K seed points (e.g. the first K patterns, or K patterns chosen randomly from the pattern matrix). Different initial partitions can lead to different final clusterings because algorithms based on square error can converge to *local minima*. One way to overcome local minima is to run the partitional algorithm with several different initial partitions. If they all lead to the same final partition, we have some confidence that the global minimum of square error has been achieved. This is shown in Figure 2.20. In the example given, there is no unique solution, though the majority of clusters detected indicate some structure even if this turned out to be weak in later analysis.

Updating the partition

Partitions are updated by reassigning patterns to clusters in an attempt to reduce the square error. The centre can be recomputed after each new assignment, or after all patterns have been examined (Forgy, 1965; McQueen, 1967).

The Euclidean metric is the most common metric for computing the distance between a pattern and a cluster centre, but the Mahalanobis distance is also used. However, the Mahalanobis distance requires computation of the inverse of the sample covariance matrix every time a pattern changes its cluster label.

Number of clusters

In the example given for $K = 2$, this number of clusters is not self-evident. At a later stage, we will discuss the proper choice of the number of clusters (K) in greater detail. In Figure 2.21, the above example is presented for $K = 3$. Clearly, the variability of the clusters detected increases with K if clustering is not very well structured.

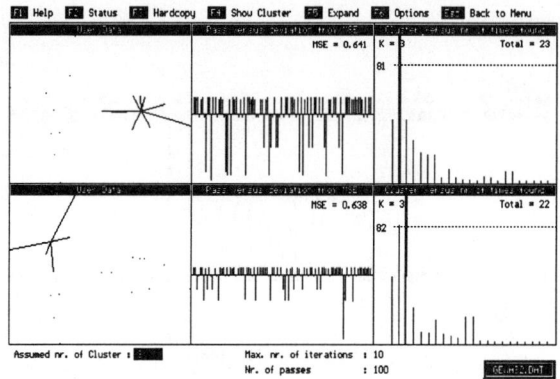

Figure 2.21: EDAPLUS example of partitional clustering for K = 3.

In one of the popular partitional clustering algorithms called ISODATA (Ball and Hall, 1964), adjustment of the number of clusters is foreseen if certain conditions are met. These conditions are specified by the user. A cluster is split if it has too many patterns and an unusually large variance along the feature with the largest spread. Two clusters are merged if their cluster centres are sufficiently close, again based on a parameter supplied by the user.

An *outlier* is a pattern that is sufficiently far removed from the rest of the data. Outliers can provide useful information (if they are not due to noise in the measurement process), but forcing an outlier to belong to a cluster distorts the shape of that cluster as well as its statistics. Thus it is best to identify an outlier and remove it from further consideration.

Convergence

Partitional algorithms terminate when the criterion function cannot be improved. There is no guarantee that an iterative algorithm will reach a global minimum. Some algorithms stop when the cluster labels for all the patterns do not change between two successive iterations. A maximum number of iterations can be specified to prevent endless oscillations. In practice, K-means type algorithms converge rapidly.

Computational complexity

The computational complexity of this algorithm is of the order $O(ndKT)$, where n is the number of patterns, d the number of features, K the number of clusters desired, and T the number of iterations.

For comparison, Figure 2.22 shows a well–separated three cluster structure, indicating that 66 out of 100 passes with random initialization yields the majority solution. Then, we assume that the global optimum has been achieved.

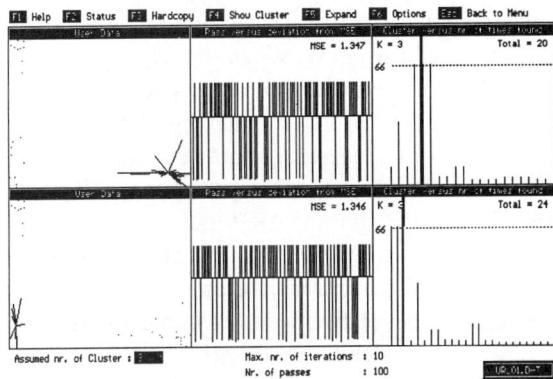

Figure 2.22: EDAPLUS example of a well-separated 3-clustering; the global optimum is likely to be achieved, though not guaranteed.

It is common practice in EDAPLUS to run the partitional algorithm 100 times with random initialization. Generally, as a result, different clusters are detected. The majority of detected – possibly slightly differing – clusters indicate the extent to which one particular clustering is superior with respect to others. The more pattern points do have questionable cluster assignments (under random starting configurations), the less convincing the majority of detected clusters will be. The effects of 100 passes under random starting configurations are shown in Figures 2.20, 2.21 and 2.22 for $K = 2$ and $K = 3$, respectively. Clearly, the URL01 data seem to be more structured than the GEWAS data. However, in both cases the number of pattern points is low to very low, and they are not necessarily generated from normal distributions.

Variability of detected clusters due to random starting configurations.

2.6 Minimum spanning tree

A graph is a mathematical structure that has a multitude of applications in cluster analysis. A graph G is the triplet $G = \langle V,E,f \rangle$ where V is the set nodes or vertices usually representing the objects or patterns, E is the set of edges representing the interaction between pairs of vertices, and f is a mapping of edges onto the product set $V \times V$. A graph is *connected* if a path exists between any two vertices in the graph. A graph is *complete* if an edge is assigned to every possible pair of nodes. A cycle is the same as a path except that vertices v_1 and v_n are the same vertex.

A *tree* is a connected graph with no cycles. If a subgraph has m vertices, it is easy to prove that a tree containing these vertices has exactly $m - 1$ edges. A *spanning* tree is a tree containing all vertices of the graph. When the edges in a graph are weighted by dissimilarities, the weight of a tree is the sum of the edge weights in the tree. Then, the *minimum spanning tree* (MST) of G is a tree having minimal weight among all other spanning trees of G. MSTs of complete graphs are especially important in cluster analysis. Prim's (1957) algorithm is generally taken to be the best computationally.

Note that by the above definition the MST is uniquely related to the single-link characteristic. Although the single-link hierarchy can be derived from the MST, the MST cannot be found from a single-link hierarchical clustering.

For present purposes, it is the relative positions of the points that are important; pairs of patterns in the same cluster should be closer than pairs of patterns belonging to different clusters. Among other graph structures, the MST reflects the 'structure' or the inherent separation among clusters. The edges correspond to small interpoint distances. Zahn (1971) demonstrated how the minimum spanning tree can be used to detect clusters.

Step 1. Construct the MST for the set of n patterns given.

Step 2. Identify inconsistent edges in the MST.

Step 3. Remove the *inconsistent edges* and call the connected components clusters.

The crucial step in the algorithm is the definition of *inconsistency*. Informally stated, an edge is said to be inconsistent if its interpoint distance is significantly larger than the average of nearby interpoint distances. Thus the inconsistent edges are related to cluster separation.

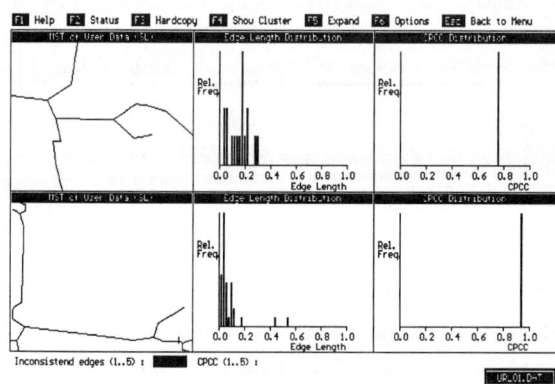

Figure 2.23: EDAPLUS examples of the minimum spanning tree; the MST of GEWAS.DAT in the upper screen; the MST of URL01.DAT in the lower screen.
Inconsistent edges can hardly be detected for GEWAS.DAT, though they are quite obvious in the case of URL01.DAT. The values of the cophenetic correlation coefficient (CPCC) are fairly high (see Section 2.7).

In Figure 2.23, the MST of the user data is displayed. As the display of the MST is on a two-dimensional plot of the data, we observe the meaning of global and local properties of a projection. Globally, previously observed projections resemble one another. However, locally they differ in the sense that Sammon's projection preserves interpoint distances globally whereas the MST projection shows preservation of the local structure.

> Note that the MST is determined by the $n \times n$ dissimilarity matrix only. Generally, the projection of the MST on a two-dimensional plot does not preserve MST characteristics. Only if the intrinsic (or true) dimensionality of the data is two, will the MST appear as a true MST in the projection plane.

Appearance of the MST in EDAPLUS.

2.7 Cluster validity

Validation of clustering results is an essential step that changes a qualitative analysis into hard evidence. Both external and internal indices are of importance. External indices compare the clustering results to what the investigator would like to see. Internal indices assess the merit of the clustering results on an objective basis. Validation often involves Monte Carlo analysis and statistical testing.

Indices
An index should make good intuitive sense, should have a basis in theory, and should be readily computable.

Baseline distribution
A baseline distribution is a null distribution derived from a population containing 'no structure'.

Baseline distributions form the heart of the assessment of the validity of suggested structure, which is assumed to be present in the user data. Figures 2.24 and 2.25 show the EDAPLUS representations of baseline distributions for hierarchical and partitional structures. In Chapter 6, we shall derive internal indices for tendency and validity based upon these baseline distributions. At this stage we discuss only the internal indices which are always taken into account in every analysis.

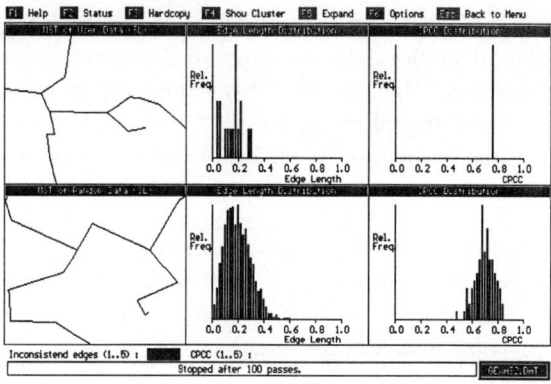

Figure 2.24: Baseline distribution for the edge length, and CPCC, given the sampling window of the user data, number of pattern points and dimensionality.

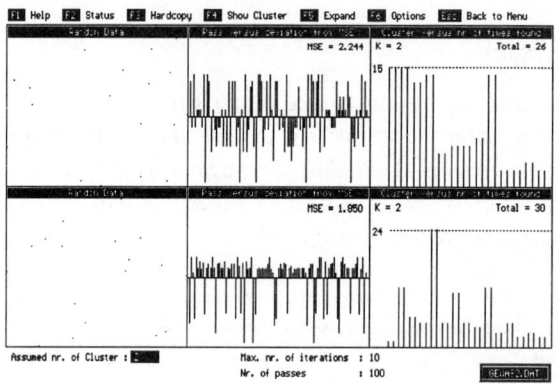

Figure 2.25: Baseline distribution for the occurring clusters, given the sampling window of the user data, number of pattern points and dimensionality.

In Figure 2.24, the characteristics of the MST of the user data (upper screen) are always compared with the baseline distributions of the random MST characteristics (within the same data window, and with the same number of pattern points and dimensionality) (lower screen). Clearly, no inconsistent edges are detected. Also, the CPCC value of the user data is not exceptionally high with respect to the CPCC baseline distribution (lower screen).

Figure 2.25 shows two partition experiments with random data. Clearly, no dominant clusters are likely to be found, even if the number of patterns is rather low.

Baseline distributions in EDAPLUS.

2.7.1 Hierarchical structures

The question is, should one be confident in the results of a hierarchical clustering, or not?

The cophenetic correlation coefficient (CPCC) has been proposed for quantitative data. The CPCC is the product-moment correlation coefficient between the entries of the input proximity matrix and the output *cophenetic matrix*. The CPCC is defined as follows:

$$\text{CPCC} = \frac{(1/M)\sum d(i,j)\, d_C(i,j) - m_D m_C}{[(1/M)\sum d^2(i,j) - m_D]^{1/2}\,[(1/M)\sum d_C^2(i,j) - m_C]^{1/2}}$$

where $m_D = (1/M)\sum d(i,j),$ (input matrix)

   $m_C = (1/M)\sum d_C(i,j),$ (output matrix)

and all summations are over the set $\{(i,j): 1 \le i \le j \le n\}$. The value of CPCC is between -1 and 1; the closer to 1, the better the match and the better the hierarchy fits the data.

The Davies–Bouldin (DB) index was originally proposed as a way of deciding when to stop clustering. The index is plotted against the number of clusters and clustering is stopped when the index is minimized. Given a partition of the n objects into K clusters, one first defines the following measure of within-to-between cluster spread for all pairs of clusters (j,k) (Davies and Bouldin, 1979):

$$R_{j,k} = \frac{e_j + e_k}{m_{j,k}}$$

where e_j is the average error for the jth cluster and $m_{j,k}$ is the Euclidean distance between the centres of the jth and kth clusters.

The index for the kth cluster is

$$R_k = \max_{j \ne k} \{R_{j,k}\}$$

and the DB index for the K-cluster clustering is

$$\text{DB}(K) = (1/K) \sum_{k=1}^{K} R_k \quad \text{for } K > 1$$

The smaller $\text{DB}(K)$, the better the clustering.

Figure 2.26 shows a dendrogram of the data to be considered, on the basis of which a nested sequence of partitions can be derived. For each of those partitions, the DB index is computed and plotted against the number of clusters corresponding to the nested sequence of partitions. A plot of the DB indices is shown in Figure 2.27. Its shape provides an indication the most likely number of clusters, and whether the clustering has to be considered to be as *strong*, *weak*, or *random*.

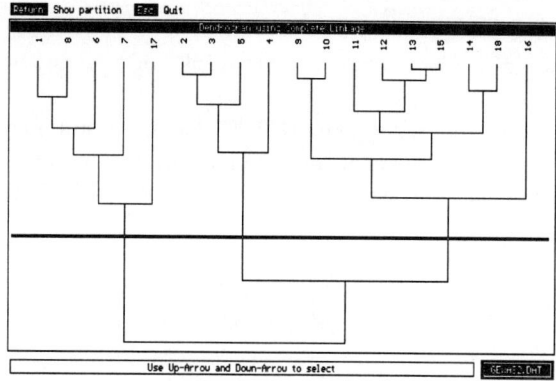

Figure 2.26: Example of a dendrogram generating a nested sequence of partitions in a K-decreasing order on the basis of which DB(K) is computed (see Figure 2.27).

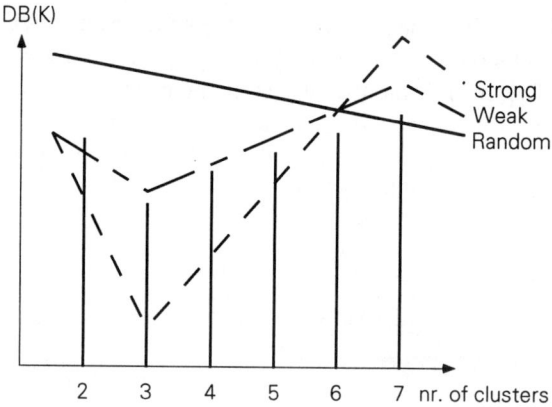

Figure 2.27: Plotting the Davies–Bouldin index gives an indication of clustering tendency.

EDAPLUS offers two options to compute the DB index.

Given a dendogram, for each level (as indicated) one may select from the menu <STATISTICS> the computed DB-index for the level as indicated; see Figure 2.28.

The second option is to determine automatically the most likely number of clusters present in the user data. Select the menu <HIERARCHICAL>. The DB index is determined for each partition in the nested sequence for the given dendogram. The number of clusters for which the minimum value of the DB index is achieved, is displayed; see Figure 2.29.

The DB index in EDAPLUS

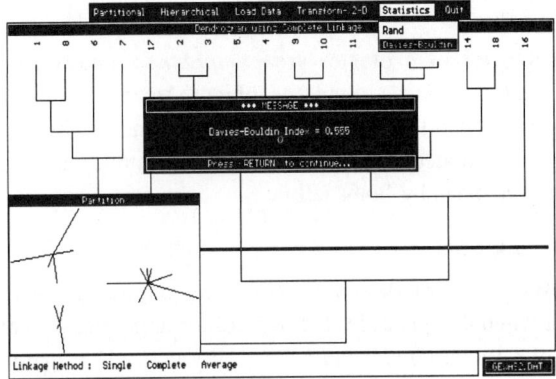

Figure 2.28: The DB index computed for the level as indicated.

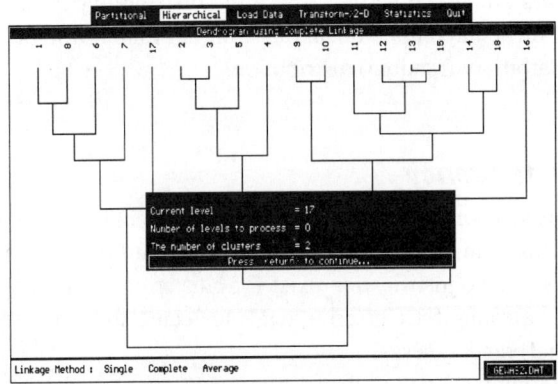

Figure 2.29: The most likely number of clusters present in the given data, determined by the DB index.

2.7.2 Partitional structures

Generally speaking, if the value of square error obtained is 'significantly' small, or small enough, the clustering is assumed to be 'valid'. This turns out to be a difficult approach, mainly because of the definition of the baseline distribution and the dependence of the problem parameters (number of patterns, features and clusters).

One could simply generate data randomly over some *sampling window*, but what sampling window is appropriate? Only careful comparison of clustering results with baseline distributions on the basis of the same number of patterns, the same dimensionality and the same number of clusters within the same sampling window may lead to reliable solutions.

2.7.3 Individual structures

The two main properties of a cluster are *compactness* and *isolation*. Compactness measures the internal cohesion among the objects in the cluster, whereas isolation measures separation between the cluster and other patterns. A valid cluster is unusually compact and unusually isolated. The problem is to obtain reference distributions so that we can specify what we mean by 'unusually'.

2.7.4 Monte Carlo analysis

Monte Carlo analysis is a method for estimating parameters and probabilities by computer sampling when the quantities are difficult or impossible to calculate directly.
The distributions of various *cluster indices* validity depend on many problem-specific parameters and can be estimated only by Monte Carlo sampling.

Monte Carlo analysis can approximate an unknown distribution if an experimental sampling procedure can be programmed on a computer that simulates the process being studied. This occurs when estimating a baseline distribution of an index for cluster validity. The most difficult computational question in Monte Carlo analysis is arranging to sample from an arbitrary (random) distribution.

2.8 Clustering tendency

The term *clustering tendency* refers to the problem of deciding whether data exhibit a predisposition to cluster into natural groups without identifying the groups themselves. In other words: does some justification exist for clustering or are the data random? As such, the problem of testing for clustering tendency can also be phrased as the problem of testing for (spatial) randomness.

A test for clustering tendency is stated in terms of an internal criterion. No category or other a priori information is brought into the analysis. The definition of randomness and the type of clustering tendency test depend on the form of the available data. A good example is the following.

MST-based test of clustering tendency (Smith and Jain, 1984)
Smith and Jain (1984) proposed an MST-based test of the random position hypothesis, H, which tests wether two sets of high-dimensional patterns arise from the same population. The test is as follows:
Step 1. Determine the convex region containing the n patterns $\{x_i\}$ being tested for clustering tendency.
Step 2. Generate m points $\{y_j\}$ uniformly over the convex region found in step 1.
Step 3. Pool $\{x_i\}$ and $\{y_j\}$ and find the MST of the $m+n$ points.
Step 4. Determine T, the number of x-y joins in the MST.
Step 5. Reject H_0 in favour of a clustered alternative if T is 'small'. Reject H_0 in favour of a regular alternative if T is 'large'.

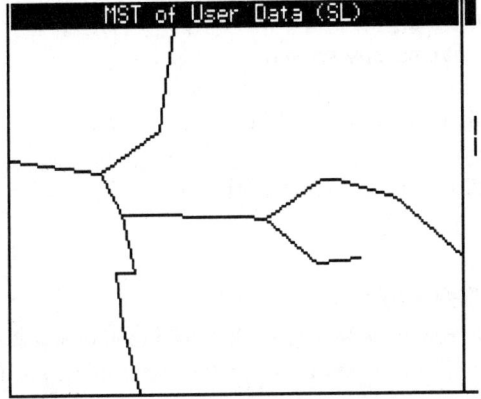

(a) Minimum spanning tree of the user data X; n = 18

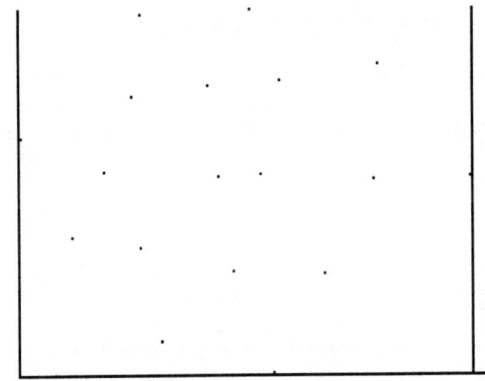

(b) Randomly generated data Y; m = 18

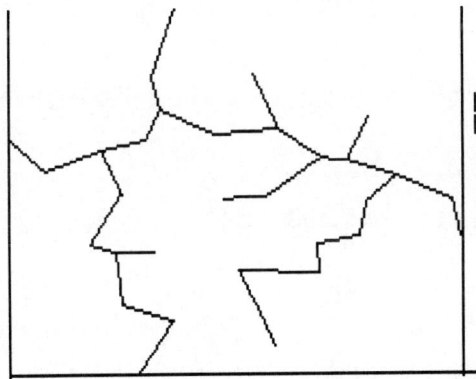

(c) Minimum spanning tree of the pooled data X + Y; m + n = 36

Figure 2.30: MST-based statistic for clustering tendency;
H_0 *is rejected in favour of a regular alternative because the number of x - y joins (T) is 'large'.*

An x-y join links a pattern in $\{\mathbf{x}_i\}$ to a generated point $\{\mathbf{y}_j\}$ by an edge in the MST of the pooled sample $\{\mathbf{x}_i,\mathbf{y}_j\}$. We usually set m to n.

Clustered data should show a higher number of x-x and y-y joins, and thus a lower number of x-y joins, than random data. The number of x-y joins under regularity should be larger than under randomness.

The above test is illustrated in Figure 2.30.

2.9 Concluding remarks

In this chapter, we have reviewed some of the basic methods and algorithms. The main objective was to provide the basic tools on the basis of which later analysis can be performed, and more specifically, useful measures for clustering tendency and validity can be derived. These measures will show their importance in computer-assisted (automatic) reasoning which will be discussed in later chapters.

Citations to be remembered

Yet despite successes, important methodological problems remain unsolved and limit the effectiveness of clustering techniques. For example, when a clustering method's input is a matrix of proximities between pairs of objects, useful information may have been lost by this transformation of the original data: might one not represent proximity relationships between objects more informatively than by a single proximity measure? And, when most clustering algorithms operate on data, they cluster objects irrespective of whether the data exhibit natural clusters: there is a need for further research about the relevance of clusters (Day, 1990).

The bewildering variety of clustering methods reflects in part the diversity of problem domains and in part the lack of rigorous basis for the problem: clustering is most applied in problems where no clear mathematical formulation exists. Indeed many algorithms require no more than the existence of a set of data points in some suitable space, although some require knowledge of the number of classes. While many of these methods can be applied to a given problem, there is in general no guarantee that any two will produce consistent answers, leading to the vexed issue of 'cluster validity'. Indeed one rule of thumb for establishing the validity of clusters produced by one method is to compare them with those produced by a second method. This is clearly an unsatisfactory state of affairs (Wilson and Spann, 1990).

A cluster can be visualized as a collection of patterns which are close to one another or which satisfy some spatial relationship. The task of a clustering algorithm is to identify such natural groupings in spaces of many dimensions, one must be careful not to think automatically of clustering problems as two- or three-dimensional. The real benefit of cluster analysis is to organize multidimensional data where visual perception fails (Jain and Dubes, 1988).

Practically all proposed clustering algorithms would perform well when presented with well-separated, compact groups. For elongated clusters, the minimum variance method would perhaps cut the clusters in two in its search for compactness. The single link method – with its chaining effect – or MST would be ideally suited instead. In the case of linked, globular clusters an estimate may be made of the density in the region of each point (e.g. the number of points within a specific radius: this point will be taken in mode analysis). ... For touching globular groups, the MST may not be of direct use. The minimum variance method should however find the clusters, and mode analysis may also be profitably applied here. Finally, in the cases of concentric groups, or groups characterized by differing densities, the MST may be used (Murtagh, 1985).

3 | Examples and discussion

Further, we believe that it is particular important in exploratory data analysis situations to fit the data without distorting the data, and our methodology eschews all preprocessing of the data by, for example, normalization, substitutions for missing point values, or elimination of outliers (Matthews and Hearne, 1991).

The purpose of this chapter is threefold. Firstly, we consider the very aim of any hierarchical cluster method. This is to impose algorithmically the ultrametric inequality (transitivity). This has to be considered in depth in order to understand of what hierarchical data grouping is all about. Secondly, we touch upon the principle of normalization, which is particularly important if the measurements have different scales. Finally, we discuss a number of problems which are associated with hierarchical and partitional clustering. The chapter ends with practical guidelines for real applications.

3.1 The transitivity property

Let d be a *cophenetic proximity measure*. Then, the *equivalence relation $R_c(a)$* on a set of objects is given by

$$R_c(a) = \{(\mathbf{x}_i, \mathbf{x}_j) : d(i,j) < a\}$$

The relation $R_c(a)$ can be shown to be an equivalence relation for any $a \geq 0$ by checking the three conditions necessary for an equivalence relation.

Since $d_c(i,i) = 0$ for all i

$$(\mathbf{x}_i, \mathbf{x}_i) \in R_c(a) \text{ for all } a \geq 0$$

so $R_c(a)$ is *reflexive*.

Since $d_c(i,j) = d_c(j,i)$ for all (i,j)

$$(\mathbf{x}_j, \mathbf{x}_i) \in R_c(a) \text{ if } (\mathbf{x}_i, \mathbf{x}_j) \in R_c(a)$$

for all $a \geq 0$. So $R_c(a)$ is *symmetric*.

The final condition, *transitivity*, requires that for all $a \geq 0$,

$$\text{if } (\mathbf{x}_i, \mathbf{x}_j) \in R_c(a) \text{ and if } (\mathbf{x}_k, \mathbf{x}_j) \in R_c(a)$$

$$\text{then } (\mathbf{x}_i, \mathbf{x}_j) \in R_c(a) \text{ for all } (i,j,k)$$

This can be rewritten as

$$d_c(i,j) \leq \max\ [d_c(i,k), d_c(k,j)]$$

for all (i,j,k). This requirement is called the *ultrametric inequality*. The very aim of any hierarchical cluster method is to impose the ultrametric inequality algorithmically.

Note that this requirement is stronger than the triangle inequality

$$d_t(i,j) \leq d_t(i,k) + d_t(k,j) \text{ for all } (i,j,k)$$

Consider the following distance matrix

$$[d] = \begin{bmatrix} 0 & 0.7 & 0.3 & 0.8 \\ & 0 & 0.5 & 0.1 \\ & & 0 & 0.9 \\ & & & 0 \end{bmatrix}$$

which is reflexive and symmetric. The transitivity property is *not* satisfied.

The SL algorithm reads as follows:

Step 1. $k = 0, L = 0$

Step 2. $\min\limits_{i,j}\ [d(\mathbf{x}_2,\mathbf{x}_4)] = d(\mathbf{x}_2,\mathbf{x}_4) = 0.1$

Step 3. $\{\mathbf{x}_2,\mathbf{x}_4\}$ are fused to form a cluster in C_1; $L_1 = 0.1$

Step 4. $\tilde{d}_1(\{\mathbf{x}_2,\mathbf{x}_4\},\mathbf{x}_1) = \min\ [d(\mathbf{x}_2,\mathbf{x}_1),d(\mathbf{x}_4,\mathbf{x}_1)] = 0.7$

$\tilde{d}_1(\{\mathbf{x}_2,\mathbf{x}_4\},\mathbf{x}_3) = \min\ [d(\mathbf{x}_2,\mathbf{x}_3),d(\mathbf{x}_4,\mathbf{x}_3)] = 0.5$

$\tilde{d}_1(\mathbf{x}_1,\mathbf{x}_3) = d(\mathbf{x}_1,\mathbf{x}_3) = 0.3$

Step 5. $k = k + 1$

Step 6. $\min\ [\tilde{d}_1(\mathbf{x}_1,\mathbf{x}_3),\tilde{d}_1(\{\mathbf{x}_2,\mathbf{x}_4\},\mathbf{x}_1),\tilde{d}_1(\{\mathbf{x}_2,\mathbf{x}_4\},\mathbf{x}_3)]$

$= d(\mathbf{x}_1,\mathbf{x}_3) = 0.3$

Step 7. $\{\mathbf{x}_1,\mathbf{x}_3\}$ are fused to form a cluster in C_2; $L_2 = 0.3$

Step 8. $\tilde{d}_2(\{\mathbf{x}_1,\mathbf{x}_3\},\{\mathbf{x}_2,\mathbf{x}_4\})$

$= \min\ [\tilde{d}_1(\mathbf{x}_1,\{\mathbf{x}_2,\mathbf{x}_4\}),\tilde{d}_1(\mathbf{x}_3,\{\mathbf{x}_2,\mathbf{x}_4\})] = 0.5$

Step 9. $k = k + 1 = 2$

Step 10. Clusters $\{\mathbf{x}_2,\mathbf{x}_4\}$ and $\{\mathbf{x}_1,\mathbf{x}_3\}$ are fused to form a cluster in C_3; $L_3 = 0.5$.

As a result, the cophenetic matrix yields

$$[d_c]_{SL} = \begin{bmatrix} 0 & 0.5 & 0.3 & 0.5 \\ & 0 & 0.5 & 0.1 \\ & & 0 & 0.5 \\ & & & 0 \end{bmatrix}$$

which satisfies the ultrametric property. Accordingly, we obtain the k-dendrogram and the L-dendrogram as depicted in Figure 3.1 with k being the step index and L being the level where fusing takes place.

Likewise, for CL (min operator replaced by max), we find:

Step 2. $d(\mathbf{x}_2,\mathbf{x}_4) = 0.1$

Step 3. $\{\mathbf{x}_2,\mathbf{x}_3\}; L_1 = 0.1$

Step 4. $\tilde{d}_1(\{\mathbf{x}_2,\mathbf{x}_4\},\mathbf{x}_1) = \max[d(\mathbf{x}_2,\mathbf{x}_1),d(\mathbf{x}_4,\mathbf{x}_1)] = 0.8$

$\tilde{d}_1(\{\mathbf{x}_2,\mathbf{x}_4\},\mathbf{x}_3) = \min[d(\mathbf{x}_2,\mathbf{x}_3),d(\mathbf{x}_4,\mathbf{x}_3)] = 0.9$

$\tilde{d}_1(\mathbf{x}_1,\mathbf{x}_3) = d(\mathbf{x}_1,\mathbf{x}_3) = 0.3$

Step 6. $\min[\tilde{d}_1(\mathbf{x}_1,\mathbf{x}_3),\tilde{d}_1(\{\mathbf{x}_2,\mathbf{x}_4\},\mathbf{x}_1),\tilde{d}_1(\{\mathbf{x}_2,\mathbf{x}_4\},\mathbf{x}_3)] = 0.3$

Step 7. $\{\mathbf{x}_1,\mathbf{x}_3\}; L_2 = 0.3$

Step 8. $\tilde{d}_2(\{\mathbf{x}_1,\mathbf{x}_3\},\{\mathbf{x}_2,\mathbf{x}_4\}) = \max[\tilde{d}_1,\tilde{d}_1] = 0.9; L_3 = 0.9$

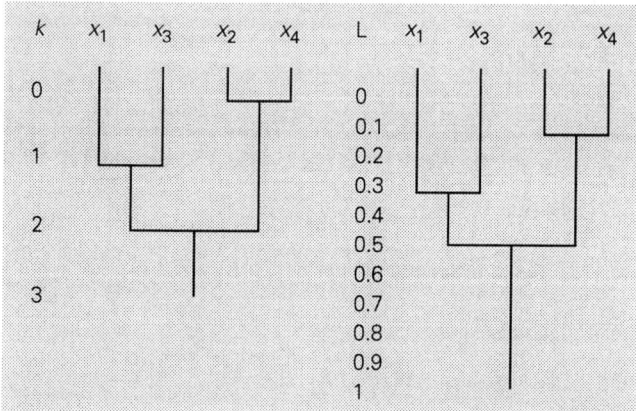

Figure 3.1: The k-dendrogram and L-dendrogram under single linkage.

This yields

$$[d_c] = \begin{bmatrix} 0 & 0.9 & 0.3 & 0.9 \\ & 0 & 0.9 & 0.1 \\ & & 0 & 0.9 \\ & & & 0 \end{bmatrix}$$

and we obtain the dendrograms shown in Figure 3.2.

As a result, we notice that we have been imposing on the input distance matrix

$$[d] = \begin{bmatrix} 0 & 0.7 & 0.3 & 0.8 \\ & 0 & 0.5 & 0.1 \\ & & 0 & 0.9 \\ & & & 0 \end{bmatrix}$$

the transitivity property, under SL:

$$[d_c]_{SL} = \begin{bmatrix} 0 & 0.5 & 0.3 & 0.5 \\ & 0 & 0.5 & 0.1 \\ & & 0 & 0.5 \\ & & & 0 \end{bmatrix}$$

and under CL:

$$[d_c]_{CL} = \begin{bmatrix} 0 & 0.9 & 0.3 & 0.9 \\ & 0 & 0.9 & 0.1 \\ & & 0 & 0.9 \\ & & & 0 \end{bmatrix}$$

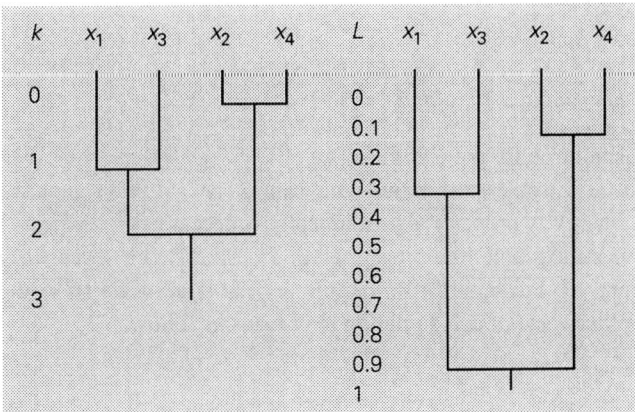

Figure 3.2: The k-dendrogram and L-dendrogram under complete linkage.

Then the question arises which result represents the original input data best.

Let us consider the following input-output matching coefficient

$$\rho(d,d_c) = \sum_{i=1}^{n-1} \sum_{j=i+1}^{n} |d(\mathbf{x}_i,\mathbf{x}_j) - d_c(\mathbf{x}_i,\mathbf{x}_j)|$$

as a measure of how well the cophenetic matrix represents the original input data. In the example above, we find

$$\rho_{SL} = 0.9 \quad \text{and} \quad \rho_{CL} = 0.7$$

from which we conclude that the CL solution represents the input data structure best. The cophenetic correlation coefficient (CPCC) is a commonly used matching coefficient.

Note that, if we impose the triangle inequality the 'closest' distance matrix is

$$[d_t] = \begin{bmatrix} 0 & 0.7 & 0.3 & 0.5 \\ & 0 & 0.5 & 0.1 \\ & & 0 & (0.6) \\ & & & 0 \end{bmatrix}$$

for which $\rho(d,d_t) = 0.3$, whereas, if we impose the ultrametric inequality, the 'closest' result is

$$[d_c] = \begin{bmatrix} 0 & (0.5) & 0.3 & (0.5) \\ & 0 & 0.5 & 0.1 \\ & & 0 & (0.5) \\ & & & 0 \end{bmatrix}$$

with $\rho = (d_t,d_c) = 0.6$.

Under SL, we find

$$\rho(d,d_c) = \rho(d,d_t) + \rho(d_t,d_c) = 0.9$$

This property is not true for the CPCC.

The transitivity property of input data cannot be assumed generally. Therefore, imposing the transitivity property algorithmically is neccessary.

Consider, as an example, the binary relation on the Dutch soccer league {Ajax (a), PSV (b), Utrecht (c), Feyenoord (d), Twente (e)}. If Ajax beats PSV, then the relation takes value 1, otherwise 0. Such a relation may reads as follows:

		a	b	c	d	e	no. of ones
Ajax	a	–	1	0	1	1	3
PSV	b	0	–	0	0	1	1
Utrecht	c	1	1	–	0	1	3
Feyenoord	d	0	1	1	–	1	3
Twente	e	0	0	0	0	–	0

It can easily be seen that this relation does not satisfy the transitivity property. However, if the same matrix were

		a	b	c	d	e	no. of ones
Ajax	a	–	1	0	0	1	2
PSV	b	0	–	0	0	0	0
Utrecht	c	1	1	–	0	1	3
Feyenoord	d	1	1	1	–	1	4
Twente	e	0	1	0	0	–	1

then the transitivity property would be satisfied. A carefull look at the matrix shows that the transitivity property is only satisfied if the number of 1's in the rows is distributed as 0,1,2,3,4 in any order. Consequently, the probability of satisfying the transitivity property yields

$$P(\text{trans}) = \frac{n!}{2^{n(n-1)}}$$

3.2 Normalization of data

Raw data, or the actual measurements, are not always used just as they are recorded. Some form of normalization may be used, particularly if measurements having different ranges of scale should be weighted equally in the measure of proximity used. However, even if this seems logical, within the context of clustering, the normalization changes the interpoint distances and consequently may influence natural clustering.

Consider the following input data: consisting of eight pattern points in a two-dimensional space by the transposed input pattern matrix is denoted X^*:

$$X^{*T} = \begin{bmatrix} 1 & 2 & 3 & 2 & 1 & 2 & 3 & 4 \\ 1 & 1 & 2 & 2 & 2 & 3 & 3 & 4 \end{bmatrix}$$

The measurement means are $m_1 = 9/4$ and $m_2 = 9/4$. If we normalize using the formula $x_{ij} = x_{ij}^* - m_j$, we obtain

$$X^T = \frac{1}{4} \begin{bmatrix} -5 & -1 & 3 & -1 & -5 & -1 & 3 & 7 \\ -5 & -5 & -1 & -1 & -1 & 3 & 3 & 7 \end{bmatrix}$$

Now, the covariance matrix yields

$$R = \frac{1}{8} X \cdot X^T = \frac{1}{16} \begin{bmatrix} 15 & 11 \\ 11 & 15 \end{bmatrix}$$

of which the eigenvalues are determined by

$$\begin{vmatrix} \frac{15}{16} - \lambda & \frac{11}{16} \\ \frac{11}{16} & \frac{15}{16} - \lambda \end{vmatrix} = 0$$

so that $32\lambda^2 - 60\lambda + 13 = 0$ from which it follows that the eigenvalues are $\lambda_1 = \frac{13}{8}$ and $\lambda_2 = \frac{1}{4}$. The rotational transformation is then given by

$$\begin{bmatrix} y_{i1} \\ y_{i2} \end{bmatrix} = \frac{1}{\sqrt{2}} \cdot \begin{bmatrix} 1 & 1 \\ -1 & 1 \end{bmatrix} \begin{bmatrix} x_{i1} \\ x_{i2} \end{bmatrix} \quad i = 1, 2, \ldots, 8$$

Thus we find

$$Y^T = \frac{1}{2\sqrt{2}} \begin{bmatrix} -5 & -3 & 1 & -1 & -3 & 1 & 3 & 7 \\ 0 & -2 & -2 & 0 & 2 & 2 & 0 & 0 \end{bmatrix}$$

The result of this transformation can be seen in Figure 3.3. The data line up with the largest eigenvalue.

If we perform the normalization

$$y_{ij}' = (y_{ij} - m_j)/S_j$$

with $m_1 = m_2 = 0$ and $S_1 = \frac{13}{8}$ and $S_2 = \frac{1}{4}$, we obtain the rotated normalized data (with $q = 1/\sqrt{13}$)

$$y'^T = \begin{bmatrix} -5/q & -3/q & 1/q & -1/q & -3/q & 1/q & 3/q & 7/q \\ 0 & -\sqrt{2} & -\sqrt{2} & 0 & \sqrt{2} & \sqrt{2} & 0 & 0 \end{bmatrix}$$

of which the resulting plot is given in Figure 3.4.

> It should be noticed that the user/random data are always normalized with respect to the mean and the variance. To enable standard plotting of the data, the range of scale is [0,1] in the x and y direction. The data window is determined by the max(x) and max(y) values. Random data are always generated within the user data window.

Normalization of data in EDAPLUS.

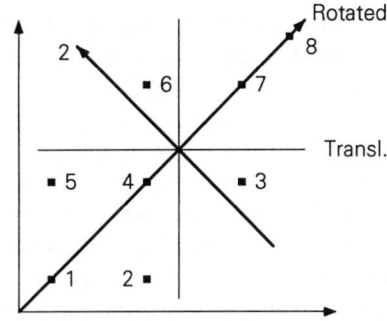

Figure 3.3: Rotational transformation based upon the eigenvectors corresponding to the eigenvalues λ_1 and λ_2.

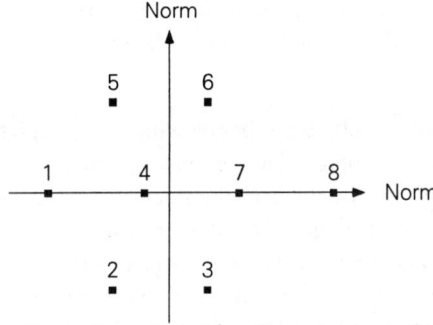

Figure 3.4: Normalized representation of the data of Figure 3.3.

3.3 Discussion of problems in cluster analysis

In the previous chapter many techniques for clustering were described. Some of the problems associated with those techniques will be discussed in the following sections. Most of them are general in the sense that they apply to the majority of techniques discussed.

The definition of a cluster

There is no universal agreement on what constitutes a cluster. In statistical terms: a cluster is a group of contiguous elements of a statistical population. However, informal statements (alike, similar, small interpoint distance, compact, homogeneous) are used and illustrate the vague, subjective nature of its meaning. It is always important (and necessary) to have an intuitive understanding of what constitutes a cluster in the domain of application and of the very goal of the analysis.

Natural clusters are very often seen as spherical clouds in the pattern space. In fact, all kinds of geometrical shapes do occur in practice. There is no technique available which is capable of finding all possible shapes of clusters. Most techniques are merely of some use in finding a particular shape.

The choice of measurements

In general, the number of variables to be measured on each object is large. Given the curse of dimensionality, it is always an issue of great importance to reduce this number of measurements. Reduction and selection of variables (measurements, features) is a well-known problem in pattern recognition where true class labels are known in advance in the training set.

In cluster analysis, this mostly turns out to be a circular problem. We need to have the clusters to be able to judge the effectiviness of the measurements of the objects to be clustered. What is left is either a priori knowledge in the domain of application (and intuition), or some sort of principal components analysis (eigenvalue analysis).

The problem of standardization scaling (normalization) has and been treated already.

Similarity and distance

The majority of clustering techniques begin with the calculation of a similarity or distance matrix. Selecting an appropriate measure and deciding wether variables should be weighted or not, can only be done in the context of the application domain. The number of possible measures is large. Each of them stresses particular details of the data. Generally, the choice should be made by the domain expert.

Even if the Euclidean distance is the distance measure most commonly used, there are many other distance measures available. In particular, if variables may be correlated, the weighted Euclidean distance or Mahalanobis distance is of great value,

$$d_M(\mathbf{x}_i, \mathbf{x}_j) = (\mathbf{x}_i - \mathbf{x}_j)R^{-1}(\mathbf{x}_i - \mathbf{x}_j)$$

where R is the pooled covariance matrix and $\mathbf{x}_i$ and $\mathbf{x}_j$ the vectors of objects i and j.

The number of clusters

A problem common to all clustering techniques is the difficulty of deciding the number of clusters present in the data. For the mean square error techniques, a plot of some square-error criterion against the number of clusters has been suggested. A sharp (in/de)crease of the value of the criterion could indicate the correct number of clusters. This procedure has in general been found to be unsatisfactory.

It has already been indicated that repeated Monte Carlo experimentation is needed to guide the investigator in deciding the 'optimal' number of clusters present in the data. However, it has also been clarified that properly windowing the random data for obtaining the baseline distributions is still a delicate problem.

Also, for hierarchical cluster methods no clear indicator is available for cutting the dendrogram 'optimally'. In many cases, a pre-indicator (like the Davies–Bouldin index) may guide the investigator, though it may suffer from outlier sensitivity. Nontheless, such an index may help to frame further experimentation.

Hierarchical techniques

The general problem here is the somewhat arbitrary choice, both of the method and of the measure of association. The only post-justification can be obtained by comparing the input data representation and the output cophenetic representation. The closer this match, the more appropriate the choise of method.

Several hierarchical clustering techniques such as the SL method or median method give rise to a property called *chaining*, which refers to the tendency of the method to cluster together at a relatively low level objects linked by chains of intermediates. This property, often seen as a defect, is simply a description of what the method does. As such, the method is appropriate if one is looking for 'optimally' connected clusters rather than for homogeneous spherical clusters.

The sensitivity of some techniques to outliers may an occassion be regarded as a drawback. In other instances, this sensitivity amy be useful, especially if one needs to be able to detect such outliers. From the investigator's point of view, there is a lot of value in knowing which object is to be considered as an outlier and why.

Optimization techniques

The problem of yielding a local optimum rather than the global optimum is the most common drawback of the optimization techniques (e.g. K-means). As it is impossible to investigate all possible suboptimal solutions, repeating the analysis with different (randomly chosen) initial partitions may guide the investigator to achieve an 'improved optimum'.

However, even if the procedure is repeated under different starting configurations and may lead to the same final partition, there is no guarantee that the global optimum has been obtained.

Empirical studies

In order to illustrate some of the properties of a particular method (as they appear in the literature) artificial data are generated so as to have a particular (and desirable) structure. For that purpose, sets of data are two-dimensional, enabling the data and the results to be plotted and examined visually. Thus, making the results obtained from that particular method more easily understood. However, in practice there will usually be many more than two variables measured on each object. Problems in proper understanding the results in higher-dimensional cases are likely to be magnified.

Figure 3.5 shows some artificial data examples illustrating the above problems.

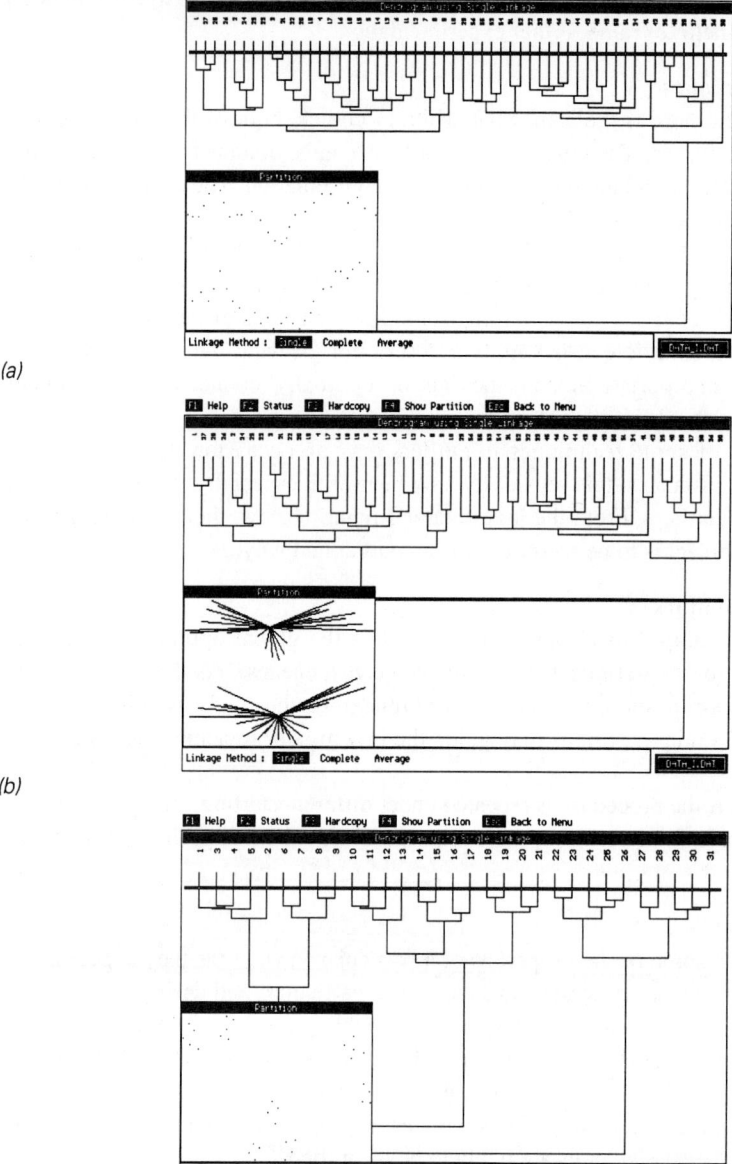

(a)

(b)

(c)

Figure 3.5: Some artificial data examples: desirable chaining (a,b); number of clusters (c,d); detected outlier (e).

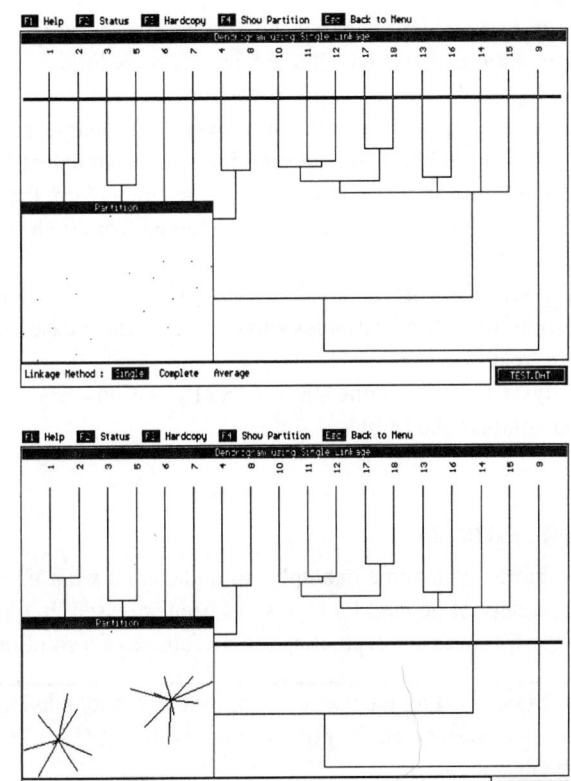

(d)

(e)

Figure 3.5 (continued)

3.4 Practical guidelines

Since most cluster analysis techniques are sensitive to the presence of outliers, some attempt should be made to remove these.

After outliers have been removed, it is generally helpful to use some projection method to obtain a two-dimensional mapping of the data (principal components, Sammon's nonlinear mapping, Kruskal's multi-dimensional scaling). Such a mapping might be used as an aid to the interpretation of clusters produced by some method at a later stage of the analysis. Before employing any formal clustering, finding the minimum spanning tree might be helpful. Its edge-length distribution informs the investigator about the presence of inconsistent edges, as well as the expected compactness of the data.

We next face the problem of selecting the variables to be used and the particular technique to employ. As already stated, although crucial, very little in general can be said about justifying the 'optimal' choices. A priori domain knowledge will be indispensable. After obtaining a set of clusters, various intuitively reasonable procedures can be followed for evaluating the stability and the usefulness of the solutions found.

1. Divide the set of entities randomly into two subsets and perform the analysis on each subset separately. Similar solutions should be obtained from both sets when the data are clearly structured.
2. Repeat the analysis with different subsets of variables and compare the results. Deletion of a small number of variables should not, in most cases, alter the clusters found if they are 'real'.
3. Repeat the analysis with different starting configurations (initial partitions, free parameters) and compare the results.

3.5 Concluding remarks

In this chapter, we have given some examples to understand what data grouping is all about. Moreover, a variety of general problems has been adressed. Intuitively reasonable procedures have to be followed for evaluating the usefulness of a solution.

> Removing outliers, dividing the user data into subsets, and deleting variables from the records cannot be accomplished within EDAPLUS. Use the editor to massage the user data.

Massaging the user data in EDAPLUS.

4 | Framing of cluster analysis studies

- *Different ends require different means and different logical structures.*
- *While techniques are important ... knowing when to use them and why to use them is more important.*
- *In the long run it does not pay a statistician to fool either himself or his clients; (Tukey, 1954).*

It has been pointed out that we should distinguish primary validation – after having applied clustering methods to the data – using all numerical results, visual interpretation, internal indices, and explicit measures of confidence (belief) in suggested evidence. This, apart from secondary validation after transforming numerical results and tree patterns into belief and feeling. How useful, informative, and valid are the results with respect to the research goal chosen? Such transformation requires the investigator to have intuition, intelligence, and knowledge about the problem to be solved; (Backer).

In this chapter, we try to identify the purpose of a cluster analysis: that is, what the researcher intends to achieve. Originating from the domain of application, the researcher has to make decisions about what has to be achieved, and how. Making such decisions and choosing the options available is called *framing* the analysis: more specifically framing the application, framing alternatives and framing decisions (Romesburg, 1984).

If we ask ourselves which is more basic, applying methods (computing) or planning, then the answer will be *planning*. In other words, after having decided *what* should be the goal of the research, then we have to decide what criteria should be adopted, and how the results of applying methods are transformed into information, into belief, or into feeling about that research goal. Framing of the analysis refers to *how* the researcher chooses specific methods to address her or his research goal.

In Figure 4.1 such planning is represented in the form of the major steps in research. There, step 3 (choosing and framing a method) will be of subjective nature; step 4 (applying the method to the data), on the other hand, is assumed to be more objective.

Step 4 and step 5 (applying the method and deciding how useful the results are) cannot be treated without an understanding of the research goal. Romesburg (1984) describes four major types of research goal: creating a question, creating a hypothesis testing a hypothesis, and performing a classification. These types will be the subject of Section 4.1.

4.1 Research framing

As mentioned already, framing includes choosing a research goal, choosing the options available in the analysis for framing the data and transforming the technical results into

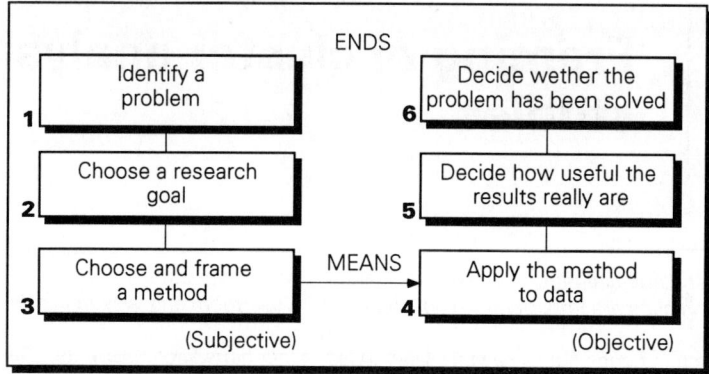

Figure 4.1: Major steps in research.

information and belief. The ordering of framing is shown in Figure 4.2.

In the foregoing, validation has been addressed to in terms of indicating how internal indices could be used to decide if the clustering results can be considered as 'hard' evidence. More generally, validation refers to how one has determined that one has attained the research goal. Therefore, we distinguish primary validation and secondary validation. In Figure 4.2 these validation steps are indicated at the appropriate level. We will return to validation in the following sections.

Following Romesburg (1984), four major research goals can be distinguished:

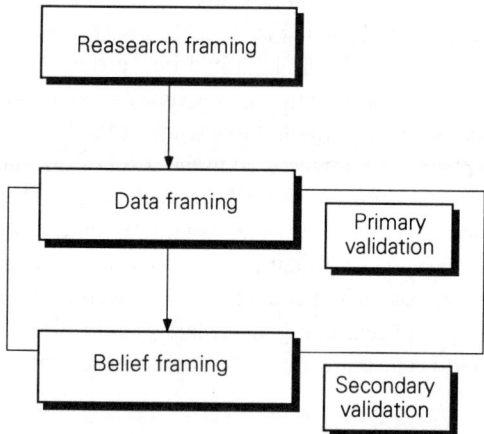

Figure 4.2: Major steps in framing.

Create a question
The first research goal, creating a question, explores the data through cluster analysis, hoping this will reveale interesting patterns of similarity that will spark a question about the processes of nature that led to the patterns of similarity, or a question about what else the results could be related to.

Create a hypothesis
Here we already have a question in mind before we perform the cluster analysis. This question helps us to frame the collection of data (helps us to collect the right kind of data). Analyzing the resulting trees may suggest a research hypothesis. This use of cluster analysis is called *retroduction* (creating a hypothetical reason that accounts for the observed facts).

Test a hypothesis
Here we already have a good question and a good hypothesis in mind. Before the analysis we are able to predict the patterns of similarity in the tree that we would expect should the hypothesis be true. Performing cluster analysis, the actual tree should – within an acceptable tolerance – match the predicted tree. This use of cluster analysis is called a *hypothetico-deductive method*.

Classification
A general-purpose *classification* serves a general scientific purpose aiming at a catalog of the entities studied. A special-purpose classification is merely focused on relating classification to specific qualitative or quantitative variables.

If we refer the framing of the cluster analysis to the choise of specific methods to address the research goal, it is evident framing and validation are unseparable. Although when performing the cluster analysis, the investigator frames first and validates last, the presentation will often be in the reverse order.

4.2 Data framing

It has been pointed out that data framing consists of a set of decisions regarding the choice of objects (patterns), attributes (features, measurements), scales of measurement, standardization, resemblance coefficients (association measures), and the clustering method.

Roughly, the following steps are always to be taken.
1. Define the *objects* used in the cluster analysis and determine why and how their nature fits the research goal.
2. Describe what sampling procedure will be used (random, systematic) and how the *sample size* has been determined.
3. Define the attributes (features, *measurements*) and explain why those chosen are appropriate for the research goal.

4.   Describe the scales of attribute measurements.
5.   Explain why the analysis includes *hierarchical* methods or *partitional* methods, or both.
6.   Should the pattern (data) matrix be *standardized* and, if so, why?
7.   State the *resemblance coefficient* (association measure) that will be used and explain why this choice fits the research goal.
8.   Justify the choice of the particular *clustering technique*.
9.   If a tree is cut into clusters to form a *classification*, explain why cuts were made where they were.
10.   State the CPCC value and other *indices of confidence*.

Next, we address the problem of deciding of what should be believed.

4.3 Belief framing

Generally speaking, belief framing goes back to background knowledge, measures of structure, baseline distributions, existing classifications, and expert intuition and expectation.

It has been pointed out that we should distinguish primary and secondary validation, after having applied clustering methods to the data. The former uses all numerical results, visual interpretation, internal indices, and explicit measures of confidence (belief) in suggested evidence. This is the very subject of Part 3, the kernel of this book. The latter includes transformation of numerical results and tree patterns into belief and feeling. How useful, informative and valid are the results with respect to the research goal chosen. Such transformation requires the investigator to have intuition, intelligence and knowledge about the problem to be solved.

Figure 4.3 illustrates the impact of primary validation and secondary validation. Clearly, as a consequence, data framing and belief framing will show some overlap.

Secondary validation refers to a complex of factors to be taken into account.

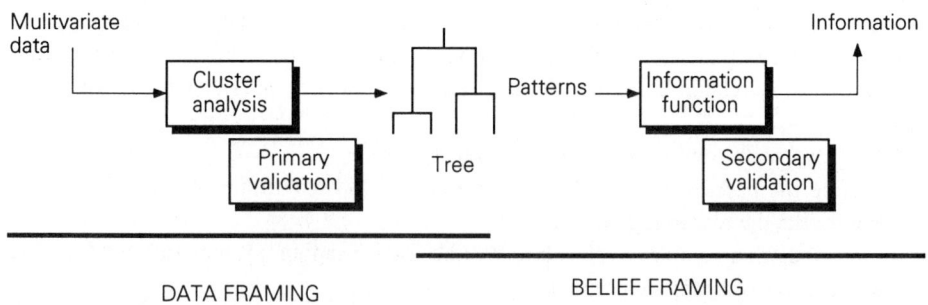

Figure 4.3: From data to information.

1. Is the resulting tree pattern well-structured?
2. Is the result in agreement with existing classifications?
3. Is the result in agreement with existing expert intuition?
4. Is the result in agreement with other data analysis methods?
5. Are the results valid for different data from the same research objective?
6. Are the results stable and robust?

At this stage, no attempt will be made to further the above questions. We shall return to the above in Chapter 10 in which a cluster analysis of delphinid sonar signals will be outlined. There, related to existing biological taxonomies and ecological habitat, a hypothesis about sonar sound signal modelling is tested and a scientific hypothesis about the mass stranding phenomenon is generated (Kamminga, 1994).

4.4 Concluding remarks

In this chapter we have discussed some of the decisions that the researcher has to make about what has to be achieved and how. Planning and framing refer to how the researcher choses specific methods to address the research goal.

Primary and secondary validation are subjects which will be re-examined in Chapter 10.

5 | Approximate reasoning in cluster analysis: a branch of fuzzy logic

In fact, most of the classes of objects encountered in the real world are of this fuzzy, not sharply defined type. If we adopt the vague expectation: 'clusters of objects are such that the degree of natural association is high among members of the same group and low between members of different groups', we recognize a prime example of imprecise description of real objects; (Backer, 1978).

5.1 The impact of this chapter

In the preceding chapters, we have been developing cluster analysis starting from a highly informal and imprecise working definition of a cluster. Thus a cluster was said to be a set of entities which are alike (similar), and entities from different clusters are not alike (dissimilar). Moreover, we have restricted ourselves to *exclusive* (non-overlapping) classifications aiming at partitioning the set of objects (patterns) such that each object belongs to exactly one cluster. Overlapping (*non-exclusive*) clustering methods have not been dealt with. Those methods are reviewed in Shepard and Arabie (1979). *Fuzzy clustering* is also a type of non-exclusive classification in which a pattern is assigned a degree of belongingness to each cluster in a partition and will be explained in this chapter in detail.

In Chapter 1, the notion of subjectivism and approximate reasoning gave rise to distinguishing formal (rigid conventional mathematics) and informal analysis (human elastic interpretation). The elastic domain tends to use linguistic variables like 'small', 'weak', 'large', 'strong', and so forth, while the rigid domain tends to use numbers.

In Chapter 2, we have identified hierarchical clustering as a sequence of partitions in which each partition is nested within the next partition in the sequence. Partitional clustering has the general objective of obtaining that partition which, for a fixed number of clusters, optimizes some criterion function (objective function), like minimization of the square error criterion. In that case, the resulting partition has also been referred to as the *minimum variance* partition. The error represents deviations of the patterns from the centroids. Gordon and Henderson (1977) wrote the above criterion function in such a way that the clustering problem can be formulated as a nonlinear programming problem.

Let $y_{ik} = 1$ if the ith pattern belongs to the kth cluster, and 0 if the ith pattern does not belong to the kth cluster. Then the centroid of the kth cluster, $\mathbf{m}_k$, is written as $\mathbf{m}_k = (m_{k1}, m_{k2}, \ldots, m_{kd})$, where

$$m_{kj} = \frac{\sum\limits_{i=1}^{n} (y_{ik}x_{jk})}{\sum\limits_{i=1}^{n} y_{ik}}$$

Then, the total within variance, S_W, can be written as

$$S_W = \sum_{i=1}^{n} \sum_{k=1}^{K} y_{ik} \sum_{j=1}^{d} (x_{ij} - m_{kj})^2$$

Minimization of S_W leaves us with the minimum variance solution.

If we minimize S_W more generally under the assumption that $y_{ik} \in [0,1]$ subject to the constraints

$$\sum_{k=1}^{K} y_{ik} = 1 \quad \text{and} \quad y_{ik} \geq 0$$

we obtain a solution where a pattern belongs to a cluster with a 'grade of membership', a concept similar to fuzzy clustering.

Even if hierarchical structures and partitional structures are philosophically similar, they are mathematically not isomorphic.

Changing the qualitative results of the above structures into 'hard' evidence has been identified as one of the most subtle problems in cluster analysis: *cluster validity*. Internal indices, like the CPCC and the DB index, were proposed to value the degree of matching of the induced hierarchy and the data, and the strength of the clustering as a whole, respectively. Baseline distributions were shown to be instrumental in the assessment of the validity of algorithmical suggested structure assumed to be present in the user data. Finally, testing the predisposition of the data to cluster in natural groups has been identified as the problem of *clustering tendency*.

In Chapter 3, the transitivity property was discussed in order to understand the very nature of hierarchical representation of the data. Furthermore, general problems and guidelines were listed in order to underline the complexity of the analysis.

Finally, in Chapter 4, making explicit of what the researcher intends to achieve was identified as the problem of framing the research, the data and the belief.

When dealing with a partition in the form of a set of compact, disjoint, well-separated clusters, there will be no ambiguity and uncertainty in the assignment of patterns to any of the clusters. However, if the clusters are touching and do have no sharp boundaries, the assignment of some patterns to any of the clusters turns out to be non-unique, ambiguous, uncertain, or vague (see Figures 2.21 and 2.22). Such a partition should be described in such a way that ordinary pattern subsets should be replaced by fuzzy pattern subsets.

Consequently, patterns can belong to two clusters simultaniously with a different *grade of membership.* In stead of an ordinary 'hard' partition, we identify the 'fuzzy' partition. The theory of fuzzy sets provides the mathematical tools to reformulate the process of clustering. As we will see, a *minimum fuzziness* partition is then the ultimate objective.

The theory of *fuzzy sets* deals with sets of events that do not have a crisply defined membership as in ordinary set theory. It allows patterns to have a grade of membership between 0 and 1. In fuzzy sets, we assign gradual transitions from membership to non-membership. Fuzzy variables like 'small', 'weak', 'large', or 'strong' are subjective and informal in nature. We are using modifiers or intensifiers such as 'not', 'very' and 'slightly', and 'weakening' and 'reinforcing' operators, in order to modify or intensify the meaning (or impact) of the fuzzy variables.

As such, fuzzy sets are likely to find increasing use in applications involving imprecise and incomplete information, common-sense reasoning, and complex concepts. Without doubt, cluster analysis is such an application.

In the broad sense, *approximate reasoning* can be viewed simply as the collection of techniques for dealing with inference (drawing conclusions) under ambiguity or uncertainty, in which the underlying mathematical framework is approximate rather than exact or deterministic. Utilizing the theory of fuzzy sets, approximate reasoning is a branch of fuzzy logic and can be viewed as a more expressive mathematical language for the representation of uncertain or vague structures.

In this chapter, cluster analysis will be reformulated as a process of fuzzy identification based on fuzzy relations, linking the fundamental structures, fuzzy relations and partitions nicely together. As will be seen at later stage, it will provide the basis for embedded simulation and approximate reasoning in cluster analysis (Chapter 6).

Figure 5.1 shows the embedding of this chapter in the approach throughout this book. Since Chapter 2 and 3 are devoted to 'classical' methods in cluster analysis, deriving algorithmically a variety of hierarchies and partitions, we are able to reformulate those methods in the setting of *fuzzy relations* and *fuzzy partitions*. The core of this chapter deals with the explication of approximation and fuzziness. Ultimately, the chapter yields a clear understanding of *fuzzy classification* rules and how the detection of 'strong' and 'weak' patterns in resulting partitions can be represented by a fuzzy relation based on repetitative trials (*embedded simulation*).

Conceptually we will then be ready to formulate an approach for approximate reasoning based on the algorithimically derived results from the methods presented in Chapter 2.

5.2 Fuzzy clustering: a historical review

The the major activity in the development of clustering techniques took place from the early 1960s on, st the same time the development of the theory of fuzzy sets was started. Zadeh (1965) introduced the concept of fuzzy sets by defining them in terms of

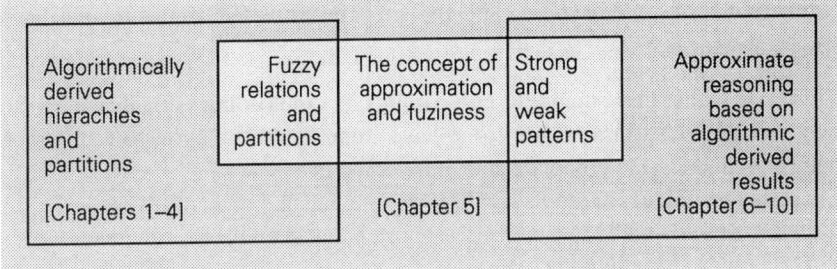

Algorithmically derived hierachies and partitions [Chapters 1–4]	Fuzzy relations and partitions	The concept of approximation and fuziness [Chapter 5]	Strong and weak patterns	Approximate reasoning based on algorithmic derived results [Chapter 6–10]

Figure 5.1: The impact of Chapter 5 in the approach adopted throughout this book.

mappings from a set into the unit interval on the real line. Fuzzy sets were introduced to provide a means of mathematically describing situations which give rise to *ill-defined classes*, i.e. collections of objects for which there are no precise criteria for membership. In real situations, especially in problems of pattern classification (cluster analysis), fuzziness is the rule rather than the exception. It is therefore believed that fuzzy sets can be applied to model these problems at least as well as, and probably better than, the methods now being used. In fact, most of the classes of objects encountered in the real world are of this fuzzy, not sharply defined type. If we adopt the vague expectation: 'clusters of objects are such that the degree of natural association is high among members of the same group and low between members of different groups' (Backer, 1978), we recognize a prime example of *imprecise description* of real objects.

Usually, imprecision and indeterminancy are considered statistical, random characteristics and are taken into account by the methods of probability theory. However, in real situations a frequent source on imprecision is not only the presence of random variables, but also the impossibility of operating with exact data, imprecision of the constraints and objectives, and intrinsic ambiguity. The theory of fuzzy sets provides a scheme for handling a number of real situations in which imprecision results in the possibility that an object may belong to a certain category (cluster) with varying grades of membership. Basic concepts and foundations of fuzzy sets can be found in Zadeh (1965; 1968; 1973) and Watanabe (1969). The basic directions for application of this theory is given in Zadeh et al. (1975).

Not surprisingly, the conceptual framework for dealing with classes in which there may be grades of membership and non-membership appears to be relevant to the clustering problem. The clustering practitioner faces the fact that any clustering technique will yield a result and he or she is left with the main question of estimating the value of the result. Therefore, a formalism, a unified theory, is desired for detecting *clustering tendency* and for scaling the value of the result (*validity*). In other words, *cluster significance* should be incorporated in the formal description of the process of cluster identification.

The vague expectations of any data analysis practitioner are best represented by the following limitations in developing and applying clustering techniques (Dubes and Jain, 1976).

-   Clustering techniques attempt to group points in a multi-dimensional space in such a way that all points in a single group have some natural relation to each other and points not in the same group are somehow different.
-   Clustering techniques are tools for discovery, not an end in themselves.
-   A cluster analysis really is a preprocessing step that should generate ideas and should help the user to form hypotheses.
-   The utility of a cluster analysis lies more in the questions raised by it than in the questions answered.
-   A cluster analysis should be supplemented by other descriptive techniques.
-   There is no true structure since we have to make *ad hoc* choices in selection of a measure of association, a criterion of clustering goodness, various parameters, and in weighing of several computational constraints.

(i)   A single criterion, a single result, cannot summarize all information available in the clustering.

From the above it is clear that we have to accept that there is no unique solution for a clustering problem, due both to the large variety of techniques available, and to the subjective choices to be made. This *non-uniqueness* occurs particularly in situations where the clusters to be detected are not very compact and not well separated. In such cases it is evident that a fuzzy set representation in cluster analysis will be convenient since fuzzy sets are sets with no sharp boundaries.

(ii)   The advantage of explicating the above representation is that various types of non-uniqueness or vagueness may be classified as such, while the corresponding classical results still remain meaningful.

Formulations like *(i)* and *(ii)* form exactly the basis for approximate reasoning based on a variety of algorithmically derived results from classical methods.

Fuzzy sets as a theoretical basis for clustering were first suggested by Bellman et al. (1966). Subsequently, a variety of fuzzy clustering was proposed.

Ruspini (1969) presented classification in fuzzy sets as the breakdown of the probability density function of the original set of data points into a weighted sum of the component fuzzy set densities.

Gitman and Levine (1970) presented an algorithm which achieves a decomposition of a multi-modal fuzzy set into a number of unimodal fuzzy sets by taking into account a coventional distance measure and the order of importance of every point given by its membership function.

Tamura et al. (1971) described for the first time a hierarchical clustering scheme generated by one parameter family of equivalence relations on a data set. The role of fuzzy hierarchical clustering in information retrieval is discussed in detail by Negoita (1973).

Dunn (1973) and Bezdek (1974) generalized the Isodata clustering process to a fuzzy-Isodata clustering technique.

In an interesting paper by Diday (1973) the notion of strong and weak patterns, in the sense of types of fuzzy sets, was developed.

Dunn (1974) used graph-theoretical arguments to show that the hierarchical clustering scheme induced by the Tamura's k-step fuzzy relation is contained in the maximal single linkage hierarchy.

Yeh and Bang (1975) extended many standard graph-theoretical concepts into fuzzy graphs and discussed their applications to cluster analysis.

Barnes (1976) developed a fuzzy set clustering containing a measure of the separation between the classes. The degree of separability can also be used to determine the number of classes into which data should be separated.

Backer (1978) proposed a general model for partitioning-optimization on the basis of a decomposition of induced fuzzy sets.

Further ideas about fuzzy relations and fuzzy optimization criteria have been suggested by Bezdek and Harris (1978) and Roubens (1978), respectively.

Experiments with a goal-directed fuzzy clustering performance measure are reported Backer and Jain (1981).

The above work has culminated in two books on fuzzy clustering by Bezdek (1981) and Backer (1978). A recent reprint volume on fuzzy models for pattern recognition (Bezdek and Pal, 1992) contains most of the prominent early contributions in the field of fuzzy cluster analysis.

While conventional clustering is making use implicitly of some degree of conceptual knowledge, contextual knowledge, background knowledge, and functional knowledge, the methodology for explicit use of such knowledge has been developed by Michalski and Stepp (1983a, 1983b), Fisher and Langley (1986), Shekar et al. (1987), Backer (1988a, 1988b) and Srivastava and Narasimhamurty (1990), leading ultimately to unifying frameworks for knowledge-based clustering.

5.3 Basic definitions

For the convenience of the reader, a brief exposition of the concept of fuzzy sets in cluster analysis is presented below, summarizing some of its definitions and illustrating its existence in some classical methods.

The *fuzzy membership function* $f: X \Rightarrow [0,1]$ can be recognized as the generalization of the classical indicator function $\chi: X \Rightarrow \{0,1\}$ where X denotes the sample space $\mathbf{x} \in X$.

Let f and h be two fuzzy sets. Then we have:

1. *complement*:

$$f^c = 1 - f \quad \Rightarrow \quad f^c(\mathbf{x}) = 1 - f(\mathbf{x}), \forall \ \mathbf{x}$$

2. *inclusion*:

$$f \subseteq h \quad \Rightarrow \quad f(\mathbf{x}) \leq h(\mathbf{x}), \forall \ \mathbf{x}$$

3. *intersection*:

$$f \cap h \quad \Rightarrow \quad (f \cap h)(\mathbf{x}) = \min \ [f(\mathbf{x}), h(\mathbf{x})], \forall \ \mathbf{x}$$

4. *union*:

$$f \cup h \quad \Rightarrow \quad (f \cup h)(\mathbf{x}) = \max \ [f(\mathbf{x}), h(\mathbf{x})], \forall \ \mathbf{x}$$

5. *level set*:

$$F(\lambda) = \{\mathbf{x} \, | \, f(\mathbf{x}) \geq \lambda\}, \lambda \in [0,1], \text{ with}$$

$$\chi_{F(\lambda)} = 1 \text{ for } \mathbf{x} \in F(\lambda)$$

$$\chi_{F(\lambda)} = 0 \text{ for } \mathbf{x} \notin F(\lambda)$$

6. *measure of fuzziness*:

$$I(f) = \frac{1}{n} \sum_{\mathbf{x}} |f(\mathbf{x}) - \chi_{F(1/2)}(\mathbf{x})|$$

5.3.1 Fuzzy relations

A second important generalization constitutes the definition of a *fuzzy relation*. A classical relation $R \subseteq X \times X \Rightarrow \{0,1\}$ is given by

$$\chi_R: X \times X \Rightarrow \{0,1\}$$

and can be generalized to

$$\mu: X \times X \Rightarrow [0,1]$$

where $\mathbf{x}, \mathbf{y} \in X \times X$.

Let μ and σ be two fuzzy relations. We have:

1. *complement*:

$$\mu^c = 1 - \mu \quad \Rightarrow \quad \mu^c(\mathbf{x}, \mathbf{y}) = 1 - \mu(x,y), \forall \ \mathbf{x}, \mathbf{y}$$

2. *inclusion*:

$$\mu \subseteq \sigma \quad \Rightarrow \quad \mu(\mathbf{x},\mathbf{y}) \leq \sigma(\mathbf{x},\mathbf{y}), \ \forall \ \mathbf{x},\mathbf{y}$$

3. *intersection*:

$$\mu \cap \sigma \quad \Rightarrow \quad (\mu \cap \sigma)(\mathbf{x},\mathbf{y}) = \min \ [\mu(\mathbf{x},\mathbf{y}),\sigma(\mathbf{x},\mathbf{y})], \ \forall \ \mathbf{x},\mathbf{y}$$

4. *union*:

$$\mu \cup \sigma \quad \Rightarrow \quad (\mu \cup \sigma)(\mathbf{x},\mathbf{y}) = \max \ [\mu(\mathbf{x},\mathbf{y}),\sigma(\mathbf{x},\mathbf{y})], \ \forall \ \mathbf{x},\mathbf{y}$$

5. *level set*:

$$F(\varepsilon) = \{(\mathbf{x},\mathbf{y}) \,|\, \mu(\mathbf{x},\mathbf{y}) \geq \varepsilon\}, \ \varepsilon \in \ [0,1]$$

If we note that for $A \subseteq X$ and R is a relation on A, with $R \subseteq X \times X$, then

$$(\mathbf{x},\mathbf{y}) \in R \ \Rightarrow \ \mathbf{x} \in A \text{ and } \mathbf{y} \in A$$

or

$$\chi_R(\mathbf{x},\mathbf{y}) = 1 \ \Rightarrow \ \chi_A(\mathbf{x}) = 1 \text{ and } \chi_A(\mathbf{y}) = 1$$

We find this equivalent to

$$\chi_R(\mathbf{x},\mathbf{y}) \leq \min \ [\chi_A(\mathbf{x}),\chi_A(\mathbf{y})]$$

Consequently, $\mu: X \times X \Rightarrow [0,1]$ leads to $\mu: f \times f \Rightarrow [0,1]$ for which

$$\mu(\mathbf{x},\mathbf{y}) \leq \min \ [f(\mathbf{x}),f(\mathbf{y})]$$

should hold.

Example fuzzy relation (Jarvis and Patrick, 1973)

Jarvis and Patrick defined a proximity measure as the number of matches in near-neighbour lists for two patterns. Their clustering algorithm can be summarized as follows. Place patterns $\mathbf{x}_i$ and $\mathbf{x}_j$ into the same cluster if $\mathbf{x}_i$ and $\mathbf{x}_j$ share at least k near neighbours and $\mathbf{x}_i$ and $\mathbf{x}_j$ are k'-near neighbours of each other with $k' \leq k$. This algorithm is non-iterative and is computationally attractive since near neighbours can be computed efficiently. The user has to specify the size of the neighbourhood k, and the similarity threshold, k_t. Large values of k bias the algorithm towards global structures, whereas small values of k favour chained or elongated structures.

The notion of proximity based on shared nearest neighbours has been modified by Gowda and Krishna (1978) to measure the 'mutual nearness' of two patterns. If $\mathbf{x}_j$ is the pth near neighbour of $\mathbf{x}_i$ and $\mathbf{x}_i$ is the qth near neighbour of $\mathbf{x}_j$, then the *mutual neighbourhood value* (MNV) between $\mathbf{x}_i$ and $\mathbf{x}_j$ is defined as $p + q$. The smaller the MNV, the more similar the patterns. This represents a stronger notion of similarity than the number of shared neighbours of Jarvis and Patrick. Again, the parameter k that controls the neighborhood depth is crucial to the performance of the algorithm. Small values of k give several 'strong' clusters and large values of k give fewer 'weak' clusters (Gowda and Krishna, 1978).

Backer, et al. (1983) have shown that in fact the Gowda and Krishna algorithm is included in the Jarvis and Patrick algorithm. Furthermore, they have shown that a nested sequence of 'reliable' solutions can be expected if the two-parameter problem, (k, k_t), is approached as a one-parameter approach for $(k, k-1)$ or $(k, k-2)$.

In the Jarvis and Patrick approach, a relation between two patterns $\mathbf{x}$ and $\mathbf{y}$ is defined as the number of neighbouring patterns they share. So, let $\Gamma_\mathbf{x}$ be the neighbourhood of $\mathbf{x}$, $\Gamma_\mathbf{y}$ be the neighbourhood of $\mathbf{y}$ and $\Gamma_\mathbf{xy} = \Gamma_\mathbf{x} \cap \Gamma_\mathbf{y}$. Further, let $|\Gamma_\mathbf{x}| = n_\mathbf{x}$, $|\Gamma_\mathbf{y}| = n_\mathbf{y}$ and $|\Gamma_\mathbf{xy}| = n_\mathbf{xy}$.

Next, let

$$\xi : X \times X \implies [0, \infty)$$

be a symmetric proximity function. Then, we define the the following fuzzy set on X

$$f(\mathbf{x}) = \sum_{\mathbf{z} \in \Gamma_\mathbf{x}} \xi(\mathbf{x}, \mathbf{z}), \ \forall \ \mathbf{x}$$

such that

$$0 \leq f(\mathbf{x}) \leq 1, \ \forall \ \mathbf{x}$$

Then,

$$\mu(\mathbf{x}, \mathbf{y}) \ = \ \sum_{\mathbf{z} \in \Gamma_\mathbf{xy}} \min [\xi(\mathbf{x}, \mathbf{z}), \xi(\mathbf{z}, \mathbf{y})], \ \forall \ \mathbf{x} \in \Gamma_\mathbf{y} \text{ and } \forall \ \mathbf{y} \in \Gamma_\mathbf{x}$$

$$= 0, \ \mathbf{x}, \mathbf{y} \notin \Gamma_\mathbf{xy}$$

is a fuzzy relation on $f \times f$. This follows immediately from the fact that

$$\mu(\mathbf{x}, \mathbf{y}) \ \leq \ \sum_{\mathbf{z} \in \Gamma_\mathbf{xy}} \min [\xi(\mathbf{x}, \mathbf{z}), \xi(\mathbf{z}, \mathbf{y})], \ \forall \ \mathbf{x}, \mathbf{y}, \mathbf{z}$$

$$\leq \min \left[\sum_{\mathbf{z} \in \Gamma_\mathbf{xy}} \xi(\mathbf{x}, \mathbf{z}), \ \sum_{\mathbf{z} \in \Gamma_\mathbf{xy}} \xi(\mathbf{z}, \mathbf{y}) \right]$$

$$< \min \left[\sum_{\mathbf{z} \in \Gamma_\mathbf{x}} \xi(\mathbf{x}, \mathbf{z}), \ \sum_{\mathbf{z} \in \Gamma_\mathbf{x}} \xi(\mathbf{z}, \mathbf{y}) \right]$$

$$= \min [f(\mathbf{x}), f(\mathbf{y})]$$

For the simple choice of

$$\xi : X \times X = \frac{1}{n}$$

it follows that

$$f(\mathbf{x}) = \sum_{\mathbf{z} \in \Gamma_{\mathbf{x}}} \frac{1}{n} = \frac{n_{\mathbf{x}}}{n}, \ \forall \ \mathbf{x}$$

and

$$\mu(\mathbf{x},\mathbf{y}) = \sum_{\mathbf{z} \in \Gamma_{\mathbf{xy}}} \min \left[\frac{1}{n},\frac{1}{n} \right] = \frac{n_{\mathbf{xy}}}{n}, \ \forall \ (\mathbf{x},\mathbf{y})$$

In conclusion, the number of shared near neighbours is a fuzzy relation on a fuzzy set whose membership function is equivalent to a *k*-nearest neighbour point density estimator in the *X*-space.

Other choices of the relation ξ lead to other already existing algorithms.

Practical application of fuzzy relations (Shapiro and Haralick, 1969)
To solve the problem of decomposition of two-dimensional shapes, Shapiro and Haralick proposed a technique which can easily be formulated in the above fuzzy relation. (This technique has been succesfully been used by Jain et al. (1980) in a problem of cell segmentation.)

In order to find the globally convex sections of *X*, 'interior line segments' are defined, LI($\mathbf{x}_i$,$\mathbf{x}_j$), being a binary relation on $X \times X$ consisting of all pairs of vertices in *X*. LI($\mathbf{x}_i$,$\mathbf{x}_j$) = 1 if *x* is 'visible' from $\mathbf{x}_j$. Then the neighbourhood of $\mathbf{x}_i$ is defined by

$$\Gamma_{\mathbf{x}_i} = \{ \mathbf{x}_j, j \neq i : \text{LI}(\mathbf{x}_i,\mathbf{x}_j) = 1 \}$$

Recalling the relation between the proximity relation ξ and the fuzzy relation μ, we obtain for

$$\xi(\mathbf{x}_i,\mathbf{x}_j) = \left(1 - \frac{d(\mathbf{x}_i,\mathbf{x}_j)}{\max[d(\mathbf{x}_i,\mathbf{x}_k)]} \right) / |\Gamma_{\mathbf{x}_i}|$$

being an appropriate choice for the proximity ξ, in the case of the two-dimensional shape, constituting $X = \{\mathbf{x}_1,...,\mathbf{x}_{16}\}$ as the ordered set of points representing the vertices of a polygonal approximation to the boundary of the planar shape, as shown in Figure 5.2.

Setting $\varepsilon = 0.40$ (level set) we find the disjunct complete subgraphs as shown in Figure 5.3. Thus,

$$C_1 = \{\mathbf{x}_1, \mathbf{x}_2, \mathbf{x}_3, \mathbf{x}_4, \mathbf{x}_5, \mathbf{x}_{15}, \mathbf{x}_{16}\}$$

and

$$C_2 = \{\mathbf{x}_6, \mathbf{x}_7, \mathbf{x}_8, \mathbf{x}_9, \mathbf{x}_{10}, \mathbf{x}_{11}, \mathbf{x}_{12}, \mathbf{x}_{13}, \mathbf{x}_{14}\}$$

and the bridge set[1]

$$\Gamma(C_1, C_2) = \{\mathbf{x}_5, \mathbf{x}_{10}, \mathbf{x}_{11}, \mathbf{x}_{14}\}$$

for which[2]

$$W(C_1, C_2) = \sum_{\mathbf{x}_i \in C_1} \sum_{\mathbf{x}_j \in C_2} \mu(\mathbf{x}_i, \mathbf{x}_j)$$

yields 0.63, being the smallest value that is attainable.

As a result, the two-dimensional shape of Figure 5.2 has been decomposed into two simple parts with the smallest amount of fuzziness W, as shown in Figure 5.3.

5.3.2 Fuzzy partitions

In Chapter 2, we described an algorithm for iterative partitional clustering; the basic idea of such an iterative algorithm is to start with an initial partition and assign patterns to clusters so as to reduce square error. Different initial partitions can lead to different final clusterings because algorithms based on square error can converge to local minima. It is important to note that each clustering result here is to be considered as a *hard partition*: cluster membership values are represented by the classical indicator function on $\{0,1\}$, such that a pattern belongs or do not belong to one of the detected clusters. The algorithm was specified by the following steps.

Step 1. Select an initial partition with K clusters. Repeat steps 2 to 4 until the cluster membership stabilizes.

Step 2. Generate a new partition by assigning each pattern to its closest cluster centre.

Step 3. Compute new cluster centres as the centroids of the clusters.

Step 4. Repeat steps 2 and 3 until an optimum value of the criterion function is found.

We next consider its generalization in the form of fuzzy clustering.

Let us recall the *hard K-means clustering* algorithm from Section 2.5. There, the algorithm was specified by the *criterion function*

$$E_K^2 = \sum_{j=1}^{K} \sum_{i=1}^{n_j} (x_i^{(j)} - m^{(j)})^{\mathrm{T}}(x_i^{(j)} - m^{(j)})$$

---

[1] The meaning of a bridge set will be subject of Section 5.4.

[2] The amount of fuzziness in a bridge set, W, will be discussed in Section 5.4.

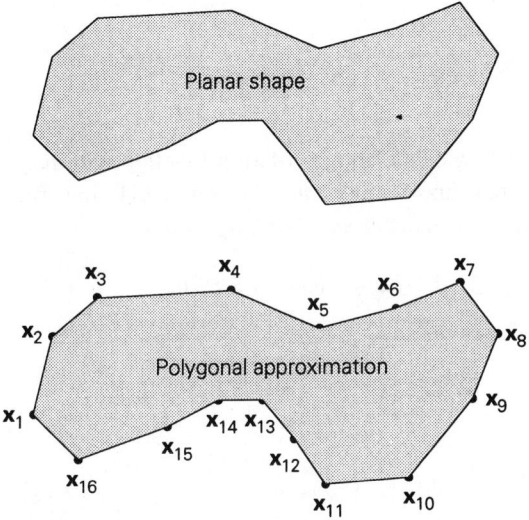

Figure 5.2: Polygonal representation of a planar shape.

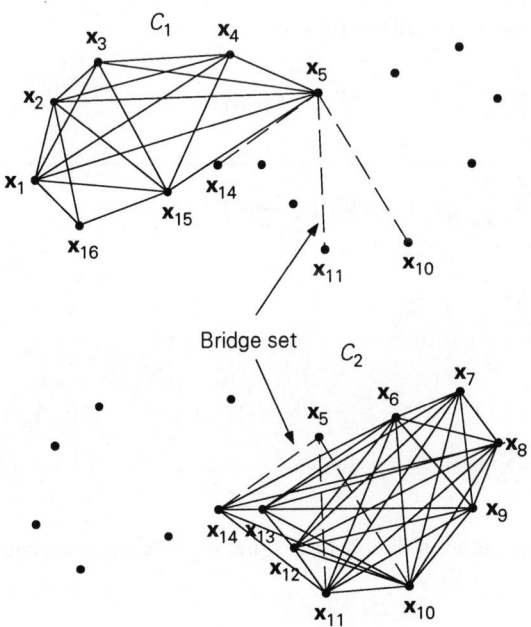

Figure 5.3: Decomposition of the shape of Figure 5.2 into two simple parts with the smallest amount of fuzziness W.

with

$$m^{(j)} = \frac{1}{n_j} \sum_{i=1}^{n_i} x_i^{(j)} \qquad \text{and} \qquad n = \sum_{j=1}^{K} n_j$$

To allow a generalization of the above criterion function with respect to 'hard' cluster membership ({0,1}) and 'fuzzy' cluster membership ([0,1]), the criterion function can be rewritten by introducing the *cluster membership function*

$$\{f_i^{(j)} = f^{(j)}(x_i), \quad 1 \leq j \leq K \quad \text{and} \quad 1 \leq i \leq n\}$$

satisfying

$$0 \leq f_i^{(j)} \leq 1, \ \forall i,j$$

$$\sum_{j=1}^{K} f_i^{(i)} = 1, \ \forall i$$

$$0 \leq \sum_{i=1}^{n} f_i^{(j)} \leq n, \ \forall j$$

Then, the criterion function reads as follows.

$$E_K^2 = \sum_{j=1}^{K} \sum_{i=1}^{n_j} (x_i^{(j)} - m^{(j)})^{\mathrm{T}}(x_i^{(j)} - m^{(j)})$$

$$= \sum_{j=1}^{K} \sum_{j=1}^{n} f_i^{(j)}(x_i - m^{(j)})^{\mathrm{T}}(x_i - m^{(j)})$$

$$= \sum_{j=1}^{K} \sum_{j=1}^{n} f_i^{(j)} D_{ij}$$

with

$$m^{(j)} = \frac{1}{n} \sum_{i=1}^{n} f_i^{(j)} x_i$$

Note that the above criterion still constitutes the 'hard' *K*-means criterion as long as

$$f_i^{(j)} = \{0,1\}$$

such that

$$f_i^{(j)} = 1 \ \text{ if } x_i \in C_j$$
$$= 0 \ \text{ otherwise}$$

Consequently, the generalization for fuzzy clustering is obtained for

$$f_i^{(j)} = [0,1], \; \forall i,j$$

and the resulting partition is called a *fuzzy partition*.

Instead of the *K*-means algorithm (or 'hard' *K*-means algorithm) we now refer to the *fuzzy K-means algorithm*. Due to Bezdek (1981), the fuzzy *K*-means algorithm has become very popular. The general format of its fuzzy objective (criterion) function became

$$E_K^2 = \sum_{j=1}^{K} \sum_{i=1}^{n} (f_i^{(j)})^\alpha D_{ij}$$

with

$$m^{(j)} = \frac{1}{n} \sum_{i=1}^{n} (f_i^{(j)})^\alpha x_i$$

where $\alpha \in [1,\infty)$ is a weighting exponent on each fuzzy membership.

The algorithm yields a stable solution for which the fuzzy partition is given by

$$f_i^{(K)} = \left(\sum_{j=1}^{K} \left(\frac{D_{iK}}{D_{ij}} \right)^{1/(\alpha-1)} \right)^{-1}, \; \forall i,K, \; \alpha > 1$$

with

$$m^{(K)} = \frac{\sum_{i=1}^{n} (f_i^{(K)})^\alpha x_i}{\sum_{i=1}^{n} (f_i^{(K)})^\alpha}$$

The above objective function is in fact a weighted sum of the distances of the feature vectors from the cluster prototype which is to be minimized. The clustering prototype is – as usual – a point (cluster centre) and this leads to detection of spherical clusters. Later investigations have shown that it is feasible to alter the objective function in such a way that other cluster shapes can be detected. One of them is the *fuzzy K-shells algorithm* (Dave, 1990; Krisnapuram et al., 1992), in which the cluster prototypes are hyperspherical shells. In 2D, the prototypes are circular boundaries, in 3D they are spherical surfaces and so on.

Now, the objective function is a weighted sum of the distances of the feature vectors from the shells. Thus, the fuzzy *K*-shells algorithm can be used to detect circles in images and more generally hyperspherical shells in higher-dimensional data.

The criterion function then reads

$$E_K{}^2 = \sum_{j=1}^{K} \sum_{i=1}^{n} (f_i^{(j)})^{\alpha} D_{ij}$$

with

$$D_{ij} = |\,(x_i - m^{(j)})^T (x_i - m^{(j)}) - r^{(j)}\,|$$

where $r^{(j)}$ denotes the radius of the jth cluster.

A recent example of the use of the fuzzy K-shell algorithm is spherical boundary detection in 2D magnetic resonance cardiac images; it can be found in Tu et al. (1994).

In order to gain more flexibility in modelling fuzziness in clustering and to enhance its interpretation by the expert, a second fuzzy partitional clustering approach emerged (Backer, 1978). The approach has also been described in Bezdek (1981) and has been compared with fuzzy K-means.

The output of this fuzzy clustering algorithm is no longer just a fuzzy partition but includes a 'hard' partition as well.

Backer (1978) and Bezdek (1981) describe Backer's fuzzy partitional clustering in detail. The output of this fuzzy clustering algorithm not only includes a 'hard' partition but also additional information in the form of membership values on [0,1], a *fuzzy partition*. The new information provided by the membership values must be interpreted by the data analyst.

Such an algorithm reads as follows:

Step 1. Select an intial partition with K clusters. Repeat steps 2 to 4 until the cluster membership stabilizes.

Step 2. Compute the membership values.

Step 3. Compute the value of the criterion function.

Step 4. Reclassify patterns to improve the criterion function.

Backer defines a variety of cluster membership functions and suggests a number of applicable fuzzy partitioning criterion functions. His program is called 'the induced fuzzy set iterative optimization program'.

Basically, the program consists of an algorithmic relationship between an inducing *object partition* $\{C_i\}_{i=1}^{k}$ on the one hand, and a collection of *induced fuzzy sets* $\{f_i(\mathbf{x}), \forall\ \mathbf{x}\}_{i=1}^{k}$ on the other hand, where

$$f_i : X \Rightarrow [0,1],\ i = 1, 2, ..., K$$

and

$$\sum_{i=1}^{K} f_i(\mathbf{x}) = 1, \forall\ \mathbf{x}$$

Usually, the initial inducing partition stems from a best guess and is subject to iterative optimization. A collection of fuzzy sets is induced by means of a point-to-subset *affinity* concept on the basis of the structural properties among objects in the representation space. This has been called *affinity decomposition* and performs in the same way as the Bayes theorem from probability theory. The appropriateness of the induced collection of fuzzy sets is made explicit by applying some performance measure (criterion function). The collection of induced fuzzy sets may cause a repartition due to a reclassification function. Then, a new inducing step follows, so we get an iterative procedure.

Definition 5.1

The affinity decomposition of each pattern $\mathbf{x}$ is given by

$$r(\mathbf{x}, C_i) \geq 0, \ \forall \ \mathbf{x}$$

and

$$r(\mathbf{x}, C) = \sum_{i=1}^{K} P_i r(\mathbf{x}, C_i), \ \forall \ \mathbf{x}$$

where P_i denotes the relative size of the ith subset given by $P_i = n_i/n$.

The (user's intuitive) notion of point-to-(sub)set affinity should be such that:
1. the affinity between a pattern and a group of patterns is not smaller when the pattern itself is contained in the group than if the pattern is not a member of the group, thus

$$r(\mathbf{x}, C_i^+) \geq r(\mathbf{x}, C_i)$$

where $C_i^+ = \{C_i, \mathbf{x}\}$;
2. the affinity between a pattern and a group of patterns is approximately zero when the pattern is distant from the group or is out of the region of interest; and
3. the affinity between a pattern and a group of patterns is equal to an absolute maximum if the group consists of just one element having the same location as the pattern under consideration.

Then, the values of the cluster membership functions for the induced fuzzy sets related to the inducing K-partition are determined by the following definition.

Definition 5.2

Given a K-partition and some affinity measure $r(\mathbf{x}, C)$, the cluster membership value $f_i(\mathbf{x})$ of pattern $\mathbf{x}$ induced by C_i is given by

$$f_i(\mathbf{x}) = P_i \frac{r(\mathbf{x}, C_i)}{r(\mathbf{x}, C)} \ \forall \ \mathbf{x}$$

Many different affinity concepts are discussed in Backer (1978), among which, the distance-oriented concept appeared to be appropriate for this discussion.

Let the subset affinity be given by

$$r(\mathbf{x}, C_i) = 1 - \frac{1}{n_i} \sum_{\mathbf{y} \in C_i} h^\beta \left[\delta(\mathbf{x}, \mathbf{y}) \right]$$

where $h^\beta[\delta(\mathbf{x}, \mathbf{y})]$ is a non-decreasing distance function on the interval [0,1], controlled by a certain parameter β, and $\delta(\mathbf{x}, \mathbf{y})$ is a distance measure satisfying reflexive and symmetric properties.

Combining the foregoing, we obtain

$$r(\mathbf{x}, C) = \sum_{i=1}^{K} \frac{n_i}{n} \left(1 - \frac{1}{n_i} \sum_{\mathbf{y} \in C_i} h^\beta[\delta(\mathbf{x}, \mathbf{y})] \right)$$

$$= 1 - \frac{1}{n} \sum_{\mathbf{y} \in C} h^\beta[\delta(\mathbf{x}, \mathbf{y})]$$

If we substitute the above in definition 5.2, we obtain

$$f_i(\mathbf{x}) = \frac{n_i - \sum_{\mathbf{y} \in C_i} h^\beta[\delta(\mathbf{x}, \mathbf{y})]}{n - \sum_{\mathbf{y} \in C} h^\beta[\delta(\mathbf{x}, \mathbf{y})]}$$

In what follows, we use for $\delta(\mathbf{x}, \mathbf{y})$ the Euclidean distance and

$$h^\beta[\delta(\mathbf{x}, \mathbf{y})] = \delta(\mathbf{x}, \mathbf{y})^2 / \beta \quad \text{for } \delta(\mathbf{x}, \mathbf{y}) \leq \beta^{1/2}$$

$$= 1 \quad \text{for } \delta(\mathbf{x}, \mathbf{y}) > \beta^{1/2}$$

Note that for $\max_{\mathbf{x}, \mathbf{y}} \delta(\mathbf{x}, \mathbf{y}) \leq \beta^{1/2}$ we find

$$f_i(\mathbf{x}) = P_i \frac{1 - \frac{1}{\beta}(\delta(\mathbf{x}, \mu_i)^2 + \frac{1}{2} D_i^2)}{1 - \frac{1}{\beta}(\delta(\mathbf{x}, \mu)^2 + \frac{1}{2} D^2)} \quad \forall\, \mathbf{x}$$

where D_i^2 is the mean square intracluster distance of the ith cluster and μ_i and μ are the mean vector of the ith cluster and the entire set, respectively.

If we represent C_i only by its sample mean μ_i, we obtain $D_i^2 = 0$, and $P_i = 1/K$, thus (after some manipulation)

$$f_i(\mathbf{x}) = \frac{1 - \frac{1}{\beta} \delta(\mathbf{x}, \mu_i)^2}{K - \frac{1}{\beta} \sum_i \delta(\mathbf{x}, \mu_i)^2}$$

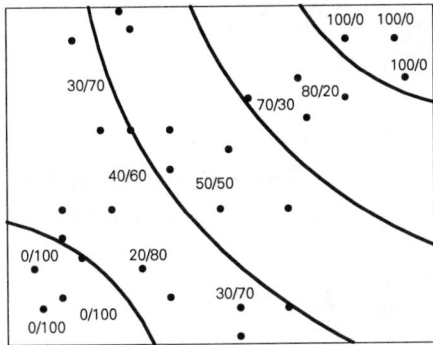

Figure 5.4: A fuzzy partitioning and its related amount of fuzzy set intersection (Backer, 1978).

which measures the ratio of the contribution of the pattern $\mathbf{x}$ to e_i^2 and E_K^2. In conclusion, there is a strong relation between the square-error hard clustering approach and the above induced fuzzy set appraoch.

Once we have established a K-collection of induced fuzzy sets we may informally characterize the partitioning as follows. If the amount of *induced fuzziness* is low it means that the K-collection of induced fuzzy sets is reasonably separable and that the inducing partition reflects the real data structure reasonably well. On the other hand, if the amount of induced fuzziness is high, it means that the inter-fuzzy set separability is low and that either the inducing partition does not reflect the real data structure well, or that almost no structure is present in the data. Consequently, our *performance measure* should measure the fuzziness in the gaps between fuzzy sets (along the fuzzy boundaries) and, therefore, should be based on the notion of intersection of fuzzy sets. Figure 5.4 illustrates a fuzzy partitioning and the related amount of intersection (Backer, 1978).

The intersection of two fuzzy sets f_i and f_j can be defined either by a min operator (as indicated before) or by a product operator. Because of its mathematical and computational attractiveness, we shall adopt here the second operator:

$$f_{i \cap j}(\mathbf{x}) = f_i(\mathbf{x}) \cdot f_j(\mathbf{x}), \ \forall \ \mathbf{x}$$

Then the clustering performance measure can be defined as follows.

Definition 5.3

Given a K-collection of fuzzy sets

$$\{f_i(\mathbf{x}), \ \forall \ \mathbf{x}\}_{i=1}^{K}$$

satisfying

$$\sum_{i=1}^{K} f_i(\mathbf{x}) = 1, \ \forall \ \mathbf{x}$$

the clustering performance measure is given by

$$\varphi = 1 - \frac{2K}{K-1} \sum_{i=1}^{K-1} \sum_{j=i+1}^{K} \frac{1}{n} \sum_{\mathbf{x}} f_i(\mathbf{x}) \cdot f_j(\mathbf{x})$$

It is easy to show that $0 \leq \varphi \leq 1$, where $\varphi = 0$ corresponds to maximum fuzziness and $\varphi = 1$ corresponds to non-fuzziness. It is also easy to show that

$$\varphi = \frac{K}{K-1} \sum_{i=1}^{K-1} \sum_{j=i+1}^{K} \frac{1}{n} \sum_{\mathbf{x}} |f_i(\mathbf{x}) - f_j(\mathbf{x})|^2$$

which shows that the performance measure can be looked upon as a cluster membership distance measure. In Bezdek (1981) a *coupling coefficient*

$$w_{ij} = \frac{1}{n} \sum_{\mathbf{x}} f_i(\mathbf{x}) \cdot f_j(\mathbf{x})$$

was suggested in order to characterize the performance of Bezdek's ISODATA program.

Bezdek defines a partition coefficient, W, as the sum of all within-cluster coupling, divided by the total coupling, or

$$W = \frac{\sum_{i=1}^{K} w_{ij}}{\sum_{i=1}^{K} \sum_{j=1}^{K} w_{ij}}$$

which results in

$$W = 1 - \frac{K-1}{K}(1-\varphi) \quad \text{or} \quad \varphi = 1 - \frac{K}{K-1}(1-W)$$

As a result, the range of variation of W becomes $[\frac{1}{K}, 1]$ which is less attractive for comparison of different clustering results since the range of variation depends on the number of clusters.

Based on this performance measure the *iterative optimization* is controlled by the following *reclassification* function. If a pattern $\mathbf{z} \in C_i$ is reclassified into $C_j, j \neq i$, then the change in the value of the performance measure φ is given by

$$\eta(\mathbf{z}, C_i) = \varphi(C_i f_i)_{i=1}^{K} - \varphi(C'_i f_i)_{i=1}^{K}$$

$$\eta(\mathbf{z}, C_j) = \varphi(C'_i f_i)_{i=1}^{K} - \varphi(C_i f_i)_{i=1}^{K}$$

$$j = 1, 2, \dots, K, j \neq i$$

where

$$\{C'_i\}_{i=1}^{K} = \{C_1, \ldots, C_i^-, \ldots, C_j^+, \ldots, C_K\}$$

with

$$C_i = \{C_i^-, \mathbf{z}\} \text{ and } C_j^+ = \{C_j, \mathbf{z}\}$$

After combining the above equations, we obtain

$$\eta(\mathbf{z}, C_i) = \frac{1}{n_j} \ [f_i^2(\mathbf{z}) - \frac{1}{n_i - 1} \sum_{\mathbf{x} \in C_i^-} f_i^2(\mathbf{x})]$$

and

$$\eta(\mathbf{z}, C_j) = \frac{1}{n_j} \ [f_j^2(\mathbf{z}) - \frac{1}{n_j + 1} \sum_{\mathbf{x} \in C_j^+} f_j^2(\mathbf{x})] \ (j \neq i)$$

Let $\eta(\mathbf{z}, C_M) = \max_j \ [\eta(\mathbf{z}, C_j)]$. Then the reclassification rule is:

if $\eta(\mathbf{z}, C_M) > \eta(\mathbf{z}, C_i)$ add pattern $\mathbf{z}$ to C_M and delete $\mathbf{z}$ from C_i, else $\mathbf{z}$ remains classified in C_i.

In Backer (1978), all concergence properties of this iterative procedure are discussed in detail. After all reclassifications (with respect to φ) have taken place, a new set of cluster membership values is calculated.

Example fuzzy partition

Here, we follow the numerical example given in Chapter 3 where the input data matrix $[d]$ was given by

$$[d] = \begin{bmatrix} 0 & 0.7 & 0.3 & 0.8 \\ & 0 & 0.5 & 0.1 \\ & & 0 & 0.9 \\ & & & 0 \end{bmatrix}$$

For $\{C_i^0\}_{i=1}^{2} = \{1,2\}, \{3,4\}$ which is the initial partition, and $\beta = 1$, we find a 2-collection of fuzzy sets $\{(f_i(\mathbf{x}), \forall \ \mathbf{x}\}$, which we shall write as

$$F_2 = \begin{bmatrix} f_1(1) & f_1(2) & f_1(3) & f_1(4) \\ f_2(1) & f_2(2) & f_2(3) & f_2(4) \end{bmatrix} = \begin{bmatrix} 0.59 & 0.48 & 0.43 & 0.50 \\ 0.41 & 0.52 & 0.57 & 0.50 \end{bmatrix}$$

characterized by $\varphi^{(0)} = 0.014$, indicating a very low separability between the two induced fuzzy sets.

After one classification step we obtain

$$\{C_i^1\}_{i=1}^{2} = \{1,3\}, \{2,4\}$$

yielding

$$F_2 = \begin{bmatrix} 0.77 & 0.30 & 0.74 & 0.14 \\ 0.23 & 0.70 & 0.26 & 0.86 \end{bmatrix}$$

and $\varphi^{(1)} = 0.60$. No further improvement is available, so F_2 is said to be the optimal (with respect to φ) fuzzy decomposition of the given data. This is in accordance with our previous observation that for $K = 2$, {1,3}, {2,4} has been indicated as an appropriate choice for cutting the dendrogram and obtaining a partition of X. However, we see that the resulting partitioning is now equipped with the cluster membership values, indicating how significant the result really is.

Evidently, the value of the performance measure φ depends on the free parameter β which controls (here) the size of the neighbourhood. The absolute value of the performance measure is obviously affected by changing β (choosing a very small value of β will lead to $\varphi \Rightarrow 1$, and a very large value of β will result in $\varphi \Rightarrow 0$. In Section 5.5, further analysis, comparison and experimental results will be discussed.

5.3.3 Fuzzy relations in fuzzy partitions

Associated with $\{C_i\}_{i=1}^K$, which is a hard K-partition, a fuzzy partitioning program provided us with

$$F_K = \begin{bmatrix} \dots f_1(\mathbf{x}) \dots \dots f_1(\mathbf{y}) \dots \\ \dots f_2(\mathbf{x}) \dots \dots f_2(\mathbf{y}) \dots \\ \vdots \qquad \vdots \\ \dots f_K(\mathbf{x}) \dots \dots f_K(\mathbf{y}) \dots \end{bmatrix}$$

Now the question arises as to whether a fuzzy relation

$$\mu : f \times f \Rightarrow [0,1]$$

for which

$$\mu(\mathbf{x},\mathbf{y}) \le \min [f(\mathbf{x}), f(\mathbf{y})]$$

does exist and can be defined.

If we define

$$\mu_f(\mathbf{x},\mathbf{y}) = 1 - \frac{1}{K} \sum_{i=1}^K |f_i(\mathbf{x}) - f_i(\mathbf{y})|$$

for which symmetry and reflexifity are easy to verify. After some manipulation (Jain, 1986), it can be shown that

$$\mu_f(\mathbf{x},\mathbf{y}) \ge \max_{\mathbf{z}} [\max [\mu_f(\mathbf{x},\mathbf{z}) + \mu_f(\mathbf{z},\mathbf{y}) - 1,0]].$$

Under the complement rule

$$\mu_f^c = 1 - \mu_f$$

it can easily be shown, [Bezdek and Harris, 1978], that

$$\mu_f^c(\mathbf{x},\mathbf{y}) \leq \mu_f^c(\mathbf{x},\mathbf{z}) + \mu_f^c(\mathbf{z},\mathbf{y})$$

which we recognize as the *triangle inequality*; in other words, we have a way to induce from every fuzzy K-partition of X a *metric* fuzzy relation on the data at hand.

In the next section, we will discuss the properties of fuzzy relations in greater detail.

5.4 More on fuzzy relations

Let $R(X)$ be the set of all fuzzy relations in X. Any fuzzy relation $\mu \in R(X)$ may satisfy some combination of the following properties:

1. A fuzzy relation is said to be *symmetric* if

$$\mu(\mathbf{x},\mathbf{y}) = \mu(\mathbf{y},\mathbf{x}), \ \forall \ (\mathbf{x},\mathbf{y}) \in X \times X$$

2. A fuzzy relation is said to be *reflexive* if

$$\mu(\mathbf{x},\mathbf{x}) = 1, \ \forall \ (\mathbf{x},\mathbf{x}) \in X \times X$$

3. A fuzzy relation is said to be *anti-reflexive* if

$$\mu(\mathbf{x},\mathbf{x}) = 0, \ \forall \ (\mathbf{x},\mathbf{x}) \in X \times X$$

4. A fuzzy relation is said to be max-min *transitive* (denoted as $\wedge$-transitivity) if

$$\mu(\mathbf{x},\mathbf{y}) \geq \max_{\mathbf{z}} \ [\min \ [\mu(\mathbf{x},\mathbf{z}),\mu(\mathbf{z},\mathbf{y})]]$$

5. A fuzzy relation is said to be max-product transitive (denoted as π-transitivity) if

$$\mu(\mathbf{x},\mathbf{y}) \geq \max_{\mathbf{z}} \ [\mu(\mathbf{x},\mathbf{z}) \cdot \mu(\mathbf{z},\mathbf{y})]$$

Two classes of fuzzy relations are particularly of interest, the class of *resemblance* relations R_r and the class of *similarity* relations R_s. The reason for this interest rests simply on the fact that in most cases a resemblance relation is assumed to be available after empirical measurement on the data to be analyzed, whereas a similarity relation is aimed at through algorithmical manipulation in order to enable mathematical data analysis.

More precisely, a resemblance relation is a fuzzy relation satisfying the symmetric and reflexive properties, whereas a similarity relation is a fuzzy relation satisfying symmetric, reflexive and transitive properties.

Clearly $R_s \subseteq R_r \subseteq R$.

As mentioned before, both classes of fuzzy relations are key concepts in cluster analysis, including their inverse relations which are usually termed *dissemblance* and *dissimilarity* relations, respectively.

Inverse relations

If we denote a similarity relation by μ_s, and a dissimilarity by μ_d, then the following property follows immediately from their *inverse relations*:

For $\mu_s, \mu_d \in [0,1]$, we consider

$$\mu_d = 1 - \mu_s \implies \mu_s(\mathbf{x,y}) \geq \max_{\mathbf{z}} \left[\min \left[\mu_s(\mathbf{x,z}), \mu_s(\mathbf{z,y}) \right] \right]$$

$$\mu_d(\mathbf{x,y}) \leq \min_{\mathbf{z}} \left[\max \left[\mu_d(\mathbf{x,z}), \mu_d(\mathbf{z,y}) \right] \right]$$

$$(\wedge\text{-transitivity})$$

and

$$\implies \mu_s(\mathbf{x,y}) \geq \max_{\mathbf{z}} \left[\mu_s(\mathbf{x,z}) \cdot \mu_s(\mathbf{z,y}) \right]$$

$$\mu_s(\mathbf{x,y}) \leq \min_{\mathbf{z}} \left[\mu_d(\mathbf{x,z}) + \mu_d(\mathbf{z,y}) - \right.$$

$$\mu_d(\mathbf{x,z}) \cdot \mu_d(\mathbf{z,y}) \right]$$

$$(\pi\text{-transitivity})$$

Likewise, for $\mu_s, \mu_d \in [0,\infty)$, we consider

$$\mu_d = 1/\mu_s \implies \mu_s(\mathbf{x,y}) \geq \max \left[\min \left[\mu_s(\mathbf{x,z}), \mu_s(\mathbf{z,y}) \right] \right]$$

$$\mu_d(\mathbf{x,y}) \leq \min_{\mathbf{z}} \left[\max \left[\mu_d(\mathbf{x,z}), \mu_d(\mathbf{z,y}) \right] \right]$$

$$\implies \mu_s(\mathbf{x,y}) \geq \max \left[\frac{\mu_s(\mathbf{x,z}) \cdot \mu_s(\mathbf{z,y})}{\mu_s(\mathbf{x,z}) + \mu_s(\mathbf{z,y})} \right]$$

$$\mu_d(\mathbf{x,y}) \leq \min_{\mathbf{z}} \left[\mu_d(\mathbf{x,z}) + \mu_d(\mathbf{z,y}) \right]$$

Now, if $d \in R_d$ (the class of dissemblance relations) satisfies

$$d(\mathbf{x,y})^{1/\lambda} \leq d(\mathbf{x,z})^{1/\lambda} + d(\mathbf{z,y})^{1/\lambda}, \lambda \in [0,1]$$

the relation d is called a *λ-metric*. For $\lambda = 1$, the metric is known as the Euclidean metric, d_e, and the inequality is known as the triangle inequality. For $\lambda = 0$, the metric is said to be an ultrametric, d^*, where the inequality can be written as

$$d^*(\mathbf{x,y}) \leq \max_{\mathbf{z}} \left[d^*(\mathbf{x,z}), d^*(\mathbf{z,y}) \right]$$

The inverse of d_e, $\mu_\Delta = 1 - d_e$, is given (Bezdek and Harris, 1978) by

$$\mu_\Delta(\mathbf{x},\mathbf{y}) \geq \max_{\mathbf{z}} \left[\max \left[\mu_\Delta(\mathbf{x},\mathbf{z}) + \mu_\Delta(\mathbf{z},\mathbf{y}) - 1, 0 \right] \right]$$

(Δ-transitivity)

As it has been shown that $\wedge$-transitivity is equivalent to the ultrametric inequality, Δ-transitivity is equivalent to the triangle inequality.

In conclusion, we observe that

$$\{\mu_\wedge\} \subseteq \{\mu_\pi\} \subseteq \{\mu_\Delta\}$$

$$\{1 - \mu_\wedge\} \subseteq \{1 - \mu_\Delta\}$$

and $$\{1/\mu_\wedge\} \subseteq \{1/\mu_\pi\}$$

5.4.1 Transitive closure

The next definitions constitute the notion of *transitive closure*. Let

$$\mu: X \times Y \quad \Rightarrow \quad [0,1]$$

and $$\sigma: Y \times Z \quad \Rightarrow \quad [0,1]$$

be two fuzzy relations. One may ask how the fuzzy relation

$$v: X \times Z \quad \Rightarrow \quad [0,1]$$

can be formalized.

Among other possibilities, two composition rules appear to be relevant:

(i) $$v_o(\mathbf{x},\mathbf{z}) = (\mu \circ \sigma)(\mathbf{x},\mathbf{z}) = \max_{\mathbf{y}} \left[\min \left[\mu(\mathbf{x},\mathbf{y}), \mu(\mathbf{y},\mathbf{z}) \right] \right]$$

(termed as max-min composition) and

(ii) $$v_\bullet(\mathbf{x},\mathbf{z}) = (\mu \bullet \sigma)(\mathbf{x},\mathbf{z}) = \max_{\mathbf{y}} \left[\mu(\mathbf{x},\mathbf{y}) \bullet \mu(\mathbf{y},\mathbf{z}) \right]$$

(termed as max-product composition).

It is easy to see that associate properties hold, thus

$$v \circ (\mu \circ \sigma) = (v \circ \mu) \circ \sigma$$

and $$v \bullet (\mu \bullet \sigma) = (v \bullet \mu) \bullet \sigma$$

Then, it follows that for any fuzzy relation $\mu \in R(X)$, we may denote

$$\forall\ (\mathbf{x},\mathbf{y})\quad \in X \times X:\ \mu_\circ^2(\mathbf{x},\mathbf{y}) = \mu(\mathbf{x},\mathbf{y})\circ\mu(\mathbf{x},\mathbf{y})$$

$$\mu_\circ^k(\mathbf{x},\mathbf{y})\quad = \mu_\circ^{k-1}(\mathbf{x},\mathbf{y})\circ\mu(\mathbf{x},\mathbf{y}) =$$

$$= \max_{\mathbf{z}_1,\mathbf{z}_2,\ldots,\mathbf{z}_{k-1}}\ [\min[\mu(\mathbf{x},\mathbf{z}_1),\mu(\mathbf{z}_1,\mathbf{z}_2),\ldots,\mu(\mathbf{z}_{k-1},\mathbf{y})]]$$

and $\quad \mu_\circ^{n+m}(\mathbf{x},\mathbf{y}) = \max\ [\min\ [\mu_\circ^n(\mathbf{x},\mathbf{z}),\mu_\circ^m(\mathbf{z},\mathbf{y})]]$

Likewise, we find

$$\mu_\bullet^k(\mathbf{x},\mathbf{y})\quad = \mu_\bullet^{k-1}(\mathbf{x},\mathbf{y})\bullet\mu(\mathbf{x},\mathbf{y})$$

$$= \max_{\mathbf{z}_1,\mathbf{z}_2,\ldots,\mathbf{z}_{k-1}}\ [\mu(\mathbf{x},\mathbf{z}_1)\bullet\mu(\mathbf{z}_1,\mathbf{z}_2)\bullet\ldots\bullet\mu(\mathbf{z}_{k-1},\mathbf{y})]$$

The max-transitive closure of μ is a fuzzy relation $\hat{\mu}$ for which

$$\hat{\mu} = \max\ [\mu_\circ^k]$$

where $\mu^k = \mu\circ\mu\circ\ldots\circ\mu$ (k times, $k > 1$).

The transitive closure forms the essential key in generating a nested family of equivalence relations on a data set. In fact, the single-linkage hierarchical structure is induced by a k-step transitive closure.

It can easily be understood that if μ and σ are two fuzzy relations on $f \times f$ it follows that

$$\max_{\mathbf{z}}\ [\min\ [\mu(\mathbf{x},\mathbf{z}),\mu(\mathbf{z},\mathbf{y})]] = (\mu\circ\sigma)(\mathbf{x},\mathbf{y}) \leq \min\ [f(\mathbf{x}),f(\mathbf{y})]$$

So, $(\mu\circ\sigma)$ is a fuzzy relation on f; the same holds for $(\mu\bullet\sigma)$.

5.4.2 Hierarchical structures

Practitioners know that intrinsic transitivity is particularly difficult to justify in many applications and turns out to be one of the most subtle of properties to generalize. This is why algorithmical linking of the input resemblance data has to be considered as the fundamental methodology in constructing clusters.

Beginning with a reflexive, symmetric fuzzy relation (resemblance relation), its transitive closure is obtained by any valid composition rule. Once transitivity is obtained, the entries of the resulting relation matrix are used to define a nested sequence of hard equivalence relations by thresholding at levels in between successive values of the fuzzy similarity relation which ultimately yields a partition tree or dendrogram.

Alternatively, beginning with an anti-reflexive, symmetric distance relation (dissemblance relation), an ultrametric relation is induced by any hierarchical clustering algorithm. Then, the entries of the ultrametric relation matrix are used to define a nested sequence of equivalence relations, yielding a partition tree or dendrogram.

The single-linkage algorithm can be phrased mathematically in the above composition rules, where the similarity relation is dual with the ultrametric relation.

In anticipation of the next section, imposing transitivity can also be phrased as follows. Beginning with a reflexive, symmetric fuzzy relation (resemblance relation), there exists a *minimum distortion approximation* applying any cost function operating on $\mu' = \mu + \Delta\mu$ such that μ' satisfies the transitivity properties. Imposing transitivity property upon a relation matrix with as little distortion (costs) as possible has been suggested in Zahn (1964).

5.5 Generalized approximation

Imposing transitivity upon a relation matrix can be phrased in terms of minimum cost transformation of the input reflexive, symmetric resemblance relation. The transformation will be phrased in graph-theoretical terms.

5.5.1 Fuzzy graphs

Any relation R on a set X can be regarded as defining a *graph* with node set X and arc set R. Following Rosenfeld (1975), similarly, any fuzzy relation μ can be regarded as defining a weighted graph, or fuzzy graph, where the arc $(\mathbf{x},\mathbf{y})$ has weight $\mu(\mathbf{x},\mathbf{y}) \in [0,1]$.

Then, formally, a fuzzy graph $G = (f,\mu)$ is a pair of functions:

$$f{:}X \quad\Rightarrow\ [0,1]$$

and $\mu{:}X \times X \ \Rightarrow\ [0,1]$

where for all $\mathbf{x}$ and $\mathbf{y}$ in X we have

$$\mu(\mathbf{x},\mathbf{y}) \leq \min [f(\mathbf{x}),f(\mathbf{y})].$$

The *fuzzy graph $H = (h,\sigma)$* is called a fuzzy subgraph of G if

$$h(\mathbf{x}) \leq f(\mathbf{x}), \ \forall\ \mathbf{x}$$

and $\sigma(\mathbf{x},\mathbf{y}) \leq \mu(\mathbf{x},\mathbf{y}), \ \forall\ (\mathbf{x},\mathbf{y})$

The basic propositions which we have encountered in Section 5.3:

- a fuzzy relation μ is a similarity relation if and only if its ε-level sets $\{\mu(\varepsilon)\}$, $\varepsilon \in [0,1]$, are equivalence relations in X;
- if a fuzzy relation μ is a similarity relation, then its ε-level sets $\{\mu(\varepsilon)\}$ are equivalence relations in $F(\varepsilon)$;

lead to the following proposition in graph-theoretical terms.

Proposition

If $H(h,\sigma)$ is a fuzzy graph of $G(f,\mu)$, then for any threshold ε, $\varepsilon \in [0,1]$, $H(\chi_{h(\varepsilon)},\sigma(\varepsilon))$ is a subgraph of $G(\chi_{h(\varepsilon)},\mu(\varepsilon))$.

A fuzzy graph is symmetric if $\mu(\mathbf{x},\mathbf{y})$ is symmetric. We shall assume from now on that fuzzy graphs are symmetric.

An ε-*path* in a fuzzy graph is a sequence of distinct nodes $\mathbf{x}_0$, $\mathbf{x}_1$, $\mathbf{x}_2$,..., $\mathbf{x}_k$ such that $(\mathbf{x}_{i-1},\mathbf{x}_i) \geq \varepsilon$ for $1 \leq i \leq k$ and $\varepsilon \in [0,1]$. The length of an ε-path is called $k \geq 0$; the *strength* τ of an ε-path is defined to be the weight of the weakest arc of the path, thus

$$\tau = \min \, [\mu(\mathbf{x}_{i-1},\mathbf{x})], \, i = 1, \, ..., \, k$$

Example

For $X = \{\mathbf{x}_1,\mathbf{x}_2,\mathbf{x}_3,\mathbf{x}_4,\mathbf{x}_5,\mathbf{x}_6,\mathbf{x}_7\}$ the following fuzzy relation matrix is assumed to be given:

$\mu(\mathbf{x}_i,\mathbf{y}_j)$	$\mathbf{x}_1$	$\mathbf{x}_2$	$\mathbf{x}_3$	$\mathbf{x}_4$	$\mathbf{x}_5$	$\mathbf{x}_6$	$\mathbf{x}_7$
$\mathbf{x}_1$	1	0.5	0	0.9	0.3	0	0
$\mathbf{x}_2$		1	0.5	0	0	0	0
$\mathbf{x}_3$			1	0.2	0	0	0
$\mathbf{x}_4$				1	0	0	0
$\mathbf{x}_5$					1	0.3	0.2
$\mathbf{x}_6$						1	0
$\mathbf{x}_7$							1

Figure 5.5 shows the resulting relational graph $G(1,\mu)$, and $H(1,\sigma)$ as a fuzzy subgraph of $G(1,\sigma)$ since

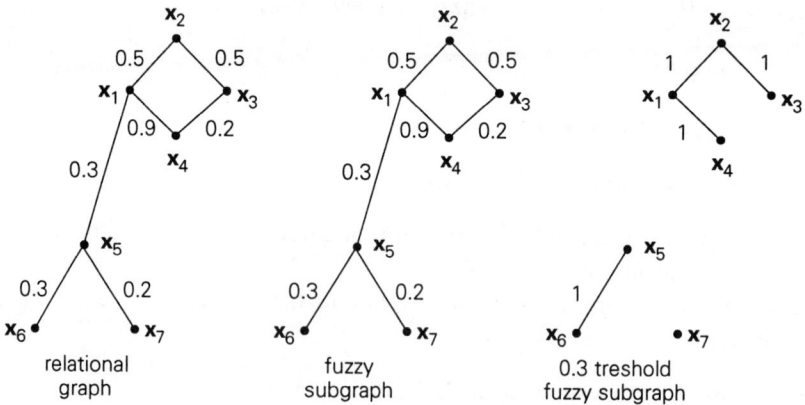

relational graph   fuzzy subgraph   0.3 treshold fuzzy subgraph

Figure 5.5: A relational graph G(1,μ) and a fuzzy subgraph H(1,σ).

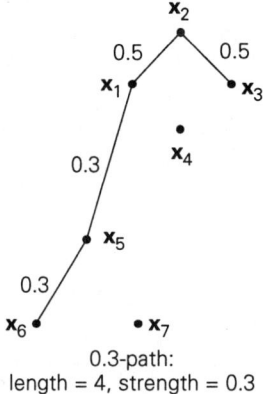

0.3-path:
length = 4, strength = 0.3

Figure 5.6: An ε-path (ε = 0.3) for G(1,μ(0.3)) and H(1,σ(0.3)).

$$\sigma(\mathbf{x},\mathbf{y}) \leq \mu(\mathbf{x},\mathbf{y}), \ \forall \ (\mathbf{x},\mathbf{y}).$$

If we select, as an example, $\varepsilon = 0.3$, it follows that $H(1,\sigma(0.3))$ is a subgraph of $G(1,\mu(0.3))$, illustrated in Figure 5.6. The strength of distinct nodes $\{\mathbf{x}_6,\mathbf{x}_5,\mathbf{x}_1,\mathbf{x}_2,\mathbf{x}_3\}$ is a 0.3-path with length $k = 4$ and strength $\tau = 0.3$

A k-step relation

$$\mu^k(\mathbf{x},\mathbf{y}) = \max_{\mathbf{z}_0,\mathbf{z}_1,\ldots,\mathbf{z}_{k-1}} \ [\min \ [\mu(\mathbf{x},\mathbf{z}_0),\mu(\mathbf{z}_0,\mathbf{z}_1),\ldots,\mu(\mathbf{z}_{k-1},\mathbf{y})]]$$

has the following meaning:

find the strength of the strongest path of all possible paths from x to y which have the same length k.

Then, *connectedness* (transitive closure):

$$\hat{\mu}(\mathbf{x},\mathbf{y}) = \max_{k} \ [\mu^k(\mathbf{x},\mathbf{y})]$$

where $\mu^k = \mu \circ \mu \circ \ldots \circ \mu$ (k times, $k \geq 1$) is defined as the strongest path of all possible paths from $\mathbf{x}$ to $\mathbf{y}$ irrespective of the length of paths.

Consequently, two elements $\mathbf{x}$ and $\mathbf{y}$ are called ε-connected, if

$$\hat{\mu}(\mathbf{x},\mathbf{y}) \geq \varepsilon$$

From the previous example we learn that $\mathbf{x}_1$ and $\mathbf{x}_2$ are 0.5-connected since

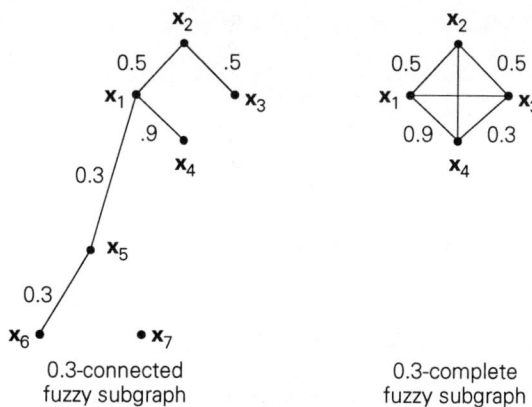

$$\mu^2(\mathbf{x}_1,\mathbf{x}_3) = \max \, [\min \, [\mu(\mathbf{x}_1,\mathbf{x}_2),\mu(\mathbf{x}_2,\mathbf{x}_3)],\min \, [\mu(\mathbf{x}_1,\mathbf{x}_4),\mu(\mathbf{x}_4,\mathbf{x}_3)]] =$$

$$= 0.5$$

$$= \hat{\mu}(\mathbf{x}_1,\mathbf{x}_3)$$

Next, a graph is called *ε-complete* if for all pairs $(\mathbf{x},\mathbf{y})$ it holds that $\mu(\mathbf{x},\mathbf{y}) \geq \varepsilon$. Figure 5.7 illustrates an *ε*-connected graph and an *ε*-complete graph for $\varepsilon \in [0,0.3]$.

In $G(f,\mu)$, we call C a *fuzzy cluster* of order k if

$$\min_{(\mathbf{x},\mathbf{y})\in C} [\mu^k(\mathbf{x},\mathbf{y}) \geq \max_{\mathbf{u}\notin C} [\min_{\mathbf{v}\in C} [\mu^k(\mathbf{u},\mathbf{v})]]$$

In fact, the fuzzy clusters of order k obtained by using the above definition are just ordanary complete subgraphs obtained by thresholding the kth power of the given fuzzy graph.

Let C_1 and C_2 be two distinct subsets of nodes of the graph $G = (f,\mu)$. Then we call the set of arcs $(\mathbf{x},\mathbf{y})$, where $\mathbf{x} \in C_1$ and $\mathbf{y} \in C_2$, a *bridge set* $\Gamma(C_1,C_2)$. The weight of the bridge set $W(C_1,C_2)$ can be given by

$$W(C_1,C_2) = \sum_{\mathbf{x}\in C_1} \sum_{\mathbf{y}\in C_2} \mu(\mathbf{x},\mathbf{y})$$

Example

Assume the following graph $G(1,\mu)$, shown in Figure 5.8, and the coresponding fuzzy clusters of order 1. The weight of the bridge set $\Gamma(C_1,C_2)$ yields

$$W(C_1,C_2) = .2$$

The foregoing provides us with the basic properties of fuzzy graphs and enables us to redefine the clustering problem as a constrained approximation problem where transitivity properties are imposed upon a relation matrix of the data to be clustered.

5.5.2 The approximation approach

As we have seen, a symmetric relation $R: X \times X \Rightarrow \{0,1\}$ can be visualized by a graph $G = (X, \chi_R)$. The question then arises of how to transform the relation R into an equivalence relation E such that we obtain a partition

$$\{C_i\}^K_{i=1}, K \in \{1,n\}$$

satisfying $\bigcup\limits_{i=1}^{K} C_i = X$, and $C_i \cap C_j = \emptyset, i \neq j$.

In terms of graphs this simply means forming K disjunct complete subgraphs. Clearly, some of the values have to be altered from 1 to 0, and vice versa. As a result, we have that

$$\chi_E(\mathbf{x},\mathbf{y}) \geq \max_{\mathbf{z}} \left[\min \left[\chi_E(\mathbf{x},\mathbf{z}), \chi_E(\mathbf{z},\mathbf{y}) \right] \right]$$

in other words: the approximation χ_E of χ_R is transitive. This approximation, suggested in Zahn (1964), is constrained by the following *cost criterion*:

$$\rho(E,R) = \sum_{i=1}^{n} \sum_{j=1}^{n} |\chi_R(\mathbf{x}_i,\mathbf{x}_j) - \chi_E(\mathbf{x}_i,\mathbf{x}_j)|$$

Then E^*, for which we can write

$$\rho(E^*,R) = \min_{E} \left[\rho(E,R) \right]$$

is assumed to be the '*best*' *approximation*.

The above approximation has been generalized to fuzzy relations in Backer (1979), as follows. Given a symmetric fuzzy relation μ, find a fuzzy equivalence relation μ_E^*, such that

$$\rho(\mu,\mu_E^*) = \min_{\varepsilon} \left[\rho(\mu,\mu_E^*)_\varepsilon \right]$$

and μ_E is transitive, where

$$\rho(\mu,\mu_E^*) = \sum_{i=1}^{n} \sum_{j=1}^{n} |\mu(\mathbf{x}_i,\mathbf{x}_j) - \mu_E(\mathbf{x}_i,\mathbf{x}_j)|$$

Figure 5.8 illustrates the approximation as a function of ε. We observe the following approximation costs:

costs	ε
$r(\mu,\mu_E)_e = 2.7$	0
2.4	0.1
2.1	0.2
2.2	0.3
2.5	0.4
$\vdots$	$\vdots$
6.5	1

Let us denote the optimal value of ε:by ε_{opt}. Then, the optimal partition $\{C_i{}^*\}^K_{i=1}$ for which the value of $\rho(\mu,\mu_E{}^*)$ is minimal.

The problem is to find ε_{opt}.

As is known, except from complete enumeration, no general algorithm can be formulated which should lead to the optimal solution. Only in terms of the iterative partitioning optimization algorithm from Section 5.3 can one obtain at least a local optimum.

The set $X_0 \subseteq X$ is called ε-semicomplete if for each partition $\{C_1,C_2\}$ of X_0, we have

$$W(C_1,C_2) \geq \frac{|C_1||C_2|}{2} \cdot \varepsilon$$

where $W(C_1,C_2)$ is the weight of the bridge set. Following Backer (1979), two theorems can be proven.

Theorem 5.1

Each class C_i of an optimal partition is ε-semicomplete.

Theorem 5.2

If C_i and C_j are two disjunct classes of an optimal partition with the smallest number of classes, then the following inequality holds:

$$W(C_1,C_2) < \frac{|C_i||C_j|}{2} \cdot \varepsilon$$

The essence of the above approach is not to be found in finding the optimal solution. On the contrary, it offers a way to compare solutions from different approaches in transforming any fuzzy relation μ into fuzzy equivalence relations μ_E, particularly in a sequence of trials. This will be the subject of the next section.

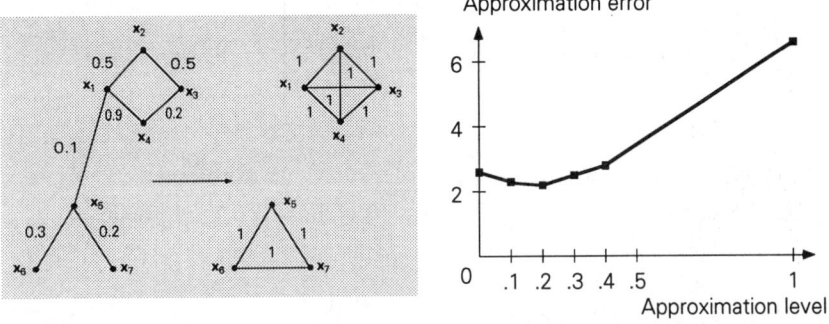

Figure 5.8: Example of approximation costs as a function of ε.

Example

Let us reconsider our example from Chapter 3. The input relation matrix was given as

$$\mu = \begin{bmatrix} 1 & 0.3 & 0.7 & 0.2 \\ & 1 & 0.5 & 0.9 \\ & & 1 & 0.1 \\ & & & 1 \end{bmatrix}$$

The above relation is symmetric and reflexive. It is easy to verify that μ does not satisfy the triangle inequality.

Let T_1 be a transform imposing metric properties. From the foregoing, we know that

$$T_1: \mu_\Delta(\mathbf{x}_i,\mathbf{x}_j) = \max_{\mathbf{x}_k} \left[\mu(\mathbf{x}_i,\mathbf{x}_k) + \mu(\mathbf{x}_k,\mathbf{x}_j) - 1,0 \right]$$

$$\mu_\Delta = T_1(\mu) = \begin{bmatrix} 1 & 0.3 & 0.7 & 0.2 \\ & 1 & 0.5 & 0.9 \\ & & 1 & 0.4 \\ & & & 1 \end{bmatrix}$$

with $\rho(\mu,\mu_\Delta) = 0.6$.

Let T_2 be a transform imposing ultrametric properties (transitivity):

$$T_2: \mu_\wedge(\mathbf{x}_i,\mathbf{x}_j) = \max_{\mathbf{x}_k} \left[\min \left[\mu(\mathbf{x}_i,\mathbf{x}_k),\mu(\mathbf{x}_k,\mathbf{x}_j) \right] \right]$$

As a result, we obtain

$$\mu_\wedge = T_2(\mu_\Delta) = T_2(T_1(\mu)) = \begin{bmatrix} 1 & 0.5 & 0.7 & 0.5 \\ & 1 & 0.5 & 0.9 \\ & & 1 & 0.5 \\ & & & 1 \end{bmatrix}$$

with $\rho(\mu,\mu_\wedge) = 1.8$ and $\rho(\mu_\Delta,\mu_\wedge) = 1.2$.

Let $T_3: \mu(\varepsilon) = \{(\mathbf{x}_i,\mathbf{x}_j)|\,\mu(\mathbf{x}_i,\mathbf{x}_j) \geq \varepsilon\}$, $\varepsilon \in [0,1]$. For $\varepsilon = 0.7$, we obtain

$$\mu' = \mu_\wedge(\varepsilon) = T_3(T_2(T_1(\mu))) = \begin{bmatrix} 1 & 0 & 1 & 0 \\ & 1 & 0 & 1 \\ & & 1 & 0 \\ & & & 1 \end{bmatrix}$$

with $\rho(\mu,\mu') = 3.0$ and $\rho(\mu_\wedge,\mu') = 4.8$.

Apparently, we observe

$$\rho(\mu,\mu_\Delta) + \rho(\mu_\Delta,\mu_\wedge) = \rho(\mu,\mu_\wedge)$$
$$(\therefore \mu_\Delta \text{ is an 'in-between' relation of } \mu \text{ and } \mu_\wedge)$$

and

$$\rho(\mu',\mu) + \rho(\mu,\mu_\wedge) = \rho(\mu',\mu_\wedge)$$
$$(\therefore \mu \text{ is an 'in-between' relation of } \mu' \text{ and } \mu_\wedge).$$

Clearly, our task is to find

$$\min_{T,\varepsilon} [\rho(\mu,\mu(\varepsilon))] \text{ for given } \mu$$

Remark

In practice, very often symmetry is not satisfied a priori. That is $\mu_{ij} \neq \mu_{ij} \; \forall \; i,j$. Then, at least the following transforms can be imagined.

$$\mu^1 = [\mu^1{}_{ij} = \mu^1{}_{ji} = \mu_{ij}]$$

$$\mu^2 = [\mu^2{}_{ij} = \mu^2{}_{ji} = \mu_{ji}]$$

$$\mu^3 = [\mu^3{}_{ij} = \mu^3{}_{ji} = (\mu_{ji} + \mu_{ij})/2]$$

$$\mu^4 = [\mu^4{}_{ij} = \mu^4{}_{ji} = \max [\mu_{ji},\mu_{ij}]]$$

$$\mu^5 = [\mu^5{}_{ij} = \mu^5{}_{ji} = \min [\mu_{ji},\mu_{ij}]]$$

Note that from the point of view of distortion, they all are equal. That is

$$\rho(\mu,\mu^1) = \rho(\mu,\mu^2) = \rho(\mu,\mu^3) = \rho(\mu,\mu^4) = \rho(\mu,\mu^5)$$

However, if we consider

$$\mu = \begin{bmatrix} 1 & 0.3 & 0.7 & 0.2 \\ 0.7 & 1 & 0.5 & 0.9 \\ 0.5 & 0.9 & 1 & 0.1 \\ 0.6 & 0.7 & 0.3 & 1 \end{bmatrix}$$

then $\rho(\mu,\mu^k) = 1.8$ for $k = 1,2,\ldots,5$.

Applying single linkage (SL), complete linkage (CL), and average linkage (AL) yields the following result

	SL	CL	AL
$[\mu^1]$	<u>2.2</u>	3.2	2.3
$[\mu^2]$	2.8	3.0	2.6
$[\mu^3]$	2.6	2.6	<u>2.2</u>
$[\mu^4]$	3.2	3.0	3.1
$[\mu^5]$	<u>2.2</u>	3.2	2.4

Not surprisingly, minimal distortion is obtained if the symmetry operator corresponds with the operator used in the algorithm for linkage.

In conclusion, one should always keep in mind the nature of operators used in cascade.

5.6 Strong and weak patterns in clustering: a fuzzy relation

The problem of finding a 'good and valid partition' of X has been stated as finding the partition $\{\widehat{C}_i\}_{i=1}^K$ which optimizes some criterion (the amount of fuzziness, approximation error, weight of bridge sets, and so on), generally denoted as φ. In fact all the working techniques optimize some criterion φ, but without guarantee of attaining the best possible solution. Many different partitions can be obtained, due to either changing thresholds, or changing the initial conditions. Then, a general study of these partitions may be quite informative while varying the conditions as mentioned before.

Let $\{\widehat{C}_i\}_{i=1}^K$ be denoted as a resulting K-partition P; then P^1, $P^2,\ldots,P^q$ are q resulting K-partitions in q trials. Further, let P^i_j be the jth cluster of the partition P^i. Thus

$$P^i = \{P^i_1, P^i_2, \ldots, P^i_j, \ldots, P^i_K\}, \ i = 1,2,\ldots,q$$

A pattern $\mathbf{x}$ has been in cluster α_1 in the first partition P^1, in cluster α_2 in $P^2,\ldots$, in cluster α_q in P^q. With X we can associate a mapping $H:X \Rightarrow n^q$ such that with each pattern $\mathbf{x} \in X$ is associated a q-vector.

Let **x** and **y** be two patterns in X, and

$$H(\mathbf{x}) = (\alpha_1, \alpha_2, \ldots, \alpha_q)$$

and $\qquad H(\mathbf{y}) = (\beta_1, \beta_2, \ldots, \beta_q)$

Then, we define $\delta(\mathbf{x},\mathbf{y})$ as the index number for which $\alpha_i = \beta_i$. It is easy to see that δ is an ultrametric distance.

Next, we define

$$F_q(\mathbf{x}) = \{\mathbf{y} \in X : \delta(\mathbf{x},\mathbf{y}) = q\}$$

and $\qquad F_p(\mathbf{x}) = \{\mathbf{y} \in X : \delta(\mathbf{x},\mathbf{y}) = p\}$

Following Diday and Simon (1976), we define the *cross-partition* $\prod$:

$$\prod = P^1 \cap P^2 \cap P^3 \cap \ldots \cap P^q$$

which is the partition defined by the connected part of the graph $G = (X, F_q)$. Intuitively, a class of $\prod$ is made up with patterns **x** which are always in the same cluster in the q K-partitions P^j.

More generally, let us consider the set γ_p of connected parts of the nodes of the graph $G_p = (X, F_p)$. The list $\gamma_1, \ldots, \gamma_p, \ldots, \gamma_q$ defines a hierarchy on X, δ being the ultrametric. Then, γ_1 constitutes the set of '*weak(est) patterns*', whereas γ_q constitutes the set of '*strong(est) patterns*'. Each class, π, of $\prod$ defines a fuzzy membership function

$$f_\pi(\mathbf{x}) = \delta(\mathbf{x},\mathbf{y})/q \text{ with } \mathbf{y} \in \pi$$

Clearly,

$$f_\pi(\mathbf{x}) = 1 \text{ if } \mathbf{x} \in \pi, \text{ and } f_\pi \text{ is constant for all } \mathbf{y} \in \pi$$

Figure 5.9 (Diday and Simon, 1976) shows an example of a resulting hierarchy defined by $0.1, \ldots, .p, \ldots, .q, .(x,y)$ being the ultrametric. The tree represents the 'weak' and 'strong' patterns nicely in the global partitioning.

Now let us recall the example of Figure 5.4 (Backer, 1978). The data are shown in Figure 5.10. Following Section 2.5, we apply partitional clustering for $K = 2$ (100 trials: $q = 100$). The dominant partitions are shown in Figure 5.11a, b and c.

In view of the foregoing, we consider – as an example – the $(k = 2)$-partition P^1 (corresponding to Figure 5.11a), P^2 (corresponding to the partition shown in Figure 5.11b), and P^3 (the partition as predicted in Figure 5.11c). Out of 100 trials, we observe that P^1, P^2, P^3 are detected 27, 20 and 14 times, respectively. The cross-partition $\prod = P^1 \cap P^2 \cap P^3$ defines the subclass π of patterns which are in the same cluster in all 61 ($k = 2$)-partitions considered. Clearly, the strongest patterns yield a membership value of 100%, while other subsets contain weaker patterns, yielding lower membership values of 75%, 70% and 55% due to lower valuesfor the index $\delta(x,y)$ with $y \in \pi$. As a result, we observe induced fuzziness, depicted in Figure 5.12, which resembles the induced fuzziness as illustrated in Figure 5.4. In Chapter 6, we use repeated experimentation as a means to represent partitional fuzziness adequately.

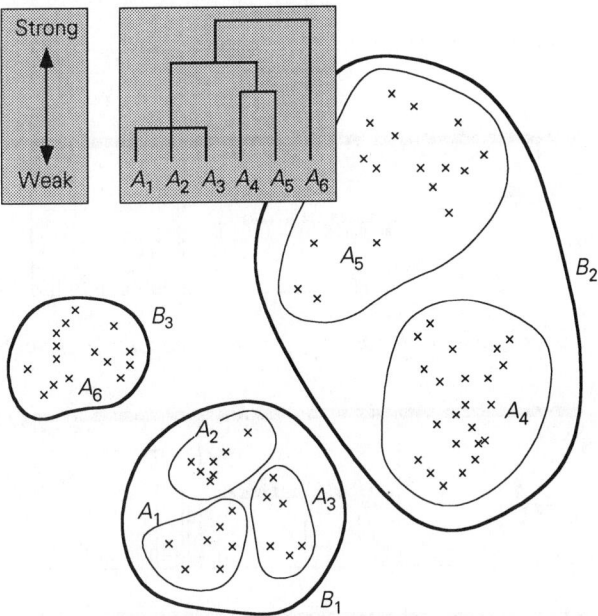

Figure 5.9: Resulting hierarchy defined by the cross-partition of q trials: A_1, A_2, A_3 being the weakest pattern, and B_1, B_2, B_3 being the strongest pattern.

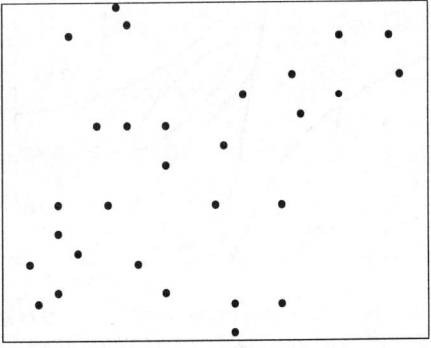

Figure 5.10: Data (identical to Figure 5.4) on which 'strong' and 'weak' patterns are to be imposed by large amount of trials of K-partitioning (K = 2).

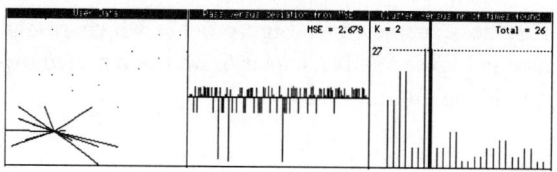

Figure 5.11a: The strongest (most dominant) clustering pattern found for (K = 2) partitional clustering.

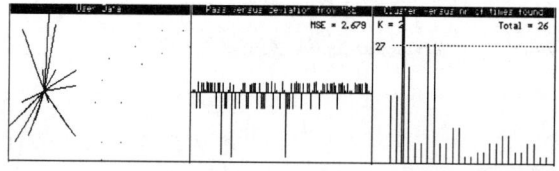

Figure 5.11b: Weaker (less dominant) clustering pattern found for (K = 2) partitional clustering.

Figure 5.11c: The weakest clustering pattern considered in this example.

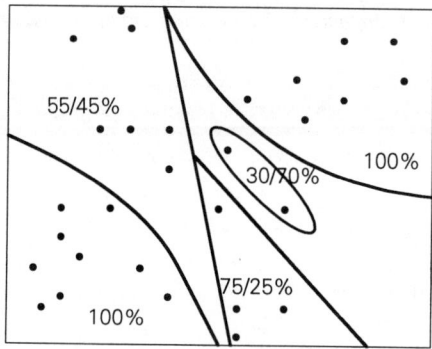

Figure 5.12: Induced fuzziness (strong and weak patterns) based on repeated experimentation with partitional clustering; compare with Figure 5.4).

5.7 Goal-directed methods of comparison

As has already been indicated, the intuitive appeal of the fuzzy set approach lies in the fact that the data analyst is now equipped with a value of the performance measure as well as the identification of bridging (overlapping) pattern points. However, a single performance measure cannot summarize all the information that can be gleaned from a clustering (Dubes and Jain, 1976). Dubes and Jain discussed different points of view in comparing clustering techniques and clustering results. No general answer exists to questions like 'which program is best', or 'what structure is most accurate'.

The method of comparison we are dealing with is to establish a *ranking* of the *utility* of clustering results obtained from different clustering programs with respect to a certain application domain where utility can be measured uniquely. So, we may call this a goal-directed comparison. To make this point clear, Backer and Jain (1981) conducted a number of experiments with real and artificial data. We will summarize these in order to emphisize their promising nature.

Experiment 1

Five partitions (Figure 5.13) achieved by five different square-error programs[3] are ranked on the basis of the value of φ resulting from their corresponding induced fuzzy sets. Clearly, the ranking based on the φ value is visually appealing to human under-standing but no formal judgement can be made.

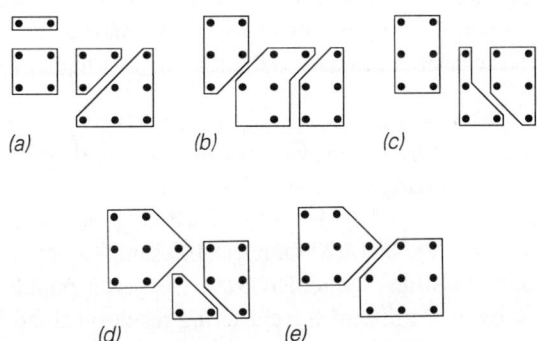

Figure 5.13: Five clusterings obtained from five different square-error programs; ranking from a to e with increasing value of φ (experiment 1).

---

3 The programs involved are
   a. FORGY   (simple *K*-means program)
   b. ISODATA   (as FORGY; heuristics employed)
   c. WISH   (sequential program)
   d. CLUSTER   (hill-climbing technique)
   e. FUZZ   (fuzzy partitioning; Backer, 1978)
   Details of these algorithms can be found in Anderberg (1973).

Experiment 2

The five different clustering programs above were applied to a particular data set, derived from Munson handprinted Fortran character set. Four characters were selected, namely *I*, *M*, *O* and *X*, for a total of 192 characters. Since the true categories are known, the number of errors obtained by comparing the true categories and the clustering result may guide us in interpreting the values of the performance measure φ.

The results are as follows:

clustering	φ value[4]	no. of errors
FUZZ	0.402	19
FORGY	0.399	21
CLUSTER	0.391	22
WISH	0.373	27
ISODATA	0.365	36
'true categories'	0.348	0

Here, we see that the the the ranking based on the φ values matches the ranking based on the number of errors. Note that the absolute value of φ depends on the free parameter β. Within a reasonable range of values for β, no changes in the ranking were observed. In fact, the absolute value of φ is not important. The choice an appropriate value for β should guarantee to induce sufficient fuzziness.

The results reported above suggest that if a classifier should be designed on the basis of the resulting clustering, the utility of this approach is made explicit by the number of errors when testing this classifier. This leads to the next experiment.

Experiment 3

Fifty observations were sampled from two bivariate normal distributions each and the total of 100 observations presented to both FUZZ and ISODATA. Each of the clustering results forms the set of design labels for a 1-nearest neighbour classifier. Each 1-NN classifier is then tested with another 200 samples (100 samples per class) – but the same 200 samples for each classifier – sampled from the given populations. As may be expected, φ values show the utility of the clustering result obtained by FUZZ over the clustering result given by ISODATA. The goal-directed comparison yields 5 against 11 errors, respectively, in the testing phase.

For these data and the choice of β the maximum value of φ is 0.701. One way to *test* the *significance* of $\varphi = 0.701$ given a certain β is to compare this result with the φ value obtained from random data with no apparent clustering at all.

---

[4] Euclidean distance; $\beta = 10$.

Experiment 4

The clustering programs FUZZ and ISODATA were applied to a data set generated from a single bivariate normal distribution. The scaling was changed so that β remained unchanged.

If, for this β, the resulting φ is also high it simply means that $\varphi = 0.701$ does not indicate well-separated clusters. On the contrary, if the resulting φ for the random data is very low, it signifies the result $\varphi = 0.701$ as representing a well-separated clustering.

The results

clustering	no. of clusters	φ value[5]
ISODATA	2	0.077
	3	0.083
	4	0.101
	5	0.097
FUZZ	2	0.087
	4	0.105

provide the φ values for the no-clustering case, indicating that $\varphi = 0.701$ should be considered as a characterization of a well-separated clustering.

5.8 Fuzzy classification inference

In anticipation of the next chapter, in this section we develop some fundamental ideas about *fuzzy classification* and *fuzzy inference* given a set of patterns having partially (or fully) 'hard' or 'soft' labels assigned to 'hard' or 'soft' classes (clusters) contained in 'hard' or 'soft' partitionings of the X-space.

5.8.1 Classification of patterns

Traditional *statistical methods* for pattern classification are based on certain assumptions regarding the attributes (features, pattern properties, pattern measurements), the patterns, and their categories of origin (classes, clusters). The categories are assumed to form disjunct subsets within the set of all patterns (objects). Each object has a class label; for some objects the class labels are known. From such a *training set* the decision rules for classification are derived which are then used to predict the class label of an unknown object. A number of statistical procedures allow the estimation of *probabilities* of the unknown (test) object belonging to the various classes, which reflect the *uncertainty* in the final allocation.

As pointed out before, the fuzzy set approach recognizes the fact that dividing the set of objects in disjunct subsets is often unnatural. It has been indicated that this is particularly true in our aim of partitioning our data in the classical sense. In the setting of

---

5   Euclidean distance; $\beta = 10$.

particularly true in our aim of partitioning our data in the classical sense. In the setting of fuzzy sets, we demonstrated that, instead of one 'hard' class label (cluster assignment), each pattern has 'soft' class labels indicating that it belongs to all classes (clusters), each to its own degree. While statistical methods use 'hard'-labelled patterns in the training set, fuzzy set methods are more general by allowing both 'hard'-labelled and 'soft'-labelled patterns.

In statistical discriminant analysis the objects (e.g. biological samples to be analyzed) are assumed to be randomly selected from the population of objects to be classified. In other words, the training set is assumed to be representative in probability. This requirement is seldom met in cluster analysis. Using the fuzzy set approach, the set of training objects has to be representative in terms of *possibility*. The objects should adequately cover the range of possible samples. Apart from this requirement the distribution of the samples is not important. In fact, no assumptions such as multivariate normality or equality of within-class covariances should have to be made about this distribution.

But even if the model assumptions are met and the classification results themselves are largely correct, we consider it unlikely that human beings, when classifying objects into different categories, reason along the same lines. Much of the human reasoning is approximate rather than exact. This is more or less inevitable since the concepts used in the human classification process (attribute assessment, definition of categories) are vaguely defined.

Fuzzy inference

Apart from Zadeh (1965), in the literature methods early methods for fuzzy reasoning have been proposed in Mamdani and Assilian (1975). A central issue of such fuzzy reasoning methods is *conditional inference* of the form:

Antecedent 1:	**if** x is A **then** y is B
Antecedent 2:	x is A'

Consequence:	y is B'

Or, as an example

Antecedent 1:	**if** $I(f) \mid \{f_i\}_{i=1}^{K}$ is *low* **then** Ct is *high*
Antecedent 2:	$I(f') \mid \{f_i\}_{i=1}^{K}$ is *very low*

Consequence:	Ct is *very high*

where $I(f)$ is the amount of detected fuzziness in a fuzzy partition, and Ct is a index for clustering tendency. If Ct is very high, this is equivalent with the fact that detected clusters are very well separated.

Fuzzy logic

As we have already seen, there is a strong relationship between classical set theory and the truth value of the statement '*x* is an element of *A*', identifying {0,1} with {*false,true*}. Similarly, the membership function of fuzzy sets links fuzzy set theory to *fuzzy logic* on [0,1] or multi-valued fuzzy logic identifying a termset like {*true, more or less true, borderline, more or less false, false*}.

In the case of a fuzzy conditional proposition '**if** *x* is *A* **then** *y* is *B*', $\mu_R(x,y)$ represents the truth value of the *implication* $\mu_A(x) \Rightarrow \mu_B(y)$ as a function of the truth values $\mu_A(x)$ and $\mu_B(y)$. Many rules have been given in the literature to obtain $\mu_R(x,y)$ from $\mu_A(x)$ and $\mu_B(y)$. The problem is to select a rule that does not seem to conflict with intuitive human reasoning within the problem domain.

We note that the implication '**if** *x* is *A* **then** *y* is *B*' does not tell us what happens if *x* is not *A*. Possible escapes are either *y* is unknown, or *y* is not *B*. Detailed discussion can be found in Mizumoto and Zimmermann (1982).

Fuzzy conditional inference

An inference procedure includes the combination of two antecedents (the first one is called the conditional proposition) to arrive at the fuzzy consequence. For this, a 'compositional rule of inference' has to be defined, formally written as

$$B' = A' \circ R_{A \Rightarrow B}$$

From the earlier analysis of fuzzy relations, it has become clear that the max-min composition is a common rule. Thus

$$\mu_{A' \Rightarrow B}(x) = \max_{} [\underset{x}{\text{MIN}} [\mu_A(x), \mu_R(x,y)]]$$

where MIN denotes taking the minimum for each value of *x* and max denotes taking the maximum of the ensuing list. As a result, the inferred membership function cannot be larger than either the membership function of *x* to *A'* or the implication relation μ_R. As we start with a conservative composition (MIN), it is intuitively permitted to choose the maximum subsequently.

Interpolation

The induced membership values can also be obtained using an interpolation procedure. Using this method we first determine the *resemblance* between the two objects

$$\mu_{AA'}(x) = \max [\underset{x}{\text{MIN}} [\mu_A(x), \mu_A(x)]]$$

This similarity and the membership of the learning object to set *B* induce the membership of the test object to set *B*

$$\mu_{B'}(y) = \mu_{AA'}(x) \circ \mu_B(y)$$

If we choose a MIN operator, it is easily understood that the interpolation is equivalent to the inference procedure.

Many objects and many attributes

We now consider the more complex situation where we have many fuzzy conditional propositions to extract information from, x_i $(i = 1,...,n)$, each sample being characterized by a manifold of attributes A_j $(j = 1,...,d)$, and belonging, in principle, to many fuzzy classes, B_L $(L = 1,...,K)$.

Then, the simple inference can be generalized to:

if	**then**
$\mu_{A_1}(x_1)\wedge\mu_{A_2}(x_1)\wedge...\wedge\mu_{A_d}(x_1)$	$\mu_{B_1}(y_1)\wedge...\wedge\mu_{B_K}(y_1)$
$\mu_{A_1}(x_2)\wedge\mu_{A_2}(x_2)\wedge...\wedge\mu_{A_d}(x_2)$	$\mu_{B_1}(y_2)\wedge...\wedge\mu_{B_K}(y_2)$
$\mu_{A_1}(x_3)\wedge\mu_{A_2}(x_3)\wedge...\wedge\mu_{A_d}(x_3)$	$\mu_{B_1}(y_3)\wedge...\wedge\mu_{B_K}(y_3)$
$\vdots$	
$\mu_{A_1}(x_n)\wedge\mu_{A_2}(x_n)\wedge...\wedge\mu_{A_d}(x_n)$	$\mu_{B_1}(y_n)\wedge...\wedge\mu_{B_K}(y_n)$

and given:
$$\mu_{A_1}(x')\wedge\mu_{A_2}(x')\wedge...\wedge\mu_{A_d}(x')$$

conclusion: $\mu_{B_1}(y')\wedge...\wedge\mu_{B_K}(y')$

For each given training sample x_i and a given category B_L we can deduce inferred membership values (or functions) of x' to B_L for each attribute A, say $\mu_{L_{ij}}(x')$, following the inference rule previously given. To arrive at the final inference, taking account of all samples and all attributes, we apply a second max-min cycle

$$\mu_L(x') = \max_i \left[\min_j [\mu_{L_{ij}}(x')] \right]$$

The rational behind this is that the object to be classified resembles a training object i not more than the attribute displaying the least similarity. This attribute determines the membership of x' in B_L as far as object i is concerned (i.e. μ_{L_i}).

Comparing object x' with all training objects x $(i = 1,...,n)$ one is free to choose the object which resembles x' best, giving rise to the largest $\mu_{L_i}(x')$.

This procedure has to be applied for each of the possible categories B_L $(L = 1,...,K)$ separately.

So, let each attribute be identified by a termset – for example, $\{high, medium, low\}$. Let us consider training sample x_i (denoted i for short), and the kth attribute. Then A_{ik} denotes the membership function $(a_{ik1}, a_{ik2}, a_{ik3})$ where

$$a_{ik1} = f_{high}(a_{ik})$$

$$a_{ik2} = f_{medium}(a_{ik})$$

$$a_{ik3} = f_{low}(a_{ik})$$

Likewise, we have for label B_L of x_i

$$b_{iL1} = f_{high}(b_{iL})$$

$$b_{iL2} = f_{medium}(b_{iL})$$

$$b_{iL3} = f_{low}(b_{iL})$$

Then, the resemblance between x_i and x' (the unknown), denoted as $\mu_k(x_i,x')$, is given by

$$\mu_k(x_i,x') = \max \ [\ \min \ [f_{high}(a_{ik}),f_{high}(b_{iL})],$$

$$\min \ [f_{medium}(a_{ik}),f_{medium}(b_{iL})],$$

$$\min \ [f_{low}(a_{ik}),f_{low}(b_{iL})] \ ]$$

Now, the inferred class membership function for the Lth label of unknown x', based on sample x_i and attribute k only, is defined as:

$$Q_{kL}(x_i,x') = (\ \min \ [\ \mu_k(x_i,x'),f_{high}(b_{iL})],$$

$$\min \ [\ \mu_k(x_i,x'),f_{medium}(b_{iL})],$$

$$\min \ [\ \mu_k(x_i,x'),f_{low}(b_{il})])$$

The resemblance between x and x' based on a set of attributes is not stronger than that for the attribute showing the least resemblance for these samples. Therefore, the resemblance on all d attributes is defined as

$$Q_L(x_i,x') = \min_k \ [Q_{kL}(x_i,x')]$$

Finally, the unknown x' is best resembled by that training sample that has the highest resemblance with x'. Thus,

$$Q(x_i,x') = \max_i \ [\ \min_k \ [Q_{kL}(x_i,x')]$$

The above fuzzy classification has been successfully applied in Lincklaen Westenberg et al. (1989).

5.9 Concluding remarks

In this chapter, we have reviewed the conceptual framework for dealing with clusters in which there may be grades of membership and which appears to be relevant to the clustering problem: i.e. scaling the value of the clustering result (validity). The theory of fuzzy sets and fuzzy relations is found to be fundamental and instrumental to explicate the various types of non-uniqueness or vagueness. In particular, the methods of induced fuzziness are of importance because they establish a formal link between 'soft' and 'hard' partitions.

Part 2
Reconsideration of the task

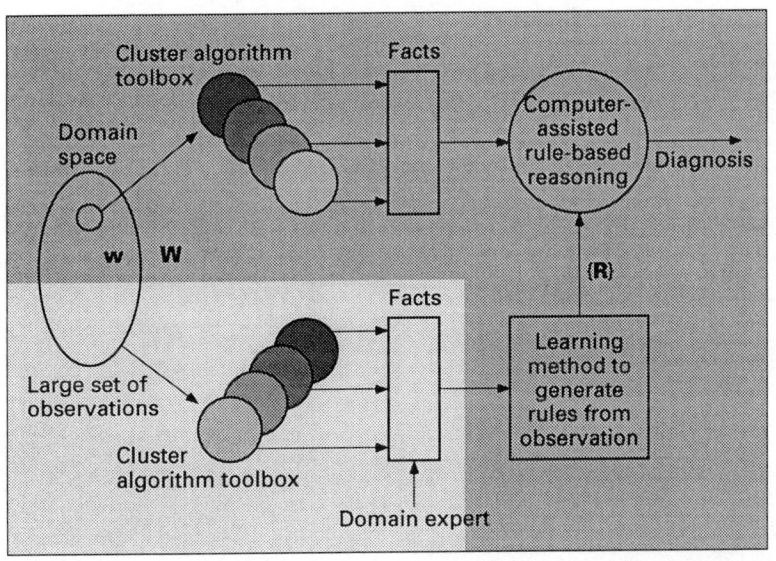

6 | Indeterminacy and uncertainty in cluster analysis

..., it is still good practice to peer into attribute spaces with as many kinds of 'eyes' as possible, with as many different multivariate methods as possible, each providing a fresh perspective. For this reason methods like principal component analysis and multidimensional scaling, and methods for detecting oddly shaped and nonlinear classes are useful adjuncts to cluster analysis (Romesburg, 1984).

There are many problems in realizing the general structure of clustering technique. They are connected with the proper choice of distance function, clustering criterion, minimization procedure and so on. Meanwhile, there is a phenomenon of a general nature that the researcher faces, regardless of how the above mentioned problems are solved. The problem is as follows. Do the clusters, obtained as a result of applying a particular clustering procedure, reflect the underlying structure of data or do they reflect artifacts and statistical irregularities of a given set of data? (Brailovsky, 1991).

This chapter focuses on intelligently exploiting the strengths of different methods, tests and algorithms over varied data domains. It has become clear that there is no single mathematical technique for evaluating the tendency to cluster, the validity of the recovered clusterings, and the performance of individual algorithms, over all types of data structures. It is therefore suggested that we use several methods and algorithms and draw conclusions from the various results. In what follows, we first try to investigate how clustering structure can be classified and what lessons can be drawn from experiments.

6.1 Preliminary remarks

From various studies (Dubes and Jain, 1976), one representative algorithm from each of the three clustering techniques, i.e. the partitional, the hierarchical, and the graph-theoretic techniques, were clearly sufficient to obtain a *rough* idea of about all the possible structures that can be imposed on the data. This helps us to find *approximately* the cluster *configuration* obtainable by these techniques and later examine which of them is natural by computation of a variety of indices and statistics for these structures.

This essentially preludes the remainder of the book. The objective of this chapter is to show, through a series of experiments, how to combine the strong points of the various techniques and algorithms.

In Balasubramandiam et al. (1990) cluster analysis is approached by first choosing appropriate algorithms, depending on the data, and combining the strong points of various methods and algorithms, using different *heuristics* and evidential inferencing.

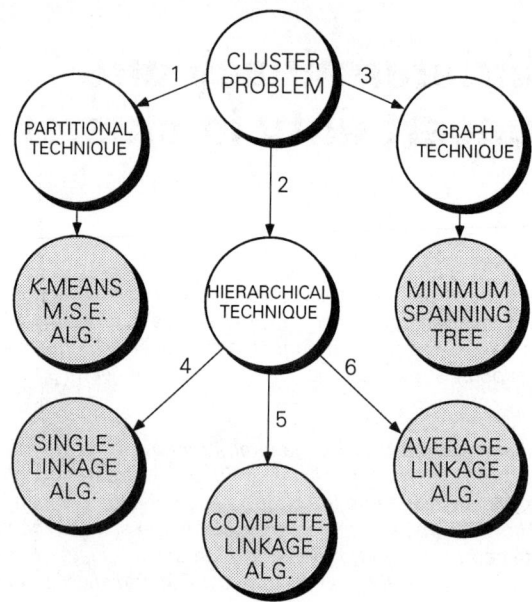

Figure 6.1: Decision tree for possible selection of clustering algorithms depending on the cluster configuration:
(1): spherical clusters;
(2): long-chained clusters; sample size < 200;
(3): concentric clusters; long-chained clusters; sample size > 200;
(4): elongated, linear non-separable, long-chained clusters;
(5): compact clusters, nearly equal sized;
(6): unequal cluster sizes.

As such, in Balasubramandiam et al. (1990) a knowledge-based approach using rules of thumb, formalized intuition and algorithmic heuristics has been outlined. We will return to those approaches in greater detail in Chapters 8 and 9. At this stage, the problem is sufficiently illustrated by considering an algorithm tree as depicted in Figure 6.1.

Almost any algorithm presupposes some structure of the data (see Figure 6.1) rather than *inferring* the structure from the data. As a result – unfortunately – all these algorithms may generate different clustering configurations of the same data, and make a final judgement very difficult as to the one that results in the *natural grouping*.

According to the study of Bayne et al. (1980), the *K-means algorithm* is the overall 'best' partition method, whereas *complete linkage* is superior among hierarchical clustering methods. More importantly, CL is also best in *not* discovering *false clusters* on random sample sets.

Intelligently exploiting different methods, techniques, tests and algorithmic results leads automatically to an *artificial intelligence* methodology: see Figure 6.2. In principle

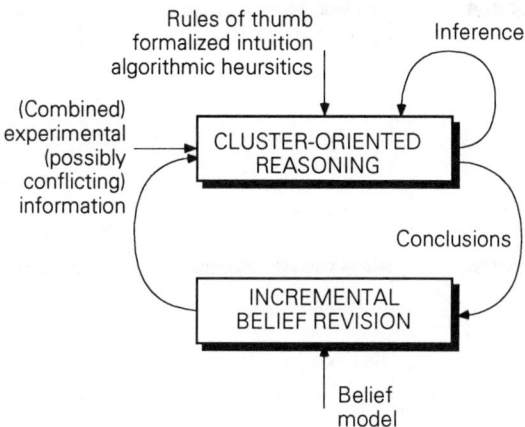

Figure 6.2: Artificial intelligence methodology: cluster-oriented reasoning and incremental belief revision.

– and within the discourse of our subject – a knowledge-based approach cluster analysis boils down to three major components: an inference engine (a cluster-oriented reasoning mechanism), a knowledge base (including formal and informal knowledge), and a belief revision mechanism (including a model of belief or certainty appropriate for the subject matter). The need for and relevance of knowledge-based systems in cluster analysis will be discussed in Chapter 7, followed by the knowledge-base methodology in Chapters 8 and 9.

As already mentioned, here our first aim is to review the potentials of
– a partitional technique: the *K*-means algorithm (see Section 2.5)
– hierarchical techniques: SL, CL, and AL (see Section 2.4)
– a graph-theoretical technique: the MST algorithm (see Section 2.6)

and the predictive value of the indices
– the Davies–Bouldin index (DB),
– the Wilks's lambda statistic (alpha),
– the inconsistency of MST-edges (i_e),
– the CPCC value,

for varied cluster configurations like spherical clusters, long-chained clusters, concentric clusters, elongated clusters and compact clusters. For that purpose, we study 12 different configurations as depicted in Figure 6.3.

For each configuration, we run each of the techniques and calculate each of the indices. The results are shown in Table 6.1. This table has to be interpreted as follows.

For each configuration, an index may be considered to be a *strong* (S) predictor for retrieving the correct cluster(s) or a *weak* (W) predictor. If nothing can be gleaned from the index value, the predictive value is to be considered as *negligible* (N). Note that for

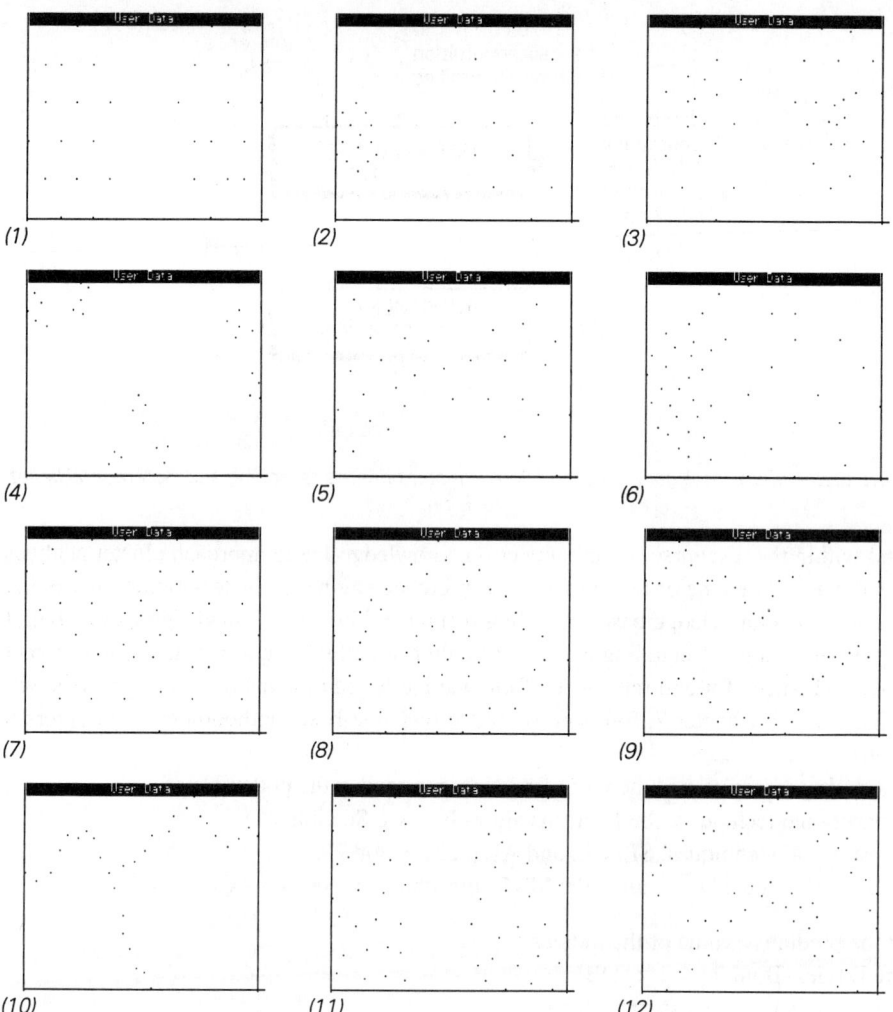

Figure 6.3: (1): randomly jittered point clusters; (2): well-separated clusters with point densities which are not equal; partly compact clusters; (3): well-separated clusters with smoothly varying point densities; (4): multi-level clustering; two levels of clustering that are apparent to most of the observers; (5): touching clusters; actually a single cluster containing a very thin 'neck' whose removal leads to two distinct clusters; (6): touching clusters; sharp gradient in point density; leads to easy detection of separating boundary; (7): non linearly separable touching clusters; sharp gradient in point density; leads to easy detection of separating boundary; (8): well-separated concentric non linearly separable clusters; (9): well-separated chain-shaped clusters; (10): cluster composed of linear pieces with definite branching structure; (11): well-separated clusters with homogeneous point densities; (12): well-separated non linearly separable chain-shaped clusters.

Table 6.1: A comparison of 12 different cluster configurations.

partitional			hierarchical		graph
K-means			SL, CL, AL		MST
repeated exp.					(incons.
(alpha)	(dom.)	(DB)	(CPCC)	edges)
1	W	W	W	W	S
2	S	S	S	S	S
3	S	S	S	S	S
4	S	S	S	S	S
5	W	S	N	N	N
6	N	N	N	N	N
7	N	N	N	N	N
8	N	N	N	N	S
9	W	W	S	W	S
10	S	S	W	W	N
11	W	W	W	W	S
12	W	W	W	N	S

(N = negligible; W = weak; S = strong)

each configuration there exist one or more strong predictors, except in cases 6 and 7: touching clusters with sharp gradient in point density. None of the indices is sensitive to difference in point density as it appears in these examples.

Following Section 1.2, we create a proximity matrix using the entries of Table 6.1. The hierarchical (AL) structure is depicted in the dendrogram of Figure 6.4. We observe two more or less distinct categories: {1,2,3,4,9,11,12} being well-separated clusters, and {5,6,7,8,10} being touching clusters.

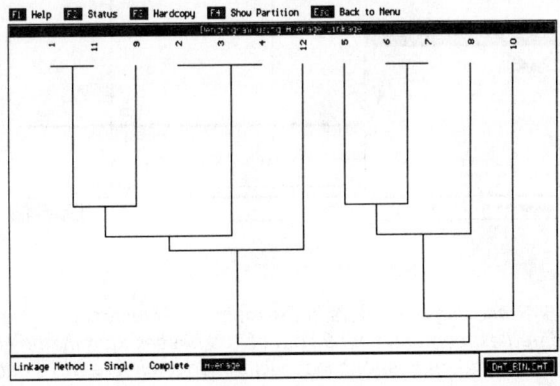

Figure 6.4: Hierarchical structure of the 12 different cluster configurations.

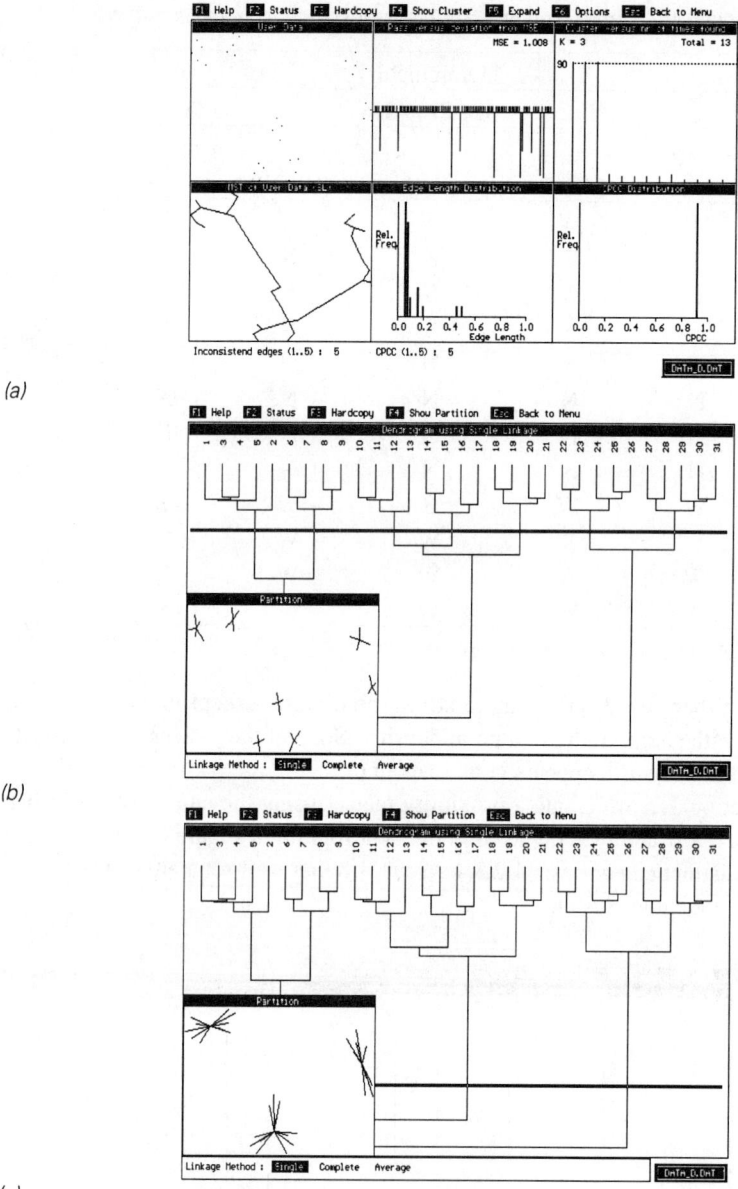

(a)

(b)

(c)

Figure 6.5: Retrieving of clusters in the multi-level clustering configuration;
(a): K-means repeated experimentation; 90 out of 100 passes identify the correct high-level
clusters (cluster dominancy (dom.)); strong inconsistent edges in the MST, and a very high CPCC
value; (b) and (c): low-level and high-level clusterings obtained from SL.

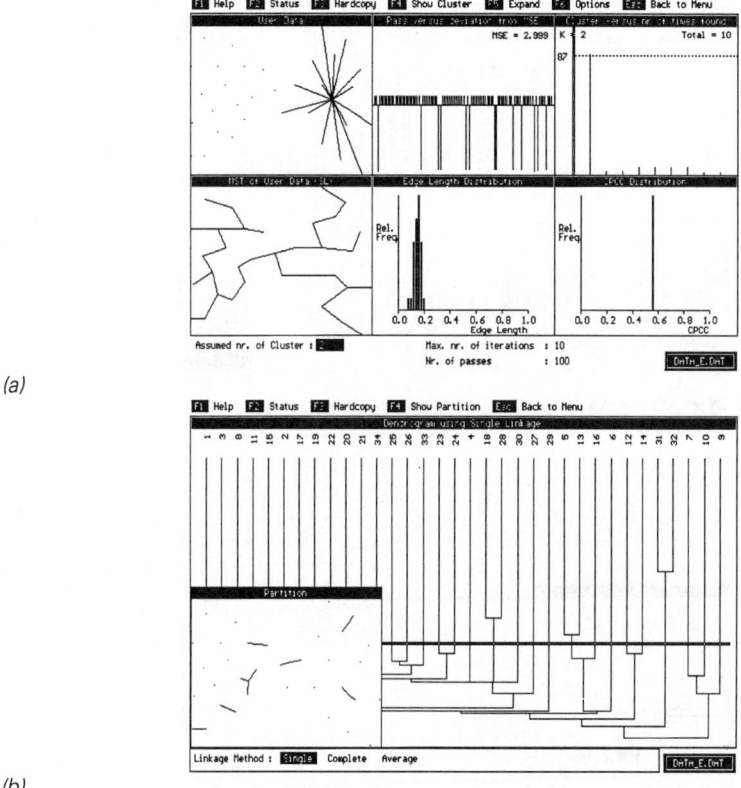

(a)

(b)

Figure 6.6: Retrieving touching clusters; (a) K-means repeated experimentation; 87 out of 100 passes identify the correct clusters (high dominancy); no inconsistent edges to be detected; a very low CPCC value; (b) no clue for cutting the SL dendrogram.

For each of these categories we show some examples to clarify the outcome. The multi-level clustering (4) is shown in Figure 6.5. The next example is a case of touching clusters (with a very thin 'neck'), (5), shown in Figure 6.6. A last example is concerned with well-separated (linearly separable) chain-shaped clusters. The results are shown in Figure 6.7.

Conclusions

We observe, not surprisingly, that no single index exists which predicts natural clusters for all possible cluster configurations. Also, there exists no general technique or algorithm which is capable of retrieving natural clusters for all possible cluster configurations. On the other hand, for each possible cluster configuration one might find an appropriate technique or algorithm to recover natural clusters, as well as one or more indices that may predict a distinct cluster tendency. So the essential message must lie in

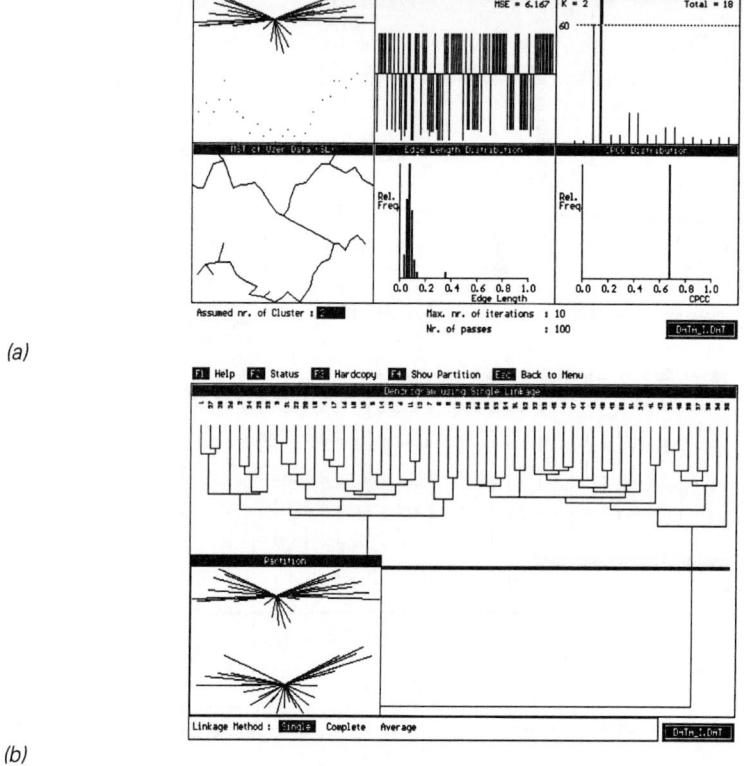

(a)

(b)

Figure 6.7: Retrieving of well-separable chain-shaped clusters; (a) K-means repeated experimentation; 60 out of 100 passes identify the correct clusters (weak dominance); a clear MST inconsistent edge, and a low CPCC value; (b) easy to detect from SL dendrogram.

finding the solution for putting data-dependent weight on the various indices that can be extracted from the cluster configuration at hand. This, and more, will be analyzed in Section 6.3.

6.2 Indeterminacy and uncertainty

In what follows, we try to justify the fact that the number of levels of structural cluster description cannot be more precise than the assessment of clustering tendency allow us to infere.

Clustering algorithms have the dubious distinction of generating clusters on the pattern space even when no clusters are really present.

This section focuses on the fact that clustering algorithms have the dubious distinction of generating clusters on the pattern sample even when no clusters are really present. Therefore, rules for deciding whether a set of points in a *d*-dimensional pattern space are arranged in a random fashion or are structured in some specific manner are very much needed and help prevent the inappropriate interpretation of the spatial arrangements of points in the pattern space. Those rules (or tests) assess *clustering tendency*. A variety of approaches have been suggested for performing tests for randomness. These use near-neighbour information and counts of patterns in local neighbourhoods and minimal spanning trees. In Section 2.8, an MST-based test of clustering tendency (Smith and Jain, 1984), was illustrated in order to convey the underlying idea of such random tests. Other tests are based on distance statistics. Comparisons of the performance of different tests and statistics have been presented in Dubes and Jain (1979), Ripley (1979), Panayirci and Dubes (1983) and Zeng and Dubes (1985).

However, not surprisingly, the performance of tests for *randomness* is greatly affected by the fact that all spatial point processes are assumed to be generated in finite *sampling windows*, and consequently no theoretical distributions are available, so that Monte Carlo simulations have to be relied on; i.e. performance estimation over different sampling windows using data with varying sample sizes and dimensionality. The edge effect of the sampling window is mostly out of control. As a result, no single test or statistic performs 'best' for varied dimensionality and sample size.

In the process of determining the degree of clustering present in the data by 'objective' means, a variety of internal indices can be used (see Section 2.7) to assess the merit of the clustering results.

The question of how many clusters are in the data is especially important when *validating* the results of a cluster analysis. Thus, internal indices are used to estimate the true number of clusters in multivariate data. On of them is the *DB-index* (see Section 2.7).

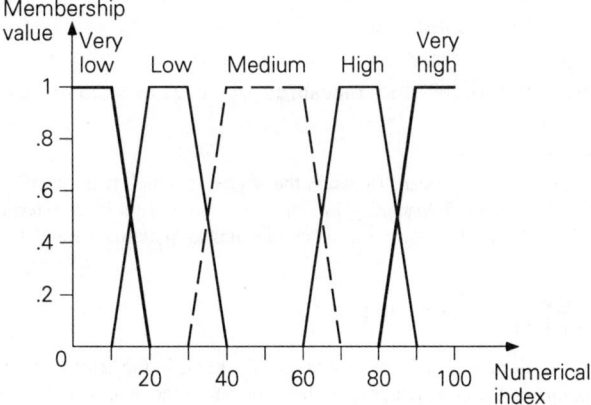

Figure 6.8: Idealized membership curves for linguistic variables; ,mapping of numerical indices on fuzzy linguistic variables.

The DB-index is a function of the ratio between the sum of within-cluster distances and a measure of the between-cluster separation, and conjectures that a *minimal value* of this measure indicates an *optimal partition*. Experimental judgements are based on plots of the indices as functions of the number of recovered clusters.

In Duda and Hart (1973) and Hartigan (1975) experimentation with the *Wilks's lambda statistic* (which measures the ratio between the total within scatter and the total scatter) has shown that in estimating the number of clusters in 'weakly clustered data', the sizes of errors[1] *(Rand coefficient)* increase dramatically compared to those with 'strongly clustered data' (but this was to be expected).

One might learn to recognize the characteristic in the plots of those indices by eye, but it is difficult to recognize automatically without any further support in the decision procedure.

Excellent surveys of Monte Carlo studies in determining the number of clusters in data can be found in Milligan (1981), Milligan and Cooper (1985) and Dubes (1987). Generally, the effects of sample size, dimensionality, cluster spread, number of true clusters and sampling window were examined. It appeared that no effect occurs due to the sampling window. Complete linkage and MSE methods recognize the true number of clusters consistently better than the single linkage clustering method.

In Sneath and Sokal (1973), the CPCC was found to be very dependent on the algorithm used. Average linkage clustering tends to result consistently in high values, and is not always useful for adjudging natural clusters. In addition, it has been proved that the CPCC is unreliable as a measure for clustering tendency.

In spite of the fact that other methods are superior over the single linkage clustering algorithm, Jardine and Sibson (1971) showed that only the SL clustering algorithm posesses the *continuity property* with reference to the proximity matrix. When a proximity matrix is changed, SL adapts in a continuous fashion while CL (for example) exhibits a sudden change in cluster identity.

In Chapter 5, we encountered non-exclusive clustering methods in the form of *fuzzy clustering*. Separation indices and measures of fuzziness have been shown to be

---

[1]   Error counting is an 'objective' means to assess the degree to which two classifications (for example *'true categories'* and *'recovered clustering'*) of the data match. This is an external index. The Rand coefficient (Rand, 1971) is the relative number of pairs of patterns treated the same under both classifications:

$$R = \frac{a+d}{a+b+c+d}, \text{ with } R \in [0,1]$$

where a is the number of pairs which are in both classifications in the same cluster, b is the number of pairs which are in one of the classifications in the same but in the other classification not in the same cluster, c, the number of pairs which are in the first classification not in the same but in the other classification in the same cluster and d is the number of pairs which are in both classifications not in the same cluster; $a + b + c + d = n(n-1)/2$.

instrumental (not, however, universal) in assessing the validity of the result and selecting the proper number of clusters present in the data. Experimentation in terms of induced fuzziness has been offered by Bezdek (1974) and Backer (1978). Its major importance lies in the fact that a natural framework could be offered in which imprecision and indeterminacy indicated by numerical indices can be mapped onto fuzzy linguistic variables – for example very high, high, medium, low, very low (see Figure 6.8).

The framework of *fuzzy logic* allows numerical updating (*fuzzy relaxation*[2]), as well as linguistic modification (intensifying and weakening) due to compatibility with external evidence or compatibility with other information.

From the above, it is crystal clear that practitioners' terminology in the process of determining the degree of clustering present in the data will never be more precise than 'these data are *strongly* clustered, are *somewhat weakly clustered*, or show *no* clustering *at all* or are *random*'. Also the numerical assessment of clustering tendency and cluster validity cannot be given in greater precision than in terms of the linguistic variables {*very high, high, medium, low, very low*}. As a result, the domain of discourse cannot be better than the cubic model as given in Figure 6.9. If so, the total number of different states of structure description is limited to 75, even if some of them are meaningless.

Before discussing the need for and relevance of knowledge-based methods in cluster analysis (Chapter 7), we conduct some simple experiments to exemplify the foregoing and to stress the relative importance of two-dimensional projection and uniform sampling of the original data in order to guide the analysis.

---

2   Fuzzy relaxation can be used, for example, when one is interested in deciding that 'cluster C_k has property α_i' where α_i is a member of a set properties α_1, α_2, ..., α_N. Although this problem includes the cluster analysis problem, it is clearly more general. A given cluster could have many properties and thus correspond to several of the α_i, or it could have none of the properties and so correspond to none of the α_i. In fuzzy relaxation the variable $f_i(k)$ represents a degree of plausibility (belief) for the correspondence $C_k \approx \alpha_i$. The value of $f_i(k)$ is high if the correspondence $C_k \approx \alpha_i$ is likely and low otherwise.

Assume that there exists a non-negative function $c_{ij}(k,l)$ whose value is large when the hypothesis '$C_k \approx \alpha_i$' is compatible with the hypothesis '$c_l \approx \alpha_j$' and small otherwise. Such a function will be refered to as a *compatibility function* and does not need to be symmetric in general.

Because of the one-to-many correspondence of the C_k and the α_i, a small value of the product $c_{ij}(k,l) \cdot f_i(l)$ does not necessarily imply that the correspondence $C_k \approx \alpha_i$ is unlikely. However, a large value for the product gives strong support to the premise $C_k \approx \alpha_i$. Therefore, an update method based on the maximum (over j) of the product terms rather than the average seems appropiate (Rosenfeld et al., 1976) suggests a relaxation formula of the form

$$f_i^{(r+1)}(k) = \frac{1}{N} \sum_{l=1}^{N} \max c_{ij}(k,l) \cdot f_j^{(r)}(l)$$

where the sum yields the average contribution for all clusters considered. Other formulae are also possible.

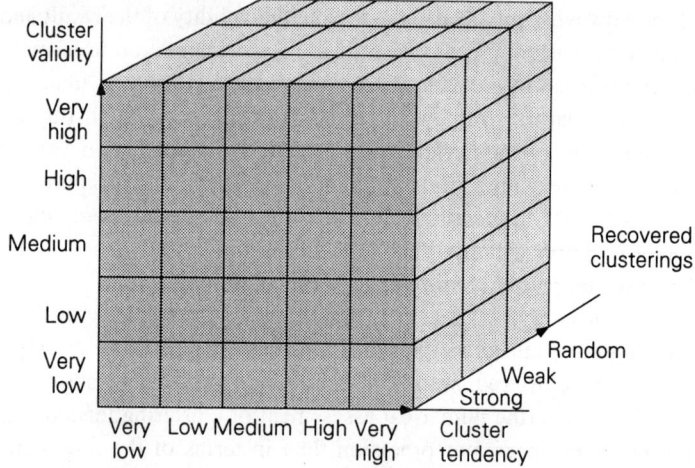

Figure 6.9: The limited number of states in clustering structures due to imprecision of the assessment of clustering tendency and validity.

Table 6.2: Unravelling multidimensional data goes with many questions to be answered and many choices to be made.

input	*output*
domain guidance?	which method first?
category labelling?	which method next?
assumptions?	which algorithm?
measurements?	what to be tested?
missing data?	what to visualize?
mixed data?	standardization?
sample size?	variable dependency
dimensionality?	clustering tendency
choice of metric?	validity of results
measure of association?	belief in results?
choice of scales?	separability?
shape of data?	how many clusters?
criterion function?	clustering strength uniqueness?
measure of compactness and isolation?	induced fuzziness?
sampling window?	intrinsic errors?

6.3 Disclosing cluster structures: some experiments

Generally speaking, if one wants to unravel multi-dimensional data by means of cluster analysis, a huge number of underlying questions has to be answered and as many choices have to be made. Even in a relatively simple practical problem, the user of clustering techniques is faced with a substantial list of input variables and output variables (see Table 6.2).

There is no way to design an experiment which is able to uncover all the effects of all variables at one time. Therefore, any reasonable experiment will have only a limited number of variables involved to keep the complexity manageable.

6.3.1 Repeated experimentation

First experiment

The first experiment is a controlled experiment, i.e. the true categories are known in advance.

The experiment is characterized as follows.
- data type: strongly clustered
- no. of dimensions: 2
- no. of true clusters: 3
- sample size: {30,30,30}
- applied uniform sampling: 30%[3]
- clustering methods: K-means, SL
- sampling window random data: cubic[4]

---

[3] Very often (particulary in the case of large samples sizes) it is preferable to attempt preliminary analysis only on a reasonably small percentage of the orignal data. The method we employ here for selecting a certain percentage of the pattern samples is similar to the approach of Murthy (1981).

Maximum and minimum feature values in each of the d dimensions are computed. Subsequently, the range in each dimension can be computed. Then a threshold in feature direction j can be defined as

$$th(j) = \text{range}(j).3.n$$

and the average threshold as

$$th_{av} = \frac{1}{dz} \sum_{j=1}^{d} th(j)$$

where z is a positive real.

The first pattern sample is taken as the first pattern sample in the sampled data set. The remaining samples are selected by examining the data in order and by accepting the data unit which differs from the previously chosen pattern in all directions more than the th_{av}. This procedure is repeated until 30% of the pattern samples are selected. The value of z is adjusted to ensure that 30% of the pattern samples are always selected.

[4] Random data consists of points generated independently and uniformly over a sampling window. Since, in general, the shape of the sampling window (cubic or spherical) has little effect on the result here, we adopt here the hypercubic sampling window for experimentation. In all cases, the sample size and the dimensionality of the random data are equal to those of the data to be analyzed.

– internal indices: DB index, Wilks's lambda statistic
– external index: Rand coefficient

In Figure 6.10, the resulting dendrogram using the CL algorithm is shown, clearly indicating the three clusters which had to be found are fully in correspondence with the aparent clusters to be seen in the two-dimensional plot of the data.

The well-structured result can then be compared with the result of generated random data within the same window as the original data (the same dimensionality, the same number of pattern points). Note that, if the sample size is relatively low, there is always local substructure in the data even if the data is generated randomly.

If we cut the dendrogram of the original data at the three-cluster level, a perfect correspondence (Rand coefficient = 1) exists with the true categories. In other words, the true clusters could be disclosed perfectly.

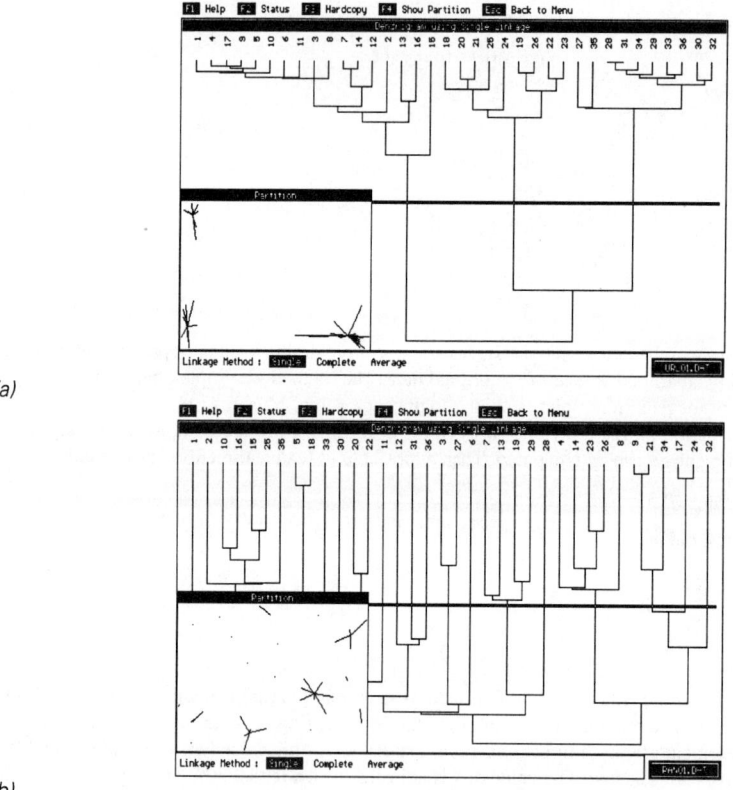

(a)

(b)

Figure 6.10: Resulting dendrograms using single linkage for the clustered data (a), and the corresponding random data (b), respecively.

If we recall the expected performance of the Davies–Bouldin index (Section 2.7, Figure 2.28), the result of the plot of the DB index for the user data given, and the random data generated accordingly, follows almost perfectly the expected behaviour of 'strongly clustered data' versus 'random data'. Undoubtedly, its minimum value is found for $k = 3$ (the number of true clusters). This is presented in Figure 6.11. A result like this will be used as 'hard' evidence: here, the ability of the DB index to recognize the true number of clusters is *very high*.

The distributions for Wilks's lambda statistic as shown in Figure 6.12 are obtained by repeated simulation (i.e. performance estimation over different realizations using the K-means algorithm with varied starting configurations chosen at random from the data points for different values of K). For the user data, 10 samplings were generated, and 100 K-means trials per sampling were carried out with randomly chosen starting configurations for $K = 2$, $K = 3$ and $K = 4$. Likewise, 10 randomly generated sets were submitted to 100 K-means trials each, with randomly chosen starting configurations for $K = 2$, $K = 3$ and $K = 4$.

In Figure 6.12, for each cluster of bars, the left triple corresponds to the minimum, the average, and the maximum value of the statistic for the user data, whereas the right triple corresponds to the random case. Clearly, a very high clustering tendency can easily be recognized. However, a clear indication for $K = 3$ cannot be assessed from this plot.

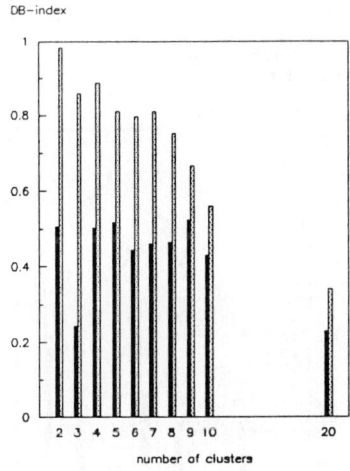

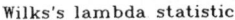

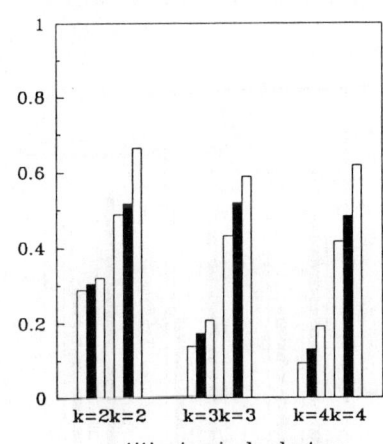

Figure 6.11: The DB index for the clustered data, and the corresponding random data, respectively; 10 samplings, 100 trials each.

Figure 6.12: The Wilks's lambda statistic for the clustered data, and the corresponding random data, respectively; 10 samplings, 100 trials each.

Two characteristics more are to be extracted from the same experiment: cluster dominance and the number of different clusters detected while performing 100 trials for $K = 2$, $K = 3$ and $K = 4$.

Cluster dominance is the maximum number of times the same cluster is detected over the 100 trials that the K-means algorithm runs with different starting configurations randomly chosen from the data set, for $K = 2$, $K = 3$ and $K = 4$, respectively. The average value is calculated over the 10 different samplings, yielding also a maximum and a minimum value. We note that 100% dominance simply implies that over all trials and samplings the same cluster is consistently detected. From Figure 6.13, we learn that cluster dominance is *high* with respect to the average value of the random data. However, experimentation with random data may yield *low* average cluster dominance, the spread tends to be high (much higher than in clustered data). Note also that the smaller the sample size, the larger the spread in cluster dominancy may be. Only in the case of $K = 2$ the distinction between the cluster dominance of the original data and random data found to be *significant*.

As may be expected, the more structure is present in the user's data, the fewer different clusters shall be detected. The average number of different clusters detected will grow more rapidly in case of random data than will be the case for well-structured data. Figure 6.14 shows the change in the number of clusters detected for $K = 2$, $K = 3$ and $K = 4$, respectively. As also may be expected, the spread in case of random data is much larger than for well-structured data.

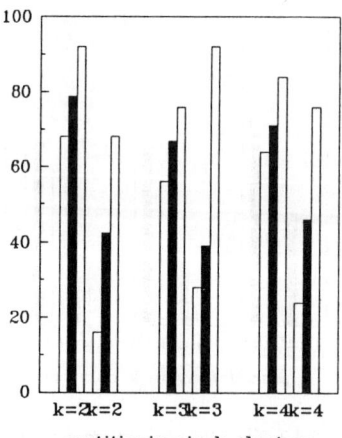

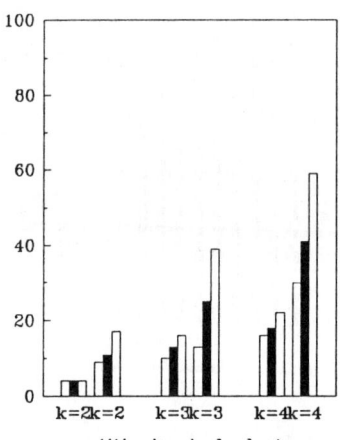

Figure 6.13: Cluster dominance for the clustered data, and the corresponding random data, respectively; 10 samplings, 100 trials each.

Figure 6.14: The number of different clusters detected in the clustered data, and the corresponding random data, respectively; 10 samplings, 100 trials each.

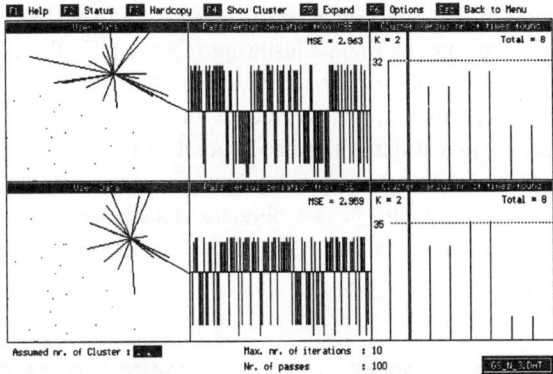

Figure 6.15: Artificially two-category generated data yielding touching clusters; low cluster dominance due to a relatively large number of patterns with inconsistent cluster assignments: a fuzzy region? Two runs, appearing in the upper and lower window, showing consistent and inconsistent cluster assignments.

Clearly, the *small* number of clusters detected corresponds with the *high* cluster dominance, observed in Figure 6.13.

Second experiment

The second experiment furthers the idea of dominance of cluster assignments in the sense of Section 5.5 (strong and weak patterns) towards induced fuzzy regions.

In Figure 6.15 two-category artificial generated data are displayed yielding touching clusters. The true categories are shown in Figure 6.16 (top left). The experiment is here characterized by:

- data type: touching clusters
- no. of dimensions: 2
- no. of true clusters: 2
- sample size: {20,20}
- clustering method: *K*-means
- internal index: mean square error, induced fuzziness

We performed 100 *K*-means trials with varied starting configurations randomly sampled from the data, for *K* = 2. Figure 6.15 shows the existence of dominant clusters (two pairs of 28% dominant clusters). One of the resulting dominant clusters has been indicated.

In comparison with the true categories, six pattern points are misclassified. The cluster dominance is relatively low due to the fact that the number of patterns with inconsistent cluster assignments is relativelly large. As such, the full region of all patterns that may have inconsistent cluster assignments over all 100 trials may be regarded as a *fuzzy region*. In Figure 6.16 (top right), this area is shown, leaving out two areas of pattern points that are consistently assigned to the recovered clusters, as indicated. In the terminology of fuzzy clustering (Chapter 5), the cluster membership value of pattern points of each of the 100% recovered clusters is {1,0} and {0,1}, respectively. The cluster membership value of the pattern points in the region of inconsistent cluster assignment lies within the interval (0,1).

If we apply a fuzzy cluster membership operator (for example a neighbourhood operator,[5] as shown in Figure 6.17), we obtain induced fuzzy regions (Figure 6.16, bottom left) for the true categories, and (Figure 6.16, bottom right) for the dominant clusters, respectively.

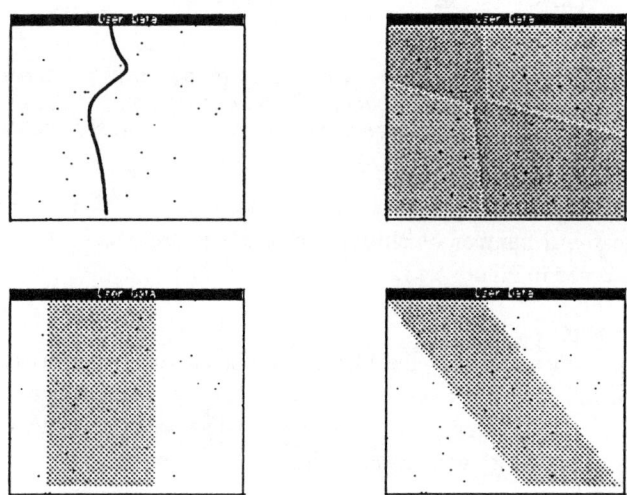

Figure 6.16: Induced fuzzy regions for the true categories (left), and the K-means partitionings (for K = 2) by taking the dominant cluster assignments into account (right).

---

[5] Following definition 5.1, Chapter 5, it can easily be shown that for

$$r(\mathbf{x}, C_i) = t_i / n_i$$

where t_i is the number of near neighbours of $\mathbf{x}$ contained in cluster C_i, and n_i is the total of pattern points of cluster C_i,

$$f_i(\mathbf{x}) = t_i / t = t_i(\mathbf{x}) / t(\mathbf{x})$$

where t is the total number of neighbours taken into account. As a result, the ammount of induced fuzziness yields

$$\varphi = \frac{1}{n} \sum_{\mathbf{x}} |(t_1(\mathbf{x}) - t_2(\mathbf{x})| / t(\mathbf{x})$$

for a two-cluster case.

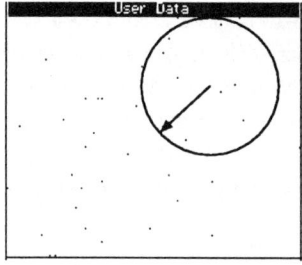

Figure 6.17: A fuzzy neighbourhood operator, as discussed in Section 5.2.

As a result, we observe that the region of inconsistent cluster assignments corresponds alsmost perfectly to the induced fuzzy region based on a fuzzy cluster membership operator. In other words, by always performing a large number of trials, say 100 (with varied starting configurations randomly chosen from the set of data points), we are able to detect a fuzzy region of which the amount of induced fuzziness is a measure of the degree of separability of the detected clusters. Note that – in the case of well-separated clusters – the region of inconsistent cluster assignments may be empty, and thus the amount of induced fuzziness is zero.

Clearly, in the search for an optimal value for K, one has to try to find the minimum value for the amount of induced fuzziness. As was the case for other internal indices, this minimum should be found by inspecting the plot of the amount of fuzziness against the number of clusters, K. In practice, the cluster dominance frequencies of detected clusters (and consequently, their related fuzzy regions due to inconsistent cluster assignments) turn out to be a reliable predictor of both cluster tendency and the number of clusters, K.

The performance of the above measure is evaluated in the third experiment.

Third experiment
The third experiment includes two-category artificial data, as shown in Figure 6.18.

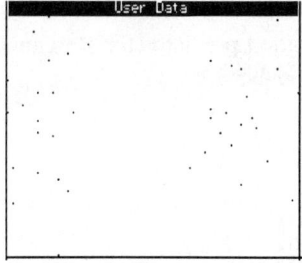

Figure 6.18: Two sets of artificial data to be used in the third experiment.

A series of three data sets (GS-L-1, GS-L-3 and GS-L-4) are constructed by shifting the true clusters towards each other.[6] Then, as a result, we obtain fixed data configurations characterized as *well-separable* clusters, *somewhat weakly separable* clusters, and *touching (nearly random?)* clusters, respectively. The experiment is thus characterized by:

- data types: {*well-clustered, weakly clustered, touching clusters*}
- no. of dimensions: 2
- no. of true clusters: 2
- sample size: {20,20}
- clustering method: AL, MST, K-means
- internal indices: CPCC (AL), Wilks's lambda statistic,
                       induced fuzziness, inconsistent edges (MST)

As before, in this experiment values of the indices are based upon 100 K-means trials with varied starting configurations randomly sampled from the respective data sets, for $K = 2$, $K = 3$, and $K = 4$.

Figure 6.19 shows, on the left, the regions of induced fuzziness in overlay with the data scatter plots, and, on the right, the corresponding minimum spanning trees. Recalling the baseline distributions for the edge length and the CPCC (see Figure 2.25), only for GS-L-1 could an inconsistent edge be detected, and a more or less significant value for the CPCC has shown up. All other connecting edges (between the true clusters) could not be recovered as being inconsistent edges. Likewise, all other values for the CPCC (even for AL) are not to be considered as significant. In conclusion, neither the inconsistent edges of the MST nor the CPCC-values are likely to be serious predictors for cluster separbility for weakly clusterings and touching clusters.

Figure 6.20 gives an overview of the Wilks's lambda statistic, cluster dominance, and number of clusters detected, for $K = 2$, $K = 3$, and $K = 4$, respectively. From the summary of the results of the Wilks's lambda statistic, the cluster dominance, and the number of clusters detected, for GS-L-1, GS-L-3 and GS-L-4, respectively, we observe that the Wilks's lambda statistic can be considered as a good predictor for the presence of structure versus randomness, but is not likely to be a good predictor for the number of clusters present in the data. Cluster dominance (and the associated amount of fuzziness) is to be considered as both a good predictor for structure and a good predictor for the number of clusters present in the data.

6

	σ_x	σ_y	μ_{1x}	μ_{2x}	μ_{1y}	μ_{2y}
GS-L-1	5	15	30	70	50	50
GS-L-3	5	15	37	62	50	50
GS-L-4	5	15	40	60	50	50

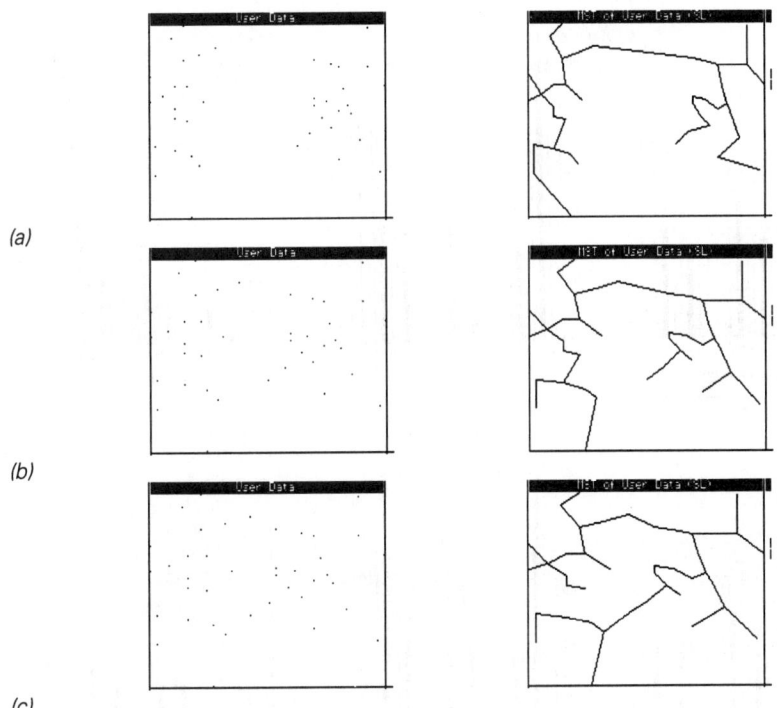

(a)

(b)

(c)

Figure 6.19: Results from experiment 3:
(a) well-clustered; (b) weakly clustered; (c) touching clusters.
Induced fuzzy regions (left) are likely to be good predictors for cluster separability; inconsistent
edges of the MST are not likely to be serious predictors for cluster separability in general.

The performance of the above internal indices is listed in Table 6.3, for GS-L-1, GS-L-3 and GS-L-4, respectively, versus their outcome from corresponding random data. Clearly, the Wilks's lambda statistic (alpha) and the distribution of clusters (cluster dominance) yielding fuzzy regions, are likely to be serious candidates for predicting the presence of structure in the data to be analyzed (clustering tendency).

6.3.2 Lessons from the experiments

(1) When using K-means partitioning, irrespective of the internal index considered, a single numerical value does not tell much. *Repeated experimentation* with varied initial configurations, chosen at random (say 100 trials over 10 samplings), with both user data and corresponding random data, may provide valid conclusions about the presence of structure in the user data at hand, and the validity of a particular partitioning obtained. Therefore, in any cluster analysis study, repeated experimentation with the user data and repeated simulation with random data should be

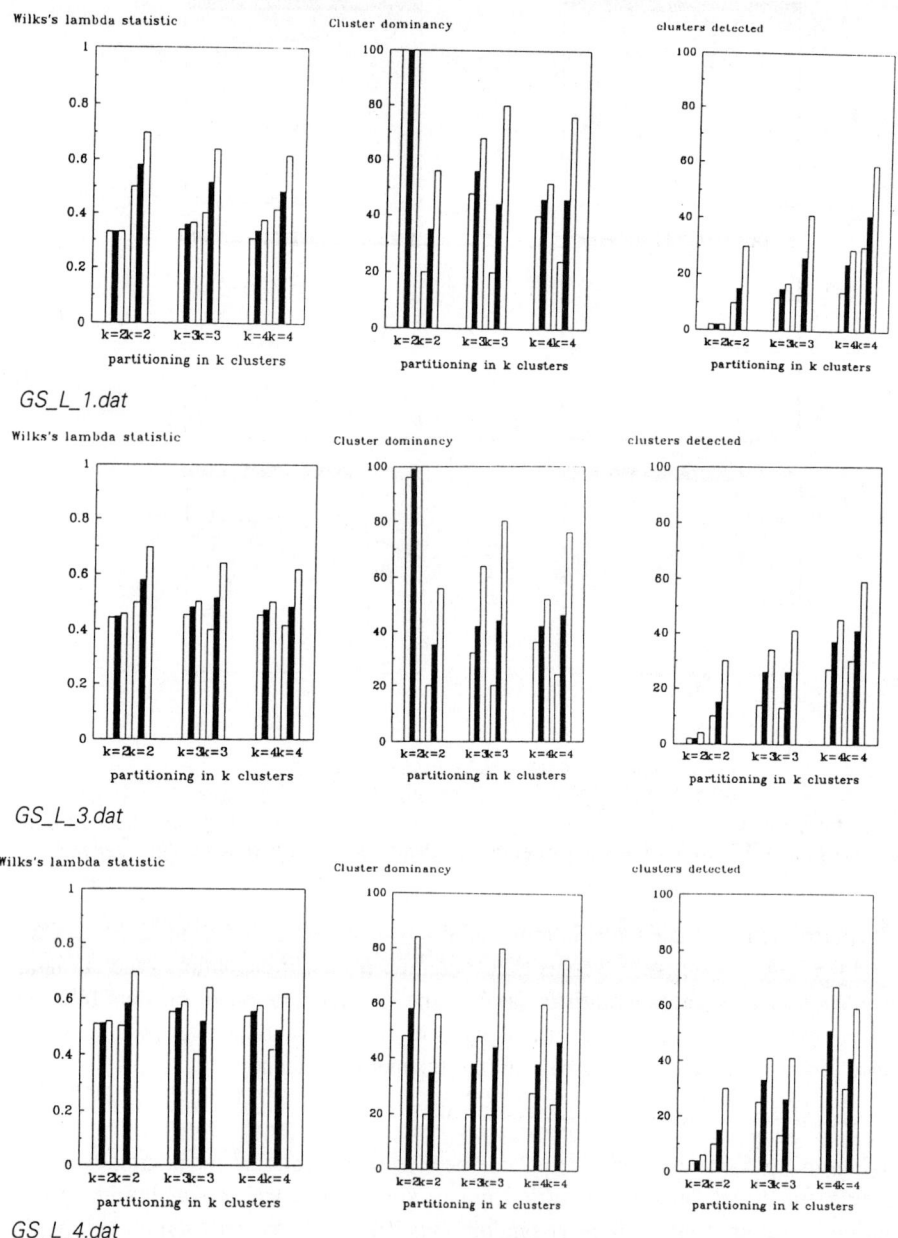

GS_L_1.dat

GS_L_3.dat

GS_L_4.dat

Figure 6.20: Summary of results for GS-L-1, GS-L-3, and GS-L-4:
from left to right, Wilks's lambda statistic, the cluster dominance, and the number of different
clusters detected, respectively. (Note that for each k, we have bars (min, average, max) for the
user data (left) and for random data (right)).

Table 6.3: Observed capabilities of some internal indices for predicting the presence of structure in the data.

	GS-L-1 *(strong)*	GS-L-3 *(weak)*	GS-L-4 *(touching)*
alpha	0.342(0.603)	0.443(0.613)	0.509(0.619)
dom.	100(18)	92(24)	55(27)
#clus.	2(28)	4(30)	6(26)

(Note that the variance of those indices over repeated experimentation with the same data may easily go over 10% (see Figure 6.20); theoretically speaking, the observed capabilities are limited to relatively low-dimensional normal distributed data).

embedded in the analysis as a matter of course. This, of course, is a rather trivial statement for any practitioner, though overlooked frequently in many practical applications.

(2) As has been pointed out before, any attempt at numerical precision is useless and overdone. Data types which we have considered, are not likely to be characterized with greater acuracy then in *approximate* terms like *well*-clustered (*strongly* clustered), *weakly*-clustered, touching clustered (or near random). Likewise, clustering tendency and cluster validity are not to be indicated more accurately than *very high*, *high*, *medium* (*marginal*), *low* and *very low*.

An attempt to formalize the results: a rule is born

As an example, we are now able to formalize experimental findings into some *heuristic* but *generalizing* rule representations which can be used in more or less automated reasoning. That will be subject of the later sections of this chapter.

If we denote the value of the Wilks's lambda statistic for the user data as $\alpha(k)$, and for random data as $\beta(k)$, and the value of the cluster dominance for the user data as $\gamma(k)$, and for random data as $v(k)$, then the following rules are easy to formalize.[7]

$$\textbf{if } \bar{\alpha}(k) < 1/k \textbf{ then } CT(1) = \{ \textit{very high, high} \}$$

and

$$\textbf{if } \bar{\alpha}(k) > 1/k \textbf{ then } CT(1) = \{ \textit{low, very low} \}.$$

Likewise, we have

$$\textbf{if } \bar{\gamma}(k) < 100/k \textbf{ then } CT(2) = \{ \textit{low, very low} \}$$

and

$$\textbf{if } \bar{\gamma}(k) > 100/k \textbf{ then } CT(2) = \{ \textit{very high, high} \}.$$

---

[7] In all bar diagrams of $\alpha(k)$, $\beta(k)$, $\gamma(k)$ and $v(k)$, so far, the values were normalized with respect to k to facilitate comparison of the values of the indices for different k.

Here, $\bar{\alpha}(k)$ denotes the average value over a large number of trials; CT denotes the *c*luster *t*endency.

We have emphasized that within the same spatial window, and with the same number of pattern points, the results of the user data should always be compared with the results of random data, thus:

user data	random data
$\bar{\alpha}(k)$	$\bar{\beta}(k)$
$\bar{\gamma}(k)$	$\bar{\nu}(k)$

Then, the initial clustering tendency can be *reinforced* or *weakened* by considering the results of the random data. So, the following rules may emerge.

if $\bar{\beta}(k) < 1/k$ **and** $\bar{\nu}(k) > 100/k$ **then** $SU(1) = \{weakening\}$

and

if $\bar{\beta}(k) > 1/k$ **and** $\bar{\nu}(k) < 100/k$ **then** $SU(1) = \{reinforcement\}$

where SU denotes the *s*upport of the initial clustering tendency found. (Note that these rules in fact imply *belief revision*.)

If we encounter *inconsistencies*, the following rules may resolve these inconsistencies:

if $\bar{\alpha}(k) < 1/k$ **and** $\bar{\gamma}(k) < 100/k$ **then** $SU(2) = \{weakening\}$

and

if $\bar{\alpha}(k) > 1/k$ **and** $\bar{\gamma}(k) > 100/k$ **then** $SU(2) = \{weakening\}$

Combining the evidence so far, may lead to rules like:

if $SU(1) = \{reinforcement\}$ **and** $SU(2) = \{weakening\}$

and $CT(1) = \{low, very\ low\}$

and $CT(2) = \{very\ high, high\}$

then $CT(3) = \{medium\ or\ marginal\}$

In the next section, we will formalize these results into rules reflecting 'facts', 'support', and 'diagnosis'.

6.3.3 Formalization from observations

In an attempt to formalize the previous results, we present an example showing how experimental findings may be formalized into heuristic but generalizing rules which may contribute to more or less automated reasoning. That will be the subject of later chapters.

Let us refer to the Wilks's lambda statistic for the user data, w, as alpha(w,K). Alpha has shown to be dependent of the number of clusters, K. Likewise, the Wilks's lambda statistic for random data, r, will be represented by beta(r,K).

We also denote the cluster dominancy for the user data and random data differently. The cluster dominancy for the user data will be refered to as %dom(w,K), and for the random data as %dom(r,K).

Note that in all bar diagrams of alpha, beta, and %dom, so far, the values were normalized with respect to K for reasons of comparison the index values for different K.

The previous observations may lead to the following rules.

> IF alpha(w,K) < $1/K$ THEN 'the clustering tendency is likely to be *high to very high*',

and

> IF alpha(w,K) > $1/K$ THEN 'the clustering tendency is likely to be *very low to low*'.

Likewise, we have

> IF %dom(w,K) < $100/K$ THEN 'the clustering tendency is likely to be *very low to low*',

and

> IF %dom(w,K) > $100/K$ THEN 'the clustering tendency is likely to be *high to very high*'.

Since we also have observed that the justification of the K-means approach for given data may be very dependent on the cluster structure present in those data, the above rules are always preceded by a so-called *certainty factor*. The certainty factor is either user-specified, or determined by other means directly from the data. Thus, a rule should read as follows.

> [*certainty factor*]IF alpha(w,K) < $1/K$ THEN 'the clustering tendency is likely to be *high to very high*'.

Moreover, it should be noted that alpha(w,K) generally denotes the average value over a large number of trials.

The reference values $1/K$ and $100/K$ do not result from theoretical considerations; they simply seem to fit in the above example.

We have emphasized that always – within the same spatial window, and with the same number of pattern points – the results of the user data should be compared with the results of random data. Thus

user data	random data
alpha(w,K)	beta(r,K)
%dom(w,K)	%dom(r,K)

Then, the established initial clustering tendency can be reinforced or weakened by considering the results of the random data, beta(r,K) and %dom(r,K). So, the following rules may emerge.

IF beta(r,K) < $1/K$ AND %dom(r,K) > $100/K$ THEN 'the *belief* in the initial clustering tendency is likely to be *decreased*'

and

IF beta(r,K) > $1/K$ AND %dom(r,K) < $100/K$ THEN 'the *belief* in the initial clustering tendency is likely to be *increased*'.

The above *belief revision* in fact is the consequence of whether or not experimentation with random data supports the initial clustering tendency.

Also, we may encounter some inconsistencies that may be resolved by the following rules.

IF alpha(w,K) < $1/K$ AND %dom(w,K) < $100/K$ THEN 'the *belief* in the initial clustering tendency is likely to be *decreased*'

and

IF alpha(w,K) > $1/K$ AND %dom(w,K) > $100/K$ THEN 'the *belief* in the initial clustering tendency is likely to be *decreased*'.

In this way, belief revision accounts for evaluation of either supporting evidences or occuring inconsistencies.

It will be clear that all kinds of basic statistics (internal or external) and their related baseline distributions may lead to rules, examplified above, for mean square error analysis, hierarchical analysis, and graph theoretical analysis, respectively.

Combining rules (or diagnostic rules) are meant to allow the system to draw final conclusions (the *diagnosis*), carefully accounting for the appropriateness of individual rules (through the certainty factors) and the various degrees of belief suggested by the rules.

The above example shows how two strongly related though different approaches line up. They are the processing paradigm and the 'data–information–knowledge' paradigm or modelling paradigm.

(i) The processing paradigm
On the basis of different *facts* deduced from the data to be analyzed (level I), we are looking to establish how well one fact *supports* the other with respect to structural variables like clustering tendency and cluster validity (level II). Combining the facts and mutual support leads to a final *diagnosis* (level III).

(ii) The 'data–information–knowledge' paradigm or modelling paradigm
Different facts deduced from the *data* to be analyzed (level I), lead to pieces of evidential *information* by implicitly associating structural variables with the data given, (level II). If

facts deduced from the data and evidential information contained in structural variables are to be generalized to rules that always hold, we end up with explicit representation of *knowledge* (level III).

In the next chapter, we will put both paradigms into perspective with respect to each other.

6.4 Concluding remarks

In this chapter we have discussed the basic indeterminacy in cluster analysis which originates from the fact that there exists no single algorithm that may uncover all possible cluster structures. This problem can be solved by using the strengths of different methods, techniques and algorithms, aiming at carefully combining weighted pieces of evidence provided by them. Repeated experimentation, simulation with random data, and approximate reasoning are offered as the keystones of a combined approach. Experiments conducted showed that numerical precision is useless and overdone. The chapter ends with a preliminary and exemplifying discussion of formalization, showing how rules might be generated from experimental observations. That and more will be the subject of later chapters.

7 | The need for and relevance of computer-assisted reasoning in cluster analysis

The complexity of the clustering problem in high-dimensional spaces, the existence of a plethora of clustering algorithms producing varied results on the same data sets and the need for choosing the right algorithms by avoiding costly exhaustive searches are the main motivations for having a knowledge based system for this problem (Balasubramaniam et al., 1990).

In the foregoing, cluster analysis and interpretation has been loosely defined as 'extracting informative patterns' from quantitative (or even qualitative) data observations – 'a bridge between data and the real world'. The process involved can be summarized as a continuous flow of:

DATA > INFORMATION > KNOWLEDGE

Consequently, the main concern is to understand and to validate the increasing level of generality.

In Chapter 6, a preliminary and exemplifying discussion was offered to show how some sort of formalization may lead to the trasnformation of facts from data into pieces of information (or even knowledge). Two strongly related, though different, approaches to be achieving this transition were refered to as the *processing paradigm* and the *modelling paradigm*.

These are the subject of the following sections.

7.1 The processing paradigm

Statistical packages, including those which carry out cluster analysis, being considered as *toolboxes*, may easily be used by inexperienced users to produce as many statistics and data representations as can be generated by the algorithmic tools as possible, without understanding the relevance of such results. Blind use of computational algorithms would not be enough to provide a good basis for interpretation of the data. Some of the deficiencies of typical toolboxes are:
– they do not inform about underlying assumptions
– they seldom warn against obvious misuse
– they do not provide guidance in the process of analysis
– they hardly assist in the interpretation of the results.

In order to provide better analysis and interpretation, a three-level architecture has been proposed by Hand (1985), which – in our context:

level I: contains numerical algorithms and forms the core of clustering techniques;

level II: contains an interface which makes it feasible for (naive) users to use a variety of different techniques;

level III: includes *clustering expertise* to interpret the results and to present them in an understandable way.

It is not the purpose of this multi-level architecture to automate the process of analysis and interpretation further but to provide a step-by-step meaningful and validated cluster-oriented reasoning process. In other words, the three-level architecture incorporates, as such, explicit clustering expertise aiming at interpreting the intermediate results, and decision-making about a variety of hypothesized conclusions about the underlying structure of the data considered. The third level, clustering expertise, which we will often refer to as the *expert system* level, can either be integrated with the clustering toolbox or reside as a separate component in the software architecture.

In Figure 7.1, we simply refer to a 'decision network' without specifying to what extent this third level is configured as an expert system (knowledge system), decision support system, or fixed hierarchical decision tree.

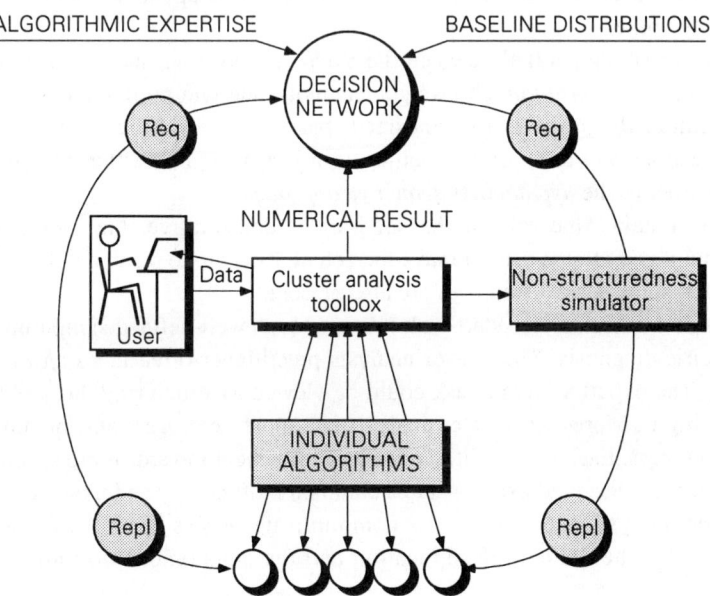

Figure 7.1: Multi-level architecture for integrated support towards meaningful and validated decision-making; the processing paradigm.

It is important that the user supplies the data to be analyzed to the toolbox – more precisely to the toolbox interface – which facilitates operation of all individual algorithms, either on request or automatically. Since ultimately, judging the presence of structure in the data is the goal, we have strongly suggested that simulation of non-structuredness (yielding corresponding baseline distributions to compare with), either on request or automatically generated, is needed. Then, at the third level, decision-making can be based upon the numerical results, appropriately chosen baseline distributions, and algorithmic and clustering expertise.

7.1.1 Knowledge-based systems (expert systems)

Barr and Feigenbaum (1982) give the follwing definition of *artificial intelligence* (AI): 'AI is the part of Computer Science concerned with designing intelligent computer systems, that is, systems that exhibit the characteristics we associate with intelligence in human behaviour – understanding language, learning, *reasoning*, solving problems, and so on'.

Research in AI is intimately related to information systems and capabilities of computers (among them computerized cluster analysis) and has resulted in systems with performance comparable to that of a human in a limited narrow domain. These types of system are often referred to as *expert systems* and are consultation programs consisting of a problem-dependent *knowledge base* and problem-independent software to exploit the knowledge base.

There is a striking parallel between the *medical consultant* making a diagnosis and the *cluster analysis consultant* choosing a technique and interpreting the results. In both cases, an initial diagnosis is hypothesized, based on very little evidence, and then proving questions and tests are used either to reject or reinforce this choice. This was described earlier as the *hypothetico-deductive approach*.

In both domains, the role of the software is consultative: the user consults the software, but it is up to the user to make the actual decision, to reach an ultimate diagnosis. However, some basic differences are also present.

– Problems in medical advice start with (more or less) well-defined symptoms, resulting in a specific diagnosis. The cluster analysis practitioner often faces an ill-structured problem. The expert system's task could be viewed as structuring the problem rather than solving it. Moreover, cluster analysis is a more heterogeneous, quantitative and iterative process. Each of these iterations is a refinement towards more specific goals.

– In medicine, the user and expert share a common terminology and medical knowledge. In cluster analysis the need is to communicate across very different areas of specialization: the domain of application and the autonomous discipline of cluster analysis.

– The choice of a particular method or technique in any cluster analysis is determined by two sources of information. One is the user, his objectives and what he can tell the system about the data. The other is the data themselves. It is the data which reveal multicollinearity and so on. In other words, there are two sources of control information in expert system support in cluster analysis, which is rare in the domain of medicine.[1]

As such, cluster analysis expert systems are still in their infancy. It would be an immature step to concentrate on the development of a general-purpose intelligent cluster analysis system as depicted in Figure 7.2, fully integrating all sources of information, the user's objectives, algorithmic experience, data expertise, application domain knowledge, and problem-structuring and problem-solving capabilities.

Therefore, we shall limit ourselves to some preliminary attempts to implement knowledge-guided reasoning within the area of low-level structural variables like clustering tendency and cluster validity.

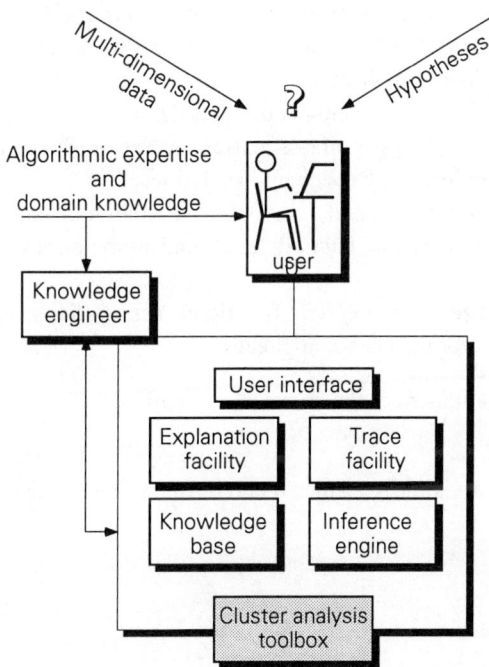

Figure 7.2: The ultimate goal: a general-purpose expert system for cluster analysis; to date, still in its infancy.

---

[1] In fact, most applications of knowledge-based approaches in cluster analysis focus on the ability to examine the data critically and use the results of this examination in determining the choices of methods to exploit (see Chapter 9).

7.2   The data–information–knowledge paradigm or modelling paradigm

In Debenham (1989), the approach to designing knowledge systems is strongly supported by strictly separate but integrated concepts and formalisms for *data, information* and *knowledge*. Designing corporate knowledge bases is thus deliberately compatible with established techniques for data-base design. The notion of *functional association* underlies the approach to knowledge systems.

Two major and fundamental issues within a given application (here cluster analysis) regarding the *structural variables* in that application (i.e. *data points, neighbourhood, regions, scatter, clusters, partitions, cluster dominance, internal* and *external indices, clustering tendency, cluster validation,* and so on) are *classification* and *representation* of those variables in that application. By 'classification' of a variable we mean deciding whether a given variable in that application should be classified as *data, information,* or *knowledge*. By 'representation' we mean constructing an actual representation of that variable in the appropriate formalism.

The notion of functional association is used to define the terms *data, information* and *knowledge*. (Note that generally classification of variables, in the above sense, is certainly non-trivial.) Following Debenham, we define:

level I: the *data* as the fundamental, indivisible variable in the application;

level II: the *information* as the *implicit* functional associations between data in the application;

level III: the *knowledge* as the *explicit* functional associations between items of information and/or data in the application.

The following example illustrates the above classification.

Example

Let us consider two indivisible variables *a* and *b*, for which the data values are given as follows (level I):

a	b
1	2
3	6
4	8
5	6
8	3
9	4

Then, $R(5,6)$, $R(8,3)$ and $R(9,4)$ are implicit functional associations between *a* and *b*, and thus constitute pieces of *information* (level II). The *explicit* functional association

$$R(a,b) \leftarrow \; \leq (a,4), \; b = 2 \times a$$

for which the general rule '**if** $a \leq 4$ **then** $b = 2 \times a$' always holds, constitutes *knowledge* (level III). This example can easily be translated into the domain of cluster analysis.

Let the data be given by

user data
$x_1 = (x_{11}, x_{12}, \ldots);$ label$_1$
$x_2 = (x_{21}, x_{22}, \ldots);$ label$_2$
$x_3 = (x_{31}, x_{32}, \ldots);$ label$_3$
$\vdots$
$x_n = (x_{n1}, x_{n2}, \ldots);$ label$_n$

and related

random data
$r_1 = (r_{11}, r_{12}, \ldots);$ label$_1$
$r_2 = (r_{21}, r_{22}, \ldots);$ label$_2$
$r_3 = (r_{31}, r_{32}, \ldots);$ label$_3$
$\vdots$
$r_n = (r_{n1}, r_{n2}, \ldots);$ label$_n$

We further consider the structural variable S (being the total scatter) and S_W (being the summed within scatter). Then,

$$R(S(x), S(r)) \leftarrow S(x) < S(r)$$

is an *implicit* functional association, and therefore constituting *information* (level II). On the other hand,

$$R(S, S_W) \leftarrow \forall(x, r): S_W < S$$

is an *explicit* functional association, representing *knowledge* (level III).

Functional associations and conceptual associations are the key issues of Section 9.4.

7.2.1 Knowledge systems versus conventional systems

The nature of knowledge systems can now be discussed and related to conventional systems. In particular, we may compare knowledge systems and expert systems. The scope of the type of changes that a system is designed to accommodate is said to be

data if the facts can be represented in the system through data values and related labels;

information if the facts can be represented in the system through values of certain tuples of variables, deduced from the data;

knowledge   if the facts can be represented in the system through certain rules which always hold.

In conventional programming, the explicit associations (*knowledge*) are encoded in a programming language (and as such are part of the program). The *information* is stored either explicitly on some storage device or implicitly in the program. The *data* are stored explicitly in a storage device. Systems of this nature are referred to as *data processing systems*.

We refer to systems in which the knowledge is encoded in a conventional programming language, and which employ a data-base management system for information and data storage, as *information-based systems* (or conventional data-base systems).

Finally, we refer to systems in which the knowledge is encoded in a knowledge language, which employ stored information and data, as knowledge-based systems (or simply knowledge systems).

We think of *expert systems* as computer systems which attempt to perform similar tasks as knowledge systems, but for an expert system the knowledge has usually been extracted (by the knowledge engineer) directly from an expert and has been deliberately represented 'as it is'. Thus, in an expert system the knowledge has not been analyzed, modelled and normalized. So, in a sense, we think of expert systems as prototype knowledge systems.

With the exception of the above, expert systems, being related to the knowledge of a particular expert, have something of a pioneering flavour, whereas knowledge systems have something of a systems architecture flavour.

Up to now, fuzzy sets were instrumental in describing cluster structures (see Chapter 5). Clearly, fuzzy sets and fuzzy logic are being used to enhance the power of knowledge systems. 'Soft' or fuzzy expert systems are valuable in those applications in which approximate reasoning is the rule rather than the exception. Medical diagnosis and clustering diagnosis belong to these applications. In what follows, fuzzy logic is considered to be the obvious choice as a language for cluster-oriented reasoning. EDAPLUS is designed along these lines. A recent reference book on the theory and applications of fuzzy expert systems is that of Kandel (1991).

7.3 Concluding remarks

In this chapter, we have identified our analysis (which we will refer to as *cluster-oriented reasoning*) as the resultant of the processing paradigm as the modelling paradigm.

(i) The *processing paradigm*: numerical algorithms producing facts deduced from data values (level I), a processing interface to direct supporting experimentation and simulation (level II), and decision-making (level III) using expertise on clustering methodology, yielding some final diagnosis about the presence of structure in the data and the validity of the ultimate findings.

(ii) The *modelling paradigm*: facts deduced from the data are associated with structural variables, implicitly constituting pieces of information (level II), directing and supporting hypotheses originating from the facts, followed by using formalized expertise (knowledge) about cluster analysis in order to reach a final diagnosis.

Figure 7.3 represents this idea of cluster-oriented reasoning (*facts*, *support*, *diagnosis*) as a combined result of a processing hierarchy (software tools) and a representative mapping of formalized functional associations (*modelling*).

As such, the development of a system for knowledge-based cluster analysis can only be viewed as the process of mapping the functional associations between data, information and knowledge onto a software architecture which is capable of processing the data in a desirable way, and which is maintainable and expandable through appropriate modularization.

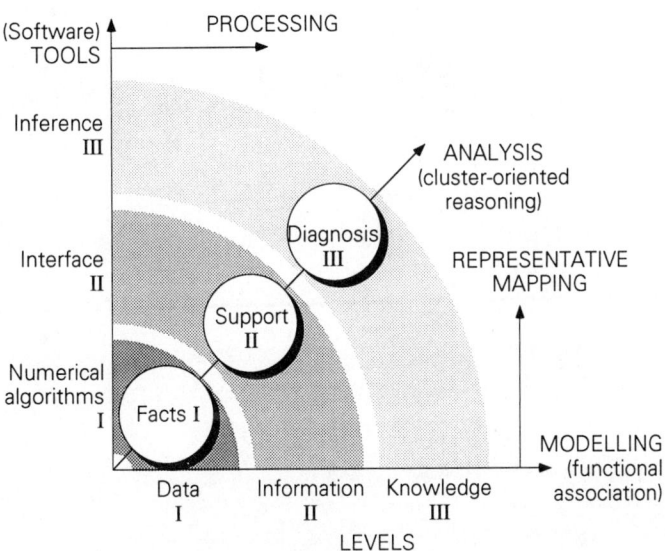

Figure 7.3: Cluster-oriented reasoning as the resultant of the processing paradigm and the modelling paradigm.

As we have developed the basic methodology of cluster analysis (Part I) and having identified the ingredients for and the underlying paradigms of the development of computer-assisted reasoning in cluster analysis (Part II), we have come to the stage where the knowledge-based methodology has to be worked out. This will be discussed in Part III.

Part 3
Computer-assisted reasoning in cluster analysis

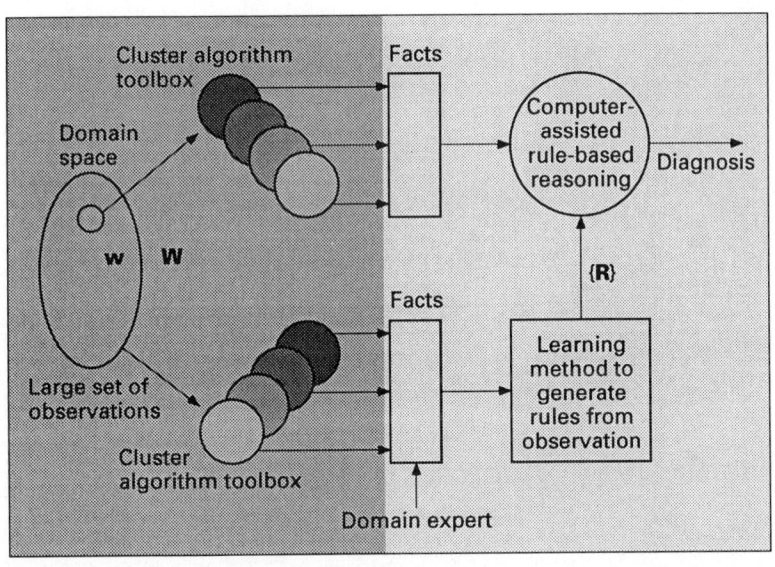

8 | Knowledge-based expert systems: an introduction

Cluster analysis is not known by a single general strategy but by a variety of possible (feasible) data dependent strategies each of which may sometimes lead to true insight, and sometimes to misleading interpretation which can only be detected and unraveled by using one's feeling and intuition, and past experience (Jain and Dubes, 1988).

One might learn to recognize the characteristics in the data views, the plots of different indices and statistics by eye, but it is difficult to recognize automatically without any further support in the decision procedure; the naive users from other disciplines should be assisted by experts in finding valid solutions (Jain and Dubes, 1988).

In view of the above, it is natural to associate such a reasoning process with expert systems:

intelligent computer programs that use knowledge and inference procedures to solve problems [here the problem of carrying out a cluster analysis] *that are difficult enough to require significant human expertise for their solution (Barr and Feigenbaum, 1981).*

In the preceding chapters we developed the basic methodology of cluster analysis, introduced some of the basic methods and showed how clustering tendency and validity indices may guide us in drawing conclusions about the true structure present in the data under consideration. Moreover, we identified the concept of fuzzy sets as being suitable to explicate the strength of the clustering patterns involved, i.e. weak and strong patterns in multi-level cluster assignments.

We have come to the stage where full understanding of the difficulties of a cluster analysis, especially in high-dimensional space, leads to a *complex network* of interrelated issues which ultimately determine *cluster-oriented conclusions*. This complex network is depicted in Figure 8.1. The input data are characterized by dimensionality, sample size, intrinsic dimensionality, cluster shapes and number of clusters. On the other hand, the representation of the data, a proximity matrix or pattern matrix, cannot be seen independent of specific choices of association measures, metrics, normalization and sampling windows. Inspection of linear and nonlinear 2D mappings may guide us in gaining insight into the intrinsic dimensionality, the cluster shapes and the number of clusters but this can hardly be justified. Even if we restrict ourselves to a limited number of algorithms (partitional and hierarchical), different or even conflicting results may be expected. Where to cut the dendrogram, how to interpret the CPCC value, what to do with outliers, how to judge the significance of inconsistent edges in the MST and what the DB index and the Wilks's lambda statistic tells us about the number of clusters, the clustering tendency, the compactness and the isolation of detected clusters. If different partitions result from different methods, what can be concluded from Rand's partition similarity coefficient? Many of these questions may be answered better if we have appropriate baseline distributions (originating from random setting of parameters or statistics). Repeated experimentation provides us with necessary baseline information and gives rise to the definition of fuzzy regions, cluster dominance and so on.

However, without a *goal*, without specifying the *intention of the classification*, our analysis is subjectively blind. We need contextual and background knowledge in order to assign appropriate weights to the various attributes (features) based on their relative importance with respect to the intended classification. We refer to this knowledge as *subjective knowledge*. On the contrary, *objective knowledge*, such as operational knowledge and functional knowledge about algorithms and procedures, and data representation, is needed to assign weights to the various attributes based on their importance in representing objects (data) appropriately. The obvious question is then how to combine this objective component (data representation) with the subjective component (the intention of the classification) in such a way that the actual representation best meets the requirements of the intention of the classification.

Evidently, Figure 8.1 clarifies the fact that the various issues like data characterization, algorithms, indices, statistics, tests, goal, etc. are interrelated and show their interdependence implicitly or explicitly. However, the complex network does not show how relevant these interrelations really are for judgement about the different data appearances (shapes and structure) which may be subject of analysis. Also, the network does not expose the extent of the contribution of each of these interrelations to the final (global) diagnosis about the true structure of the data at hand in all cases.

In other words, *rules of thumb, formalized intuition* and *experience*, algorithmic *heuristics* and so on are needed to overcome the complexity. They have to be applied in order to combine pieces of local evidence in a flexible way aiming at one global diagnosis.

We may consider this process of combining facts, views, algorithmic results and sometimes even conflicting results, and various index values, as a 'weak' reasoning process on the basis of formalized intuition and experience, and heuristics. At the same time, it can be expected that within this reasoning process one has to deal with all kinds of uncertainty and imprecision. These are due to the fact that almost all concepts used are intrinsically not sharply defined and do bear an amount of imprecision and uncertainty with respect to their relevance to the final judgement.

Why a knowledge-based expert system?
Summarizing the above:
1. Cluster analysis is a very complex process in which intuition and past experience play a dominant role.
2. The intention of cluster analysis is not to produce numerical precision but rather global approximate qualification of detected structures; numerical precision is useless and overdone.
3. Cluster analysis is not characterized by a single general strategy but by a variety of

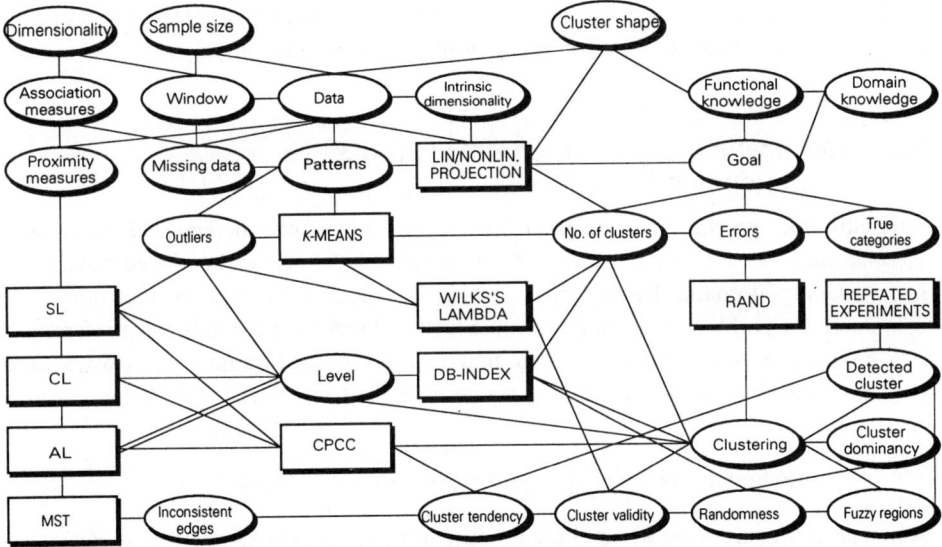

Figure 8.1: Interrelation pattern in cluster analysis.

possible (feasible) data-dependent strategies each of which may lead sometimes to true insight, and sometimes to misleading interpretation which can only be detected and unravelled by using one's feeling, intuition, and past experience, Jain and Dubes (1988).

4. One might learn to recognize the characteristics in the data views, the plots of different indices and statistics by eye, but it is difficult to recognize automatically without any further support in the decision procedure; naive users from other disciplines should be assisted by experts in finding valid solutions, Jain and Dubes (1988).

In view of the above, it is natural to associate such a reasoning process with expert systems: intelligent computer programs that use knowledge and inference procedures to solve problems (here the problem of carrying out a cluster analysis) that are difficult enough to require significant human expertise for their solution (Barr and Feigenbaum, 1981, 1982). This definition means that an expert system is intended to be a computer system that emulates the decision-making ability of a human expert.

The terms expert system, knowledge-based system and knowledge-based expert system are often used synonymously. In engineering, knowledge-based systems refer to systems that improve the performance of conventional processing by using knowledge specific to the problem domain. Those systems do not have a specific diagnostic task and generally act like a 'black box'. On the other hand, knowledge-based expert systems are specifically characterized by their diagnostic task and should be able to explain the steps of their reasoning.

In this chapter, we will cover the general concepts of expert systems, representation of knowledge, methods of inference and reasoning under uncertainty.

8.1 General concepts of expert systems
(Giarratano and Riley, 1989)

Internally, the expert system consists of two main components: the *knowledge base* and the *inference engine*. The knowledge base includes the expert's knowledge specific to some problem domain. The inference engine is a decision procedure that infers the solution of the problem. That is, given some facts, a conclusion that follows is inferred.

The knowledge of an expert system may be represented in a number of ways. One common method of representing knowledge is in the form of *IF ... THEN type rules*, such as

IF dimensionality is too high **THEN** stop

Although this method has severe limitations, many significant expert systems have been built by expressing expert knowledge in rules. Large systems may contain thousands of rules.

Expert systems are different from conventional programs because the problems to be dealt with usually have no algorithmic solution and rely on inferences to achieve a reasonable solution. Since the expert system relies on inference, it must be able to explain its reasoning so that its reasoning can be checked. However, the design of adequate explanation facilities is still undeveloped. Also, capability of the system to learn rules by examples, through rule induction, in which the system creates rules from observed data, is still in its infancy.

In many cases (and also in our case), due to the lack of deep knowledge (a full understanding of the basic structure, function and behaviour of objects in the problem domain) most systems are designed to rely on shallow knowledge. One type of *shallow knowledge* is *heuristic knowledge* ('heuristic' comes from the Greek *heuriskein* which means to find). Heuristics are not guaranteed to succeed in the same way that an algorithm is a guaranteed solution to a problem. Instead, heuristics are rules of thumb or empirical knowledge gained from experience which may aid in the solution but are not guaranteed to work.

In spite of its present limitations, a knowledge-based expert system approach to a real-world problem like cluster analysis may be very successful, especially if the system is capable of handling uncertain or incomplete information.

8.2 Elements of an expert system

The elements of a typical expert system are shown in Figure 8.2. In a rule-based system, the knowledge base contains the domain knowledge needed to solve the problem coded in the form of rules. Other representations of knowledge will be discussed in later sections.

A sophisticated expert system includes the following components:
- *user interface*: the mechanism by which the user and the expert system communicate;
- *explanation facility*: explains the reasoning of the system to the user;
- *working memory*: a global data base of facts used by the rules;
- *inference engine*: makes inferences by deciding which rules are satsified by facts, prioritizes the satisfied rules, and executes the rule with the highest priority;
- *agenda*: a prioritized list of rules created by the inference engine, whose patterns are satisfied by facts in working memory;
- *knowledge acquisition facility*: an automatic way for the user to enter knowledge into the system rather than by having the knowledge engineer explicitly code the knowledge.

Note that the knowledge acquisition facility is in fact an optional feature on many systems.

Depending on the implementation of the system, the user interface may be a simple text-oriented display or a sophisticated high-resolution, bit-mapped display.

The knowledge base is also called the production memory in a rule-based expert system. *Production rules* are stated as follows:

$$\text{the sample size is high} \rightarrow \text{use partitional method}$$

$$\text{the sample size is low} \rightarrow \text{use hierarchical method}$$

These production rules can be expressed in an equivalent pseudocode IF ... THEN format as:

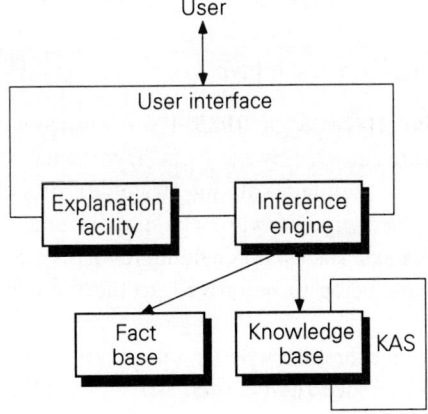

Figure 8.2: Elements of a typical expert system.

Rule: high-sample size
IF
>   the sample size is high
THEN
>   use partitional method

Rule: low-sample size
IF
>   the sample size is low
THEN
>   use hierarchical method

Each rule is identified by a name. The individual condition

>   the sample size is high

is called by various names such as *antecedent*, *conditional part*/element, *pattern*. A conditional part may consist of several conditional elements. As an example,

IF
>   the Wilks's lambda statistic is low,
>   AND
>   the sample size is high,
>   AND
>   the number of clusters is low,
>   AND
>   the intrinsic dimensionality is low
THEN
>   there is suggestive evidence that the clustering tendency is high,
>   AND
>   the validity of this evidence is high.

In a rule-based system (Hayes-Roth, 1985) the inference engine determines which rule antecedents, if any, are satisfied by the facts. Two general methods of inferencing are commonly used as the problem-solving strategies of expert systems: *forward chaining* and *backward chaining*. Forward chaining is reasoning from facts to the conclusion resulting from facts. Backward chaining involves reasoning in reverse from a hypothesis, a potential conclusion to be proved, to the facts which support the hypothesis.

Very often, an inference engine will do either forward or backward chaining. However, some types of inference engine, such as ART or KEE, offer both. The choice of inference engine depends on the type of problem. Diagnostic problems are better solved with backward chaining while prognosis, monitoring and control are better done by forward chaining.

The *fact base* contains facts regarding the *current status* of the problem at hand. The facts do not interact with one another. Instead, the knowledge base includes possible interactions (or functional associations) between facts. If a rule has multiple conditions or patterns, then all of them must be simultaneously satisfied for the rule to be placed on the agenda. Some patterns may even be satisfied by specifying the absence of certain facts in working memory. A rule whose patterns are all satisfied is said to be *activated* or *instantiated*. Activated rules are put on the agenda and selected for firing by the inference engine. Those rules that have the highest priority are executed first. Generally, all system shells let the knowledge engineer define the priorities of rules.

Expert systems are often designed to deal with *uncertainty* because reasoning is one of the best tools that we have discovered for dealing with uncertainty. The uncertainty may arise in the input data to the expert system and even the knowledge base itself. Much of human knowledge is heuristic, which means that it may only work correctly part of the time. In addition, the input data may be incorrect, incomplete, inconsistent and have other errors. Depending on the input data and the knowledge base, an expert system may come up with the correct answer, a good answer, a bad answer or no answer at all.

8.3 Representation of knowledge

Knowledge representation is of major importance in knowledge-based expert systems for two reasons. First, expert system shells are designed for a certain type of knowledge representation such as rules or logic. Second, the way in which an expert system represents knowledge affects the development, efficiency, speed and maintenance of the system.

8.3.1 Taxonomy of knowledge (an example)

Knowledge itself is hard to define exactly. Within a *domain* – a topical area or region of 'knowledge' – we may describe *knowledge* as an integrated collection of facts and relationships which, when exercised, produces competent performance. The quantity and quality of knowledge possessed by a person or a computer can be judged by the variety of situations in which the person or program can obtain successful results.

An epistemological distinction is made between a priori knowledge and a posteriori knowledge. *A priori knowledge* comes before and is independent of knowledge from the senses. It is considered to be universally true and cannot be denied without contradiction (like mathematical laws and logic). On the other hand, *a posteriori knowledge* is derived from senses. The truth or falsity of a posteriori knowledge can be verified using sense experience. Since sensory experience may not always be reliable, a posteriori knowledge can be denied on the basis of new knowledge without the necessity of contradiction.

Less philosophical is the distinction between deep knowledge and surface (shallow) knowledge. *Deep knowledge* is said to be knowledge from basic theories, first principles, axioms, and facts about a domain. In contrast with deep knowledge, *surface knowledge* (experiental or heuristic knowledge) is knowledge that is acquired from experience and is used to solve practical problems. Surface knowledge usually involves specific facts and theories about a particular domain and a large number of rules of thumb.

If we acquire and organize knowledge into a network (like Figure 8.1), the knowledge becomes *compiled* , sometimes partly in abstract patterns (deep knowledge), and partly as knowledge as a result of practical experience (surface knowledge). Expertise consists of large amounts of compiled knowledge.

Within our discourse, the clustering phenomenon is to be considered as the particular domain for which expertise is the major source of knowledge consisting of large amounts of information combined with rules of thumb, simplifications, rare facts and smart procedures in such a way that specific types of problems can be analyzed in an efficient manner. In view of this, our knowledge can be classified further into a subjective and an objective component. The *subjective knowledge* corresponds to the *intention of the classification* as specified by the user. The *objective knowledge* is associated with the *objects* and their representation.

Clustering is a process of partitioning objects into meaningful groups. It is a process that is controlled by a meaningful interaction of both the data-dependent (objective) knowledge and the user-dependent (subjective) knowledge.

Three types of knowledge affect the above process. One is *functional knowledge*: the knowledge about the functions of the objects; any physical object is created with a primary function in mind. The second is *contextual knowledge*: the knowledge about how the context affects the object representation and the intention of classification. The third is *operational knowledge*: the knowledge about the operational environment, the collection of operators on the set of objects, which yields the mapping of the object representations on a classification labelling.

The above knowledge sources determine the weights to various attributes (features) based on their importance in representing the object. An instantiation of these weights is called the *object representation*. Likewise, an instantiation of the weights to various features based on their importance to the intended classification, is called the *intention of classification*. The question is how to combine them.

Knowledge can be further classified into procedural knowledge and declarative knowledge. Both knowledge types correspond to two complementary programming implementation paradigms. *Procedural programming* is distinguished by the fact that the programmer must specify exactly how a problem solution must be coded (procedures tell a system *what to do*). Non-procedural paradigms, on the other hand, place the emphasis on specifying what is to be accomplished and letting the system determine how to accomplish it. In *declarative programming* the *goal* is separated from the methods used to achieve the goal. Examples are object-oriented programming and logic programming.

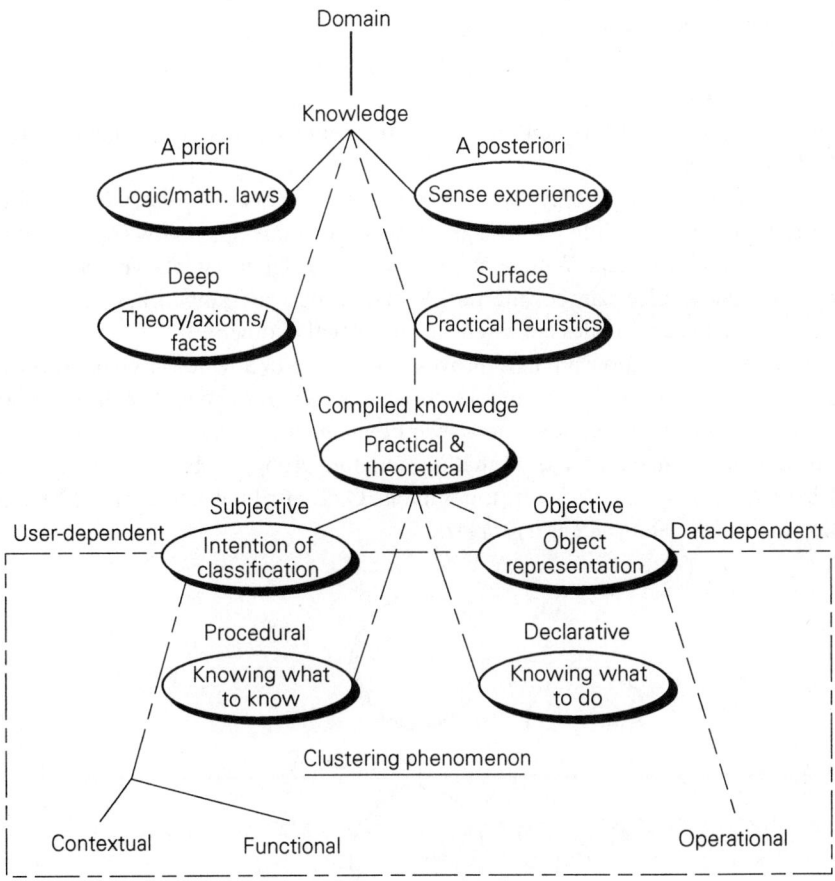

Figure 8.3: Taxonomy of knowledge.

In Figure 8.3, the above knowledge taxonomy is outlined. In the chapters that follow operational (data-dependent) practical heuristics are emphisized. In Chapter 10, a case study will exemplify the interaction between user-dependent knowledge and data dependent knowledge.

In order to accomodate all types of knowledge, a number of different knowledge representation techniques have been developed. These include rules, semantic nets, frames and conceptual graphs. As already mentioned, production rules are very commonly used as the knowledge base in knowledge-based systems since their advantages greatly outweigh their disadvantages.

8.3.2 Semantic nets (Stillings, 1987)

In short, a *semantic network* is a type of knowledge representation that formalizes objects and values as nodes and connects the nodes with arcs or links that indicate the relationships between the various nodes. In mathematical terms, a semantic net is a labelled, directed graph.

Relationships are of primary importance in semantic nets because they provide the basic structure for organizing knowledge. Without relationships, knowledge is simply a collection of unrelated facts. With relationships, knowledge is a cohesive structure about which other knowledge can be inferred. Semantic nets are sometimes referred to as *associative nets* because nodes are associated or related to others.

Certain types of relationship have proven very useful in a wide variety of knowledge representations. Two types of commonly used links are *IS_A* and *A_KIND_OF*. IS_A means 'is an instance of' and refers to a specific member of a class. A *class* is related to the mathematical concept of a set in that it refers to a group of objects. The objects in a class have one or more *attributes* in common. Each attribute has a *value*. The combination of attribute and value is a *property*.

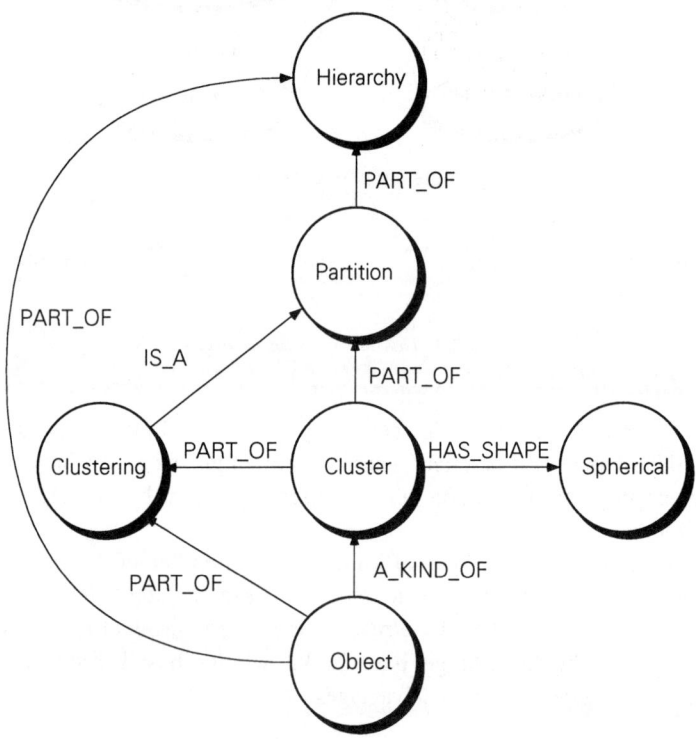

Figure 8.4: Example of a semantic net.

Other common links are *HAS_A* and *PART_OF*. Figure 8.4 is an example of a semantic net using these links. The duplication of one node's characteristics by a descendent is called *inheritance*. For example, since a clustering is a partition and partitions have the property that they divide the set of all objects into non-overlapping subsets encompassing all objects, it follows that a clustering has the same property. Inheretance is a very useful tool in knowledge representation because it eliminates the necessity of repeating common characteristics.

8.3.3 Object–attribute–value triples (Barr and Feigenbaum, 1981; Staugaard, 1987)

One problem with using semantic nets is that there is no standard definition of link names. If the IS_A link (relating a value to an attribute) or the HAS_A link (relating an object to an attribute) does occur frequently, it is possible to build a simplified semantic net using just three items: object, attribute and value. As such, the *object–attribute–value triple* (or triplet) can be used to represent all the knowledge and is convenient for listing knowledge in the form of a table. Rule induction translates the table into computer code. Table 8.1 shows an example of an object–attribute–value triple table. Triples are especially useful for representing facts, and the patterns to match the facts in the antecedent of a rule.

Table 8.1: Example of an object–attribute–value triple table.

Object	Attribute	Value
cluster	shape	spherical
cluster	density	smooth
clustering	tendency	high
clustering	No. of clusters	3
partition	m.s.e.	15.7
partition	*k*	4

A semantic net is an example of a shallow knowledge structure; many types of real-world knowledge cannot be represented by the simple structure of a semantic net. In AI, the term *schema* is used to describe a more complex knowledge structure than the semantic net. One type of schema that has been used in many AI applications is the frame.

8.3.4 Frames (Minsky, 1975)

A *frame* is a knowledge representation scheme that associates an object with a collection of features (e.g. facts, rules, defaults and active values). Each feature is stored in a *slot*. A frame is the set of slots related to a specific object, and as such, similar to a property list, schema or record, as these terms are used in conventional programming. With respect to semantic nets, frames add a third dimension by allowing nodes to have structures. Just as with semantic nets, there are no standards for defining frame-based systems. An example of a frame for a clustering is shown in Table 8.2.

Table 8.2: Example of a frame for a clustering.

Slots	Fillers
name	clustering X
type	weakly clustered
method	average linkage
data	GEWAS.DAT
no. of dimensions	4
no. of clusters	3
sample size	20,20,20
CPCC index	high
DB index	0.519
fuzz index	medium
alpha statistic	0.435
dominance	medium
tendency	medium
validity	low

Very often, frames are classified by their application. A *situational frame* contains knowledge about what to expect in a given situation. An *action frame* contains slots that specify the actions to be performed in a given situation. The frame paradigm has an intuitive appeal because the frame's organized representation of knowledge is generally easier to understand than logic, or production systems with many rules.

8.3.5 Predicate logic

In addition to rules, frames and semantic nets, knowledge can also be represented by the symbols of *logic*. A logic prescribes the rules for manipulating symbols (rules of *exact reasoning*). The application of computers to perform reasoning has resulted in *logic programming*.

One of the oldest and simplest types of formal logic is the syllogism. Any syllogism of the form

> *premise*:   All X are Y
>
> *premise*: Z is a X
>
> *conclusion*: Z is a Y

is valid no matter what is substituted for X, Y and Z.

A sentence 'Z is a X', whose truth value can be determined, is called a statement or *proposition*. Propositional logic, or propositional calculus, is a symbolic logic for manipulating propositions. Compound propositions are formed by using logical connectives on individual propositions like

> AND (*conjunction*)
>
> OR (*disjunction*)
>
> NOT (*negation*)
>
> IF ... THEN (*conditional*)
>
> IF and only IF (*biconditional*)

A *tautology* is a compound proposition that is always true, whether its individual statements are true or false. A *contradiction* is a compound proposition which is always false. If a conditional (IF...THEN) is always true, it is called an *implication*.

A basic rule of logic that asserts that if we know that *A implies B* (implication) and we know for a fact that *A* is the case, we can assume *B*, is called *modus ponens*.

Although propositional logic is useful, it does have limitations. The major problem is that propositional logic can only deal with complete statements. That is, it cannot examine the internal structure of a statement. More general cases can be studied by *predicate logic*. In particular, predicate logic is concerned with the use of special words called *quantifiers*, such as 'all', 'some' and 'no'. These words are very important because they explicitly quantify other words and so make sentences more exact. Quantifiers are concerned with 'how many' and so permit a wider scope of expression than propositional logic.

A serious limitation of predicate logic is expressing things that are sometimes but not always true. This problem might be solved by fuzzy logic, as we shall see at a later stage.

8.4 Methods of inference

If no algorithmic solution can be obtained and reasoning offers the only possibility of a solution, a method of reasoning or inference becomes particularly important. In this section, we will review a few topics like trees, deductive logic, inference rules and chaining.

8.4.1 Decision trees

A tree is a hierarchical data structure consisting of *nodes* (vertices) which store information or knowledge and *branches* (links, edges) which connect the nodes. A tree can be considered as a special type of semantic net. If a tree is used to make decisions, then it is called a *decision tree*. Figure 8.5 shows a decision tree with information and knowledge about the selection of appropriate clustering algorithms. The nodes contain

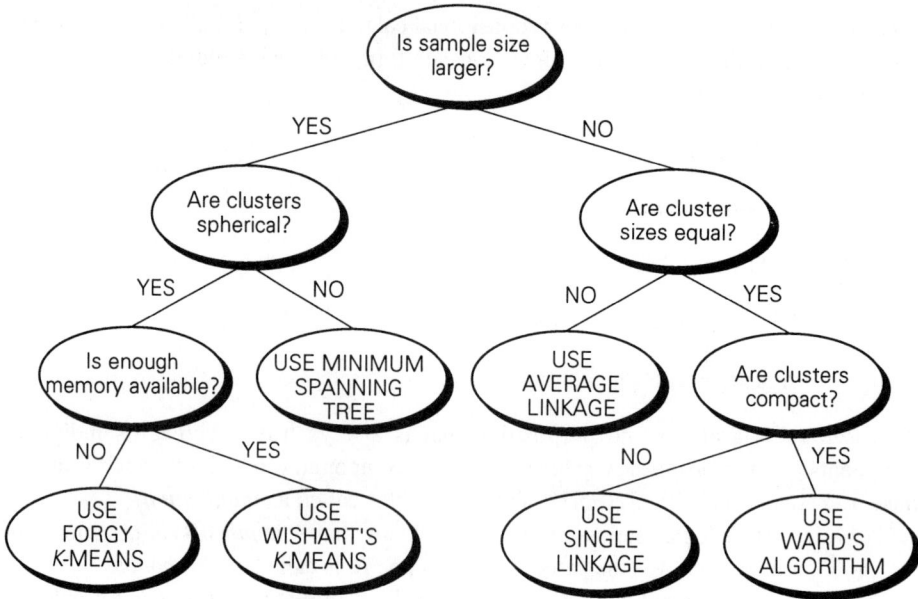

Figure 8.5: Decision tree for the selection of clustering algorithm.

questions, the branches 'YES' or 'NO' responses to the questions and the leaves contain the best guesses of what algorithm should be used. If the decisions are binary, then the decision tree is easy to construct and very efficient. If N questions are posed, a maximum of 2^N answers are possible. $N = 20$ questions can classify 1,048,576 possible answers. Decision structures can easily be translated into production rules. For example, the decision tree of Figure 8.5 could be translated into rules as follows

> **IF** QUESTION = 'is sample size large?' **AND** RESPONSE = 'NO'
> **THEN** QUESTION := 'are cluster sizes equal?'

> **IF** Question = 'is sample size large?' **AND** RESPONSE = 'YES'
> **THEN** QUESTION := 'are cluster spherical?'

and so forth for the other nodes. Leaf nodes are generating ANSWERS rather than QUESTIONS. Decision structures are limited because they cannot deal with variables as expert systems can.

8.4.2 Rules of inference

As we have identified inference as the process by which new facts are derived from known facts, it is important to define various methods of reasoning and to discuss some of the underlying schemata of logic. A brief summary of methods of inference is the

following:
- *Deduction.* Logical reasoning in which conclusions must follow from their premises; the essential characteristic of deductive logic is that the true conclusion *must* follow from true premises.
- *Induction.* Reasoning from the special case to the general (example-driven).
- *Abduction.* Reasoning back from a true conclusion to the premises that may have caused the conclusion.
- Informal methods of inference are based on *intuition, heuristics,* and *trial and error.*

Commonly, the above methods are considered in reasoning based on the assumption that once a fact is determined it cannot be altered during the course of the reasoning process. This is called *monotonic reasoning.* Reasoning that can be revised if some value changes during a reasoning session (previous knowledge may be incorrect when new evidence is obtained) is called *non-monotonic reasoning.*

Common-sense reasoning is the type that people use in ordinary situations; it may be a combination of any of the above methods. The application of *fuzzy logic* to common-sense reasoning will be discussed in Section 8.7.2.

One of the most often used methods of drawing inferences is *deductive propositional logic.* Inference schemata of this propositional form have two basic formats:
- Direct reasoning or *modus ponens,* and
- Indirect reasoning or *modus tollens.*

A compound proposition $p \rightarrow q$, corresponds to a rule as used in a rule-based expert system, while the p corresponds to the pattern that must match the antecedent for the rule to be satisfied.

A general schema for representing *modus ponens* is

$$p \rightarrow q$$
$$\underline{p}$$
$$q$$

or, in words, if we know that p implies q and we know for a fact that p is the case, we can assume q.

Table 8.3: Truth table for modus ponens.

		premises		conclusion
p	q	$p \rightarrow q$	p	q
T	T	T	T	T
T	F	F	T	F
F	T	T	F	T
F	F	T	F	F

The truth table for *modus ponens* is shown in Table 8.3. Note that only the first row has true premises and a true conclusion, and there are no other rows which have true premises and a false conclusion.

In contrast to direct reasoning, we have indirect reasoning or *modus tollens*. If we know that p implies q and we know for a fact that –q is the case, we can assume –p, or as an argument schema

$$\begin{array}{c} p \;\rightarrow\; q \\ \underline{\;–q\;} \\ –p \end{array}$$

The truth table for modus tollens is shown in Table 8.4. Note that the conclusion is true only when the premises are true.

Table 8.4: Truth table for modus tollens.

		premises		conclusion
p	q	p → q	–q	–p
T	T	T	F	F
T	F	F	T	F
F	T	T	F	T
F	F	T	T	T

Modus ponens and modus tollens are rules of inference, sometimes called *laws of inference*. A few other laws of inference for propositional logic are
– *law of contrapositive*

$$\begin{array}{c} \underline{\;p \;\rightarrow\; q\;} \\ –p \;\rightarrow\; –q \end{array}$$

– *law of the syllogism* (chain rule)

$$\begin{array}{c} p \rightarrow q \\ \underline{\;q \rightarrow r\;} \\ p \rightarrow r \end{array}$$

– *De Morgan's law*

$$\begin{array}{c} \underline{\;–(p \wedge q)\;} \\ –p \vee –q \end{array}$$

8.4.3 Forward and backward chaining

The order in which inferences are drawn is determined by a so-called control strategy.

In a rule-based system, *forward chaining* begins by asserting all of the rules whose IF clauses are true. It then checks to determine what additional rules might be true, given the facts it has already established. This process is repeated until the program runs out of new possibilities. As such, a chain that is searched or traversed from a problem to its solution is called a *forward chain*. Thus, forward chaining is reasoning from facts to the conclusions which follow from the facts.

In a rule-based system, *backward chaining* is initiated by a goal rule. The system attempts to determine if the goal rule is correct. It backs up to the IF clauses of the rule and tries to determine if they are correct. This, in turn, leads the system to consider other rules that would confirm the IF clauses. In this way the system backs into its rules. Eventually, the backward-chaining sequence ends when a question is asked or a previously stored result is found. Thus, a chain that is traversed from a hypothesis back to the facts which support the hypothesis is a *backward chain*.

As an example,

$$\frac{\begin{array}{l} p \rightarrow q \\ p \end{array}}{q}$$

$$\frac{\begin{array}{l} q \rightarrow r \\ q \end{array}}{r}$$

$$\frac{\begin{array}{l} r \rightarrow s \\ r \end{array}}{s}$$

configures a causal chain of forward inferences which deduces that s can be assumed if we know for a fact that p is the case.

Forward chaining works forward to find what solutions follow from the facts and is therefore called *data-driven, bottom-up reasoning* (from present to future) and is said to be particularly suited for planning, monitoring and control. On the other hand, backward chaining works backward to find facts that support the hypothesis and is therefore called *goal-driven, top-down reasoning* (from present to past) and is said to be particularly suited for diagnosing tasks. However, it should be remarked that it is certainly possible to do diagnosis in a forward chaining system and planning in a backward chaining one.

Two other methods of inference are of interest related to *diagnosis*. They are not as general as deduction but very useful. Related to induction is *analogical reasoning*. The basic idea of reasoning by analogy is to try and relate old situations as a guide to new ones. Rather than treating every new situation as unique, it is often helpful to try and see the *similarities* of the new situation to old ones that we know how to deal with. Analogy cannot make formal proofs; instead, it is a heuristic reasoning tool that may sometimes work. Prime examples are found in medical diagnosis.

Notice that we have encountered analogical reasoning in the form of *classification inference* in Chapter 5. See also Backer et al. (1988).

Another method that is commonly used in diagnostic problem-solving is *inference by abduction*. Abduction is sometimes referred to as reasoning from observed facts to the best explanation (Reggia, 1985).

As an example, consider the following:

> **IF** cluster C belongs to a 'strong' clustering
> **THEN** C has a relatively low scatter value
>
> **IF** C has a relatively low scatter value
> **THEN** C has a unimodal high point density

If we know that C has a unimodal high point density, can we conclude that C belongs to a 'strong' clustering? The answer to this question depends on whether we are talking about the real world or our expert system. In the real world, we could not make this conclusion with any degree of certainty. C could have a low scatter value and posess a unimodal high point density, while not being part of a 'strong' clustering. In fact, the probability of C belonging to a 'strong' clustering is rather low in the absence of any more information about the clustering. However, in a system with only the preceding rules, we could say by abduction with 100% certainty that if C has a unimodal high point density, then C belongs to a 'strong' clustering.

8.5 Reasoning under uncertainty

In this section, the issue of reasoning under uncertainty will be discussed. *Uncertainty –* to be considered as the lack of precision in information to make a decision – turns out to be a major topic in many expert system applications. Dealing with uncertainty requires reasoning under uncertainty. A number of theories have been developed to deal with uncertainty, among them Bayesian probability, Dempster–Shafer theory, and Zadeh's fuzzy theory.

8.5.1 About uncertainty

The deductive method of reasoning described in the previous sections is called *exact reasoning* because it deals with exact facts and the exact conclusions that follow from those facts. However, many expert system applications like the one we are dealing with cannot be done with exact reasoning, and require *inexact reasoning* involving uncertain facts, rules or both. Classic examples of successful systems which deal with uncertainty are MYCIN, for medical diagnosis (Davis et al., 1977), and PROSPECTOR, for mineral exploration (Duda et al., 1979). In EDAPLUS, numerical weights are given to facts or relationships to indicate the degree of confidence in the facts or relationships. In all cases, these numerical weights behave differently than *probabilities*. In general, methods for manipulating these weights (*certainty factors*) are more *informal* than approaches to manipulating probabilities.

Different types of *error* can contribute to uncertainty:
- *Ambiguity*. Something may be interpreted in more than one way.
- *Incompleteness*. Some information is simply missing.
- *Incorrectness*. Some information might be wrong.
- *Measurement errors* (precision and accuracy). Accuracy corresponds to the truth of a specific value, while precision corresponds to how well the truth is known.
- *Unreliability*. Some information may sometimes be correct and sometimes not.
- *Errors of reasoning*. Errors due to incorrect formulation of the rules. Although it is one of the interesting characteristics of human experts that they reason well under uncertainty, it should be noted that even experts are not immune to making mistakes, especially under uncertainty.

Probability is the oldest quantitative way of dealing with uncertainty, originating in the 17th century (Pascal and Fermat). The fundamental formula of 'classical' probability is defined as the probability

$$P = w/n$$

where w is the number of 'wins' and n is the number of equally possible *events*, which are the possible outcomes of an experiment or *trial*. P is said to be an *a priori* probability because the probability is calculated before the experiment is performed. When repeated trials give exactly the same result, the system is *deterministic*. If it is not deterministic, it is *non-deterministic* or *random*. Probability is fundamental to *statistics*, which is concerned with collecting and analyzing data about *populations*, a set from which samples are drawn.

Deduction and induction are the basis of statistical inference (reasoning) about populations and samples. Given a known population, deduction enables us to make inferences about the unknown sample. Correspondingly, given a known sample, induction enables us to make inferences about the unknown population.

Notice that supervised classification (pattern recognition) is concerned with inductive inference while non-supervised classification (cluster analysis) is concerned with deductive inference.

The theory of probability is based on three axioms:

1. $0 \leq P(e) \leq 1$

2. $\sum_i P(e_i) = 1$

3. $P(e_1 \cup e_2) = P(e_1) + P(e_2)$

where e_1 and e_2 are mutually exclusive events. These axioms put probability on a sound theoretical basis.

As already mentioned, a priori probabilities are not based upon experiment; all possible events and outcomes are known a priori. In practical problems, this ideal situation never occurs. So, *experimental probability* defines the probability of an event, $P(e)$, as the limit of a frequency distribution

$$P(e) = \lim_{n \to \infty} \frac{f(e)}{n}$$

where $f(e)$ is the frequency of outcomes of an event for n observed total outcomes. This type of probability is called *a posteriori probability*. Clearly, in practice our probabilities are approximated by a finite number of experiments.

Finally, the third type of probability is *subjective probability*. Subjective probability deals with events that are not reproducible and have no historical basis on which to extrapolate. A subjective probability is actually a *belief* or opinion expressed as a probability rather than a probability based on axioms or empirical measurement.

Note that a subjective probability by an expert is better than no estimate at all and is usually very accurate. Beliefs play an important role in expert systems, as we shall see in the following sections.

If events are not mutually exclusive, they influence one another. That is, knowing that one event has occurred may cause us to revise the probability that another event will occur. The probability of an event A, given that event B occurred, is called a *conditional probability*, $P(A|B)$, and is defined as

$$P(A|B) = P(A \cap B)/P(B) \quad \text{for } P(B) \neq 0$$

The solution to the inverse problem (to find the inverse probability) is *Bayes' theorem*. The general form reads as follows

$$P(H|e) = P(e|H)P(H)/P(e)$$

where e denotes an event, and H represents an hypothesis. Bayes' theorem is commonly used for decision tree analysis and is called Bayesian decision making under uncertainty. Note that $P(A|B)$ can be interpreted as the *degree of belief* that A is true, given B (*likelihood*).

8.5.2 Uncertainty in inference

Uncertainty may be present in rules, evidence used by the rules, or both. Problems may occur and sometimes probability provides an answer.

Expert inconsistency

It is not uncommon for experts to say that the observation of evidence is important but the absence of the evidence is unimportant. In this case the likelihood theory based on Bayesian probability theory is incomplete. That is, the theory is only satisfied if

$$P(e|H') = 1 - P(e|H)$$

which appears to be a rigorous constraint.

Uncertain evidence
Assume that the degree of belief in the complete evidence, E, is dependent on the partial evidence, e, given by

$$P(E|e)$$

The partial evidence, e, is the portion of E that we know. If we know all the evidence, then $E = e$ and

$$P(E|e) = P(E)$$

where $P(E)$ is the prior likelihood of the evidence E. The likelihood $P(E|e)$ is our belief in E given our imperfect knowledge, e, of the complete evidence E. If the hypothesis, H, depends on E and E is based on some partial evidence e, then $P(H|e)$ is the likelihood that H depends on e.

In real applications, experience has shown that human experts give subjective probabilities that are almost certain to be inconsistent. Then, we may easily reach the situation that

$$P(H|e) > P(H)$$

yielding a probability that is greater than it should be and may become magnified further as the inference from one rule is used by another in the chain of inferences.

This can be corrected by assuming that $P(H|e)$ is a piecewise linear function. This is an *ad hoc* assumption that worked well in various applications but is not based on traditional probability theory. An appropriate formula for $P(H|e)$ is calculated using linear interpolation as follows:

$$P(H|e) = P(H|E') + (P(H) - P(H|E')) (P(E|e)/P(E))$$
$$\text{for } 0 \leq P(E|e) < P(E)$$

$$= P(H) + (P(H|E) - P(H)) (P(E|e) - P(E))/(1 - P(E))$$
$$\text{for } P(E) \leq P(E|e) \leq 1$$

Now the value of $P(H|e)$ remains the same if $P(E|e) < P(E)$, and increases if $P(E|e) \geq P(E)$. The piecewise function ensures that when $P(E|e) = P(E)$, then $P(H|e) = P(H)$.

Combining evidence
Suppose a rule

      IF E THEN H

has E as the *conjunction of evidence*, as in

$$\text{IF } E_1 \text{ AND } E_2 \text{ AND } E_3 \dots \text{ THEN } H$$

All the E_i must be true with some probability for the antecedent to be true. In the general case, each piece of evidence is based on partial evidence e.

If we assume all E_i conditionally independent, we obtain

$$P(E_1 \cap E_2 \cap E_3 \dots |e) = \prod_i P(E_i|e)$$

This formula is correct in a theoretical sense. However, there are two difficulties in practice. First, the individual $P(E_i|e)$ are usually not independent in the real world. Second, the multiplication of many factors leads to a product that is generally far too small for $P(E|e)$.

An approximate solution to this problem is use of *fuzzy logic* to calculate $P(E|e)$ as follows:

$$P(E|e) = \min_i \, [P(E_i|e)]$$

where the min function returns the minimum value of all $P(E_i|e)$. The main problem with the fuzzy logic formula is that it makes $P(E|e)$ insensitive to any $P(E_i|e)$ except the minimum. An advantage of the fuzzy logic formula is, of course, that it is computationally simple.

Note that a *disjunction of evidence* corresponds to a max function if the formula is based on fuzzy logic.

8.5.3 Degrees of belief

Following Giarratano and Riley (1989), Figure 8.6 – purposely drawn fuzzy – shows the vague nature of the terms
– impossible
– possible
– plausible
– probable
– certain

and the vagueness in progression from one to another.

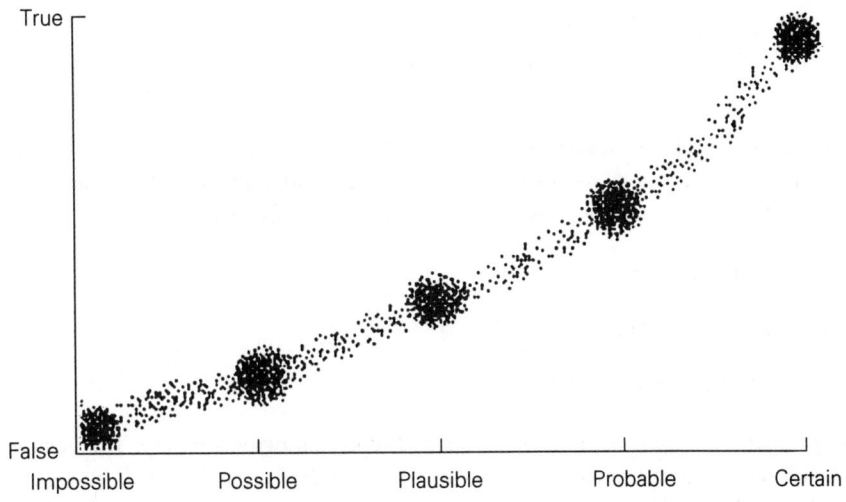

True

False

Impossible   Possible   Plausible   Probable   Certain

Figure 8.6 (From Giarratano and Riley, 1989)

A *certain belief* means that the belief is true, while *impossible belief* means that it is false. *Certain evidence* is logically true or false. This means that certain evidence corresponds to either certain belief (logically true) or impossible belief (logically false). A *possible belief* means that, no matter how remote, the hypothesis cannot be ruled out. A *plausible belief* means that more than a possibility exists. A *probable belief* means that there is some evidence favouring the hypothesis but not enough definitely to prove it.

As evidence accumulates, the plausible relations may become *probable* and then *definite*. This will be further discussed in Chapter 9.

8.6 Inexact reasoning

This section deals briefly with alternative theories that were specifically developed to deal with human belief rather than the classic frequency interpretation of probability. All these theories are examples of *inexact reasoning* in which the antecedent, the conclusion, and even the meaning of the rule itself are uncertain to some extent.

The accurate use of Bayes' theorem depends on knowing many probabilities, i.e. all prior probabilities of individual hypotheses, prior probabilities of pieces of evidence, and all conditional probabilities of a piece of evidence given a particular hypothesis. It is usually impossible to determine consistent and complete values for all these probabilities for the general population. In practice, evidence tends to accumulate piece by piece.

Let E_1 denote the existing piece of evidence, and E_2 the new piece of evidence. Then

$$P(H_i|E) = \frac{P(E_2|H_i \cap E_1)P(H_i|E_1)}{\sum_j P(E_2|H_j \cap E_1)P(H_j|E_1)}$$

However, these probabilities are not all generally known. The expression becomes even more complicated as more pieces of evidence accumulate and so more probabilities are required. Moreover, in practice, experts might be very uneasy if they believe

$$P(H|E_1 \cap E_2) = 0.8$$

since they also have to believe

$$P(-H|E_1 \cap E_2) = 0.2$$

These problems led to investigate other ways of representing uncertainty. This resulted in the development of the concept of *belief* and *disbelief*, and the *certainty factor*, as originally used in MYCIN.

8.6.1 Certainty factors

In MYCIN, the certainty factor (CF) was defined as the difference between belief and disbelief,

$$CF(H,E) = MB(H,E) - MD(H,E)$$

where

CF is the certainty factor in the hypothesis H due to evidence E;
MB is the measure of *increased belief* in H due to E;
MD is the measure of *increased disbelief* in H due to E.

The measures of belief and disbelief were defined in terms of probability by

$$MB(H,E) = 1 \qquad \text{(if } P(H) = 1\text{)}$$

$$= \frac{\max(P(H|E),P(H)) - P(H)}{1 - P(H)} \qquad \text{(otherwise)}$$

and

$$MD(H,E) = 1 \qquad \text{(if } P(H) = 0\text{)}$$

$$= \frac{\min(P(H|E),P(H)) - P(H)}{-P(H)} \qquad \text{(otherwise)}$$

Combining the measures of belief and disbelief into a single number has two uses. First, the certainty factor can be used to rank hypotheses in order of importance. Second, certainty factors allow an expert to express a belief without committing a value to the disbelief; it simply indicates the net belief in a hypothesis based on some evidence.

The definition of CF was changed in MYCIN in 1977 to

$$CF = \frac{MB - MD}{1 - \min(MB, MD)}$$

to soften the effects of a single piece of non-confirming evidence on many confirming pieces of evidence.

If two rules have the same conclusion, but with a different certainty factor, the combining function for certainty factors, given by

$$CF(CF_1, CF_2) = CF_1 + CF_2(1 - CF_1) \qquad \text{(both} > 0)$$

$$= (CF_1 + CF_2)/(1 - \min(|CF_1|, |CF_2|)) \quad \text{(one} < 0)$$

$$= CF_1 + CF_2(1 + CF_1) \qquad \text{(both} < 0)$$

yields a value for the combined certainty factor determined by the values of CF_1 and CF_2.

Although certainty factors have some basis in probability theory, the CF were also partly *ad hoc*. The major advantage is computational simplicity. CF is particularly succesful in cases of short inference chains and simple hypotheses.

We will show that in our case the use of certainty factors is demonstrated to be justified. Moreover, their use is easy to understand.

8.6.2 Dempster–Shafer theory (Shafer, 1976)

Another method of inexact reasoning is the *Dempster–Shafer theory*, which attempts to model uncertainty by a range of probabilities rather than as a single probabilistic number. The Dempster–Shafer theory has a good theoretical foundation. Certainty factors can be shown to be a special case of the Dempster–Shafer theory.

The Dempster–Shafer theory assumes that there is a fixed set of mutually exclusive and exhaustive elements called the *environment*, symbolized by θ:

$$\theta = \{\theta_1, \theta_2, \ldots, \theta_N\}.$$

Suppose

$$\theta = \{well \ separated \ compact \ spherical,$$
$$well \ separated \ elongated, \ concentric\}$$

and the question is, 'What cluster shape is linear separable?' The answer is the subset of θ

$$\{\theta_1, \theta_2\} = \{ \ well \ separated \ compact \ spherical,$$
$$well \ separated \ elongated\}.$$

Likewise, the question, 'What cluster shape is linear non-separable?' is the set

$$\{\theta_3\} = \{concentric\}$$

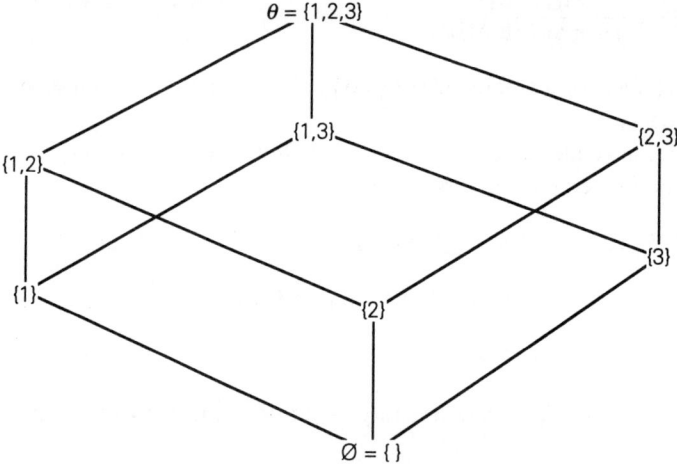

Figure 8.7: All subsets of P(θ)

which is called a *singleton* because it has only one element.

Each subset of θ can be interpreted as a possible answer to a question. If there is no correct answer, the answer is the *null set* Ø = { }.

An environment is called a *frame of discernment* when its elements may be interpreted as possible answers, and only one answer is correct. Commonly, the environment is depicted as a *hierarchical lattice*, as shown in Figure 8.7.

A set of size N has exactly 2^N subsets, including itself, and these subsets define the *power set*, denoted as $P(\theta)$,

$$P(\theta) = \{\emptyset, \{1\}, \{2\}, \{3\}, \{1,2\}, \{1,3\}, \{2,3\}, \{1,2,3\}\}.$$

In Dempster–Shafer theory, the *degree of belief* in evidence is expressed in terms of *basic probability assignment* (BPA). Every bpa can formally be expressed as a function that maps each element of the power set into a real number in the interval 0 to 1. This mapping is stated as

$$m: P(\theta) \to [0,1].$$

By convention, the empty set is usually defined as zero. Thus

$$m(\emptyset) = 0$$

and the sum of all BPAs for every subset of the power set is 1,

$$\sum_{X \in P(\theta)} m(X) = 1$$

We now examine how evidences can be combined. This has been formalized by *the rule of combination*.

$$m_1 \oplus m_2(Z) = \sum_{X \cap Y = Z}^{n} m_1(X)m_2(Y)$$

where $m_1 \oplus m_2$ represents the *consensus* of the original, possibly conflicting evidence.

If we consider a belief in the evidence of 0.7 that the cluster shapes are linear separable, then the BPA to the subset $\{1,2\}$ = {*well separated compact spherical, well separated elongated*}, is

$$m_1(\{1,2\}) = 0.7.$$

The rest of the belief is left with the environment, θ, as non-belief

$$m_1(\theta) = 1 - 0.7 = 0.3.$$

Assume next, $m_2(\{2\}) = 0.9$. Then, following the rule of combination, we arrive at

$$m_3(\{2\}) = m_1 \oplus m_2(\{2\})$$

$$= m_1(\{1,2\})m_2(\{2\}) + m_1(\{1,2,3\})m_2(\{2\})$$

$$= 0.9$$

$$m_3(\{1,2\}) = m_1 \oplus m_2(\{1,2\})$$

$$= m_1(\{1,2\})m_2(\{1,2,3\})$$

$$= 0.07$$

and

$$m_3(\{1,2,3\}) = m_1 \oplus m_2(\{1,2,3\})$$

$$= m_1(\{1,2,3\})m_2(\{1,2,3\})$$

$$= 0.03.$$

The $m_3(\{2\})$ represents the belief that the cluster shape is *well separated elongated* and only *well separated elongated*. However, $m_3(\{1,2\})$ and $m_3(\{1,2,3\})$ imply additional information.

Since $\{1,2\}$ and $\{1,2,3\}$ include $\{2\}$, it is plausible that they may contribute to the belief in $\{2\}$. Consequently, the true belief in $\{2\}$ is assumed to be in the range of belief of 0.9 to 1.

If we define the *belief function*, BEL, as the total belief of a set and all its subsets

$$BEL(X) = \sum_{X \subset Y} m(Y)$$

we obtain, as an example,

$$BEL_1(\{1,2\}) = m_1(\{1,2\}) + m_1(\{2\}) + m_1(\{1\})$$

$$= 0.7 + 0 + 0 = 0.7.$$

When more evidence becomes available, that might be conflicting, the sum of all BPAs is not guaranteed to be equal to 1, but might be less than 1. The solution to this problem is found in a normalization of the focal elements by dividing each focal element by

$$1 - \sum_{X \cap Y=0} m_1(X)m_2(Y)$$

being the amount of evidential conflict.

In Chapter 9, we present the use of the Dempster–Shafer theory of evidence as discussed above. As outlined in Part 1, it might be difficult to find a cluster algorithm which is uniformly better than other algorithms throughout the pattern space. Following Garvey et al. (1981) and Balasubramaniam et al. (1990), we will discuss a general strategy for dealing with this sort of situation in which use is made of several available heuristics and suitable ways are offered to combine them.

8.7 Approximate reasoning

In Chapter 5, cluster analysis was reformulated as a process of fuzzy identification based on fuzzy relations, linking fundamental structures, fuzzy relations and fuzzy partitions nicely together. A 'hard' cluster (ordinary crisp set) was identified as a special case of a 'fuzzy' cluster (a fuzzy set with varying membership values on [0,1]) with membership function {0,1}. We discussed a variety of fuzzy set theoretical operations in order to reformulate the process of clustering, to represent the degree of cluster ambiguity (minimum fuzziness partition), and to interpret the ultimate findings.

In anticipation of later chapters, Chapter 5 also included some fundamental ideas about approximate reasoning – as a branch of fuzzy logic – fuzzy classification, and fuzzy inference. In the context of this chapter, at this stage, we will further discuss a theory of uncertainty based on fuzzy logic, emphasizing its concern with quantifying and reasoning using *natural language.*

In natural language, many words may have ambiguous meaning, such as *high, low, very much, little,* and so on. Fuzzy logic is a development that can deal with these terms.

8.7.1 Fuzzy sets and natural language

Consider the following.

'The CPCC value is *more or less medium*'

'*Several* inconsistent edges are detected'

'The cluster tendency is *high*'

'The cluster validity is *low*'

where the words in italics refer to fuzzy sets and quantifiers. All these fuzzy sets and quantifiers can be represented and operated on in fuzzy theory.

A *fuzzy proposition* contains words such as *high* which is the identifier of a fuzzy set HIGH. In contrast to a classic proposition such as 'The CPCC value is 0.63', which represents a proposition which is either true or false, a fuzzy proposition may have degrees of truth. For example, the fuzzy proposition 'The cluster validity is *low*', may be true to some degree: *somewhat true, fairly true, very true*, and so on.

A fuzzy truth value is called a *fuzzy qualifier*, and may be used as a fuzzy set or to modify a fuzzy set. Fuzzy propositions may have fuzzy quantifiers, such as *most, many, usually*, and so on, with no dsitinction between statements and propositions as in the classical case.

Many fuzzy words are used in natural language, such as

low	*medium*	*high*
very	*not*	*little*
several	*more or less*	*few*
many	*more*	*most*
about	*approximately*	*sort of*

There are also compound terms, such as

very low	*more or less low*
not very low	*medium to high*
approximately low	*somewhat low*

which can also be defined and manipulated in fuzzy theory.

To illustrate the concept of fuzzy sets, consider the following example.

'The CPCC value is *high*'

One possible membership function is shown in Figure 8.8.

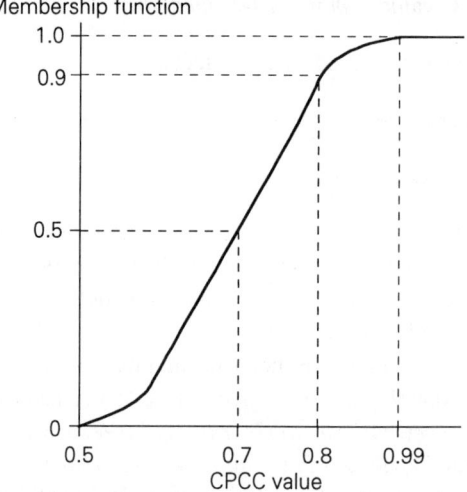

Membership function

Figure 8.8 : A membership function for the fuzzy set HIGH.

If the CPCC value is about 0.9 or above (including the maximum value of 1), the CPCC value is considered to have a membership function of 1.0 in fuzzy set HIGH. On the other hand, if the CPCC value is below 0.5, the CPCC value is considered to be in the fuzzy set HIGH and so the membership function is 0 or close to 0. In between, the membership function increases monotonically with the CPCC value.

Depending on the application, for each attribute a membership function may be constructed from one expert's opinion or from a group of experts. (See Duin and Backer, 1988.) It is important to note that although the membership function might be the result of an opinion poll of an expert panel, the membership function is really not a frequency distribution!

A mathematical function that is often used in fuzzy sets to model a membership function is called the *S-function*, and is defined as follows:

$$S(x;a,b,c) = 0 \qquad\qquad (x \leq a)$$

$$= 2[(x - a)/(c - a)]^2 \qquad (a \leq x \leq b)$$

$$= 1 - 2[(x - c)/(c - a)]^2 \quad (b \leq x \leq c)$$

$$= 1 \qquad\qquad\qquad (x \geq c)$$

A plot of this function is given in Figure 8.9.

S-function

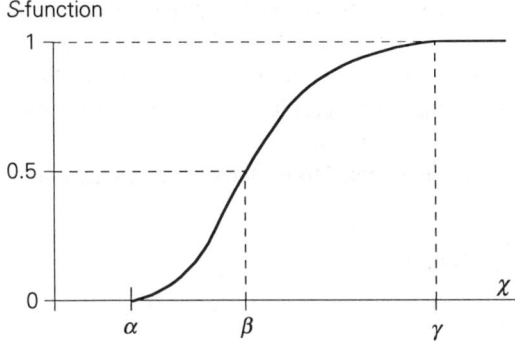

Figure 8.9: The S-function.

A second function is necessary to model for example the fuzzy set MEDIUM, called the *Π-function*

$$\Pi(x;b,c) = S(x;c-b,c-b/2,c) \qquad (x \le c)$$

$$= 1 - S(x;c,c+b/2,c+b) \qquad (x \ge c)$$

Figure 8.10 shows the shape of the *Π*-function.

In Chapter 5, we encountered the major fuzzy set operators. With their definitions, the laws for classic sets of commutativity, associativity and so forth, hold for fuzzy sets with the exception of idempotentce.

The following operations have no counterpart in ordinary set operations.
– *concentration* (CON): to be used to approximate roughly the effect of the linguistic modifier *very*;

Π-function

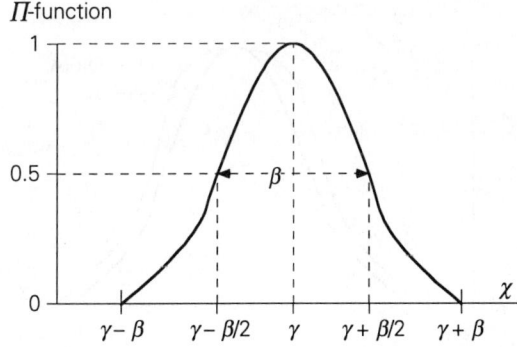

Figure 8.10: The Π-function.

– *dilation* (DIL): to be used to approximate roughly the effect of the linguistic modifier *more or less*;
– *intensification* (INT): to be used to approximate roughly the effect of articulation towards the corresponding (classical) level set based on the cross-over points of the fuzzy set.
– *normalization* (NORM): to be used to set the maximum membership value, taken over all elements *x*, to 1.

The CON operation is defined as

$$\mu_{CON}(x) = \mu(x)^2$$

and is illustrated in Figure 8.11. The DIL operation is defined as

$$\mu_{DIL}(x) = \sqrt{\mu(x)}$$

and is illustrated in Figure 8.12. The INT operation is defined as

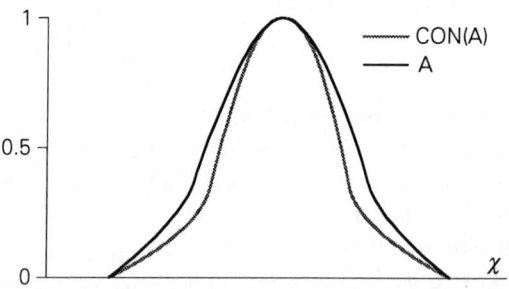

Figure 8.11: Concentration of a fuzzy set.

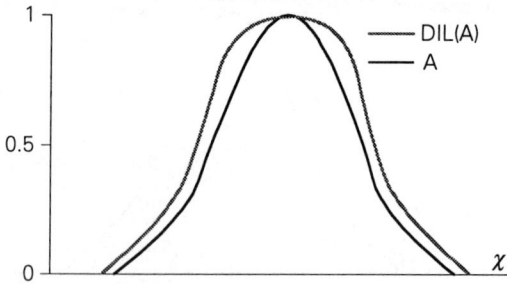

Figure 8.12: Dilation of a fuzzy set.

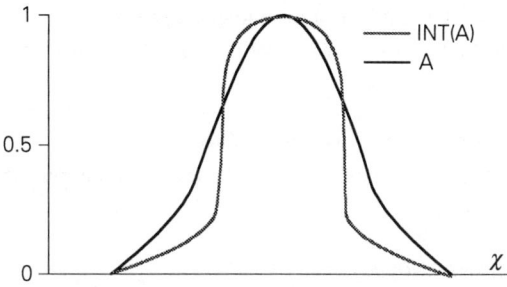

Figure 8.13: Intensification of a fuzzy set.

$$\mu_{\text{INT}}(x) = 2(\mu(x))^2 \qquad (0 \le u(x) \le 0.5)$$

$$= 1 - 2(1 - \mu(x))^2 \qquad (0.5 \le u(x) \le 1)$$

and is illustrated in Figure 8.13. These operations are of importance performing the effect of linguistic modifiers on fuzzy propositions.

In Chapter 5, we also introduced the concept of fuzzy relations. These have important applications to approximate reasoning, as shown later.

The composition of relations is the net effect of applying one relation after the other.

A common relational composition was defined (see Chapter 5) as the *max min rule*

$$u(a,c) = \max_{b}[\min(\mu(a,b),u(b,c))].$$

This rule will be used in fuzzy logic as the *max min composition rule of inference*.

8.7.2 Fuzzy logic and fuzzy rules

Just as classical logic forms the basis of conventional logic reasoning in expert systems, so *fuzzy logic* forms the basis of fuzzy expert systems. Because very little in the real world is really two-valued (true or false), common-sense reasoning is very difficult to achieve with conventional systems. To overcome the serious limitations of two-valued logic, a number of different logic theories based on multiple values of truth have been formulated, such as those of Lukasiewicz, Bochvar, Kleene, Heyting, and Reichenbach. These are commonly referred to as *multi-valued logic*.

Fuzzy logic may be considered as an extension of multi-valued logic. However, they are different in goal and use. While multi-valued logic is still a logic of exact reasoning, fuzzy logic is the logic of approximate reasoning.

A rule of approximate reasoning like

IF the dimensionality is *too high*
THEN increase the data sample size

is neither exact, nor totally inexact as a pure guess might be.

As a simple example, we consider the following set of rules for tuning the number of clusters.

R_1: IF cluster tendency is *too weak*
THEN lower the number of clusters by changing the threshold level of the dendrogram

R_2: IF cluster tendency is *strong*
THEN keep the number of clusters unchanged

The fuzzy sets for the rule antecedents and consequents are depicted in Figure 8.14.

Now assume that the tendency index yields the value 8. It follows that

$$\mu_{too\ weak}(8) = 0$$

$$\mu_{strong}(8) = 1$$

So rule R_2 is activated and fires with the resulting fuzzy consequent *NO CHANGE*. Applying the compositional rule of inference gives

$$\mu_{NO\ CHANGE} = \max[\min(\mu_{strong}(8))] = 1$$

This translates into a 0% change in the threshold value.

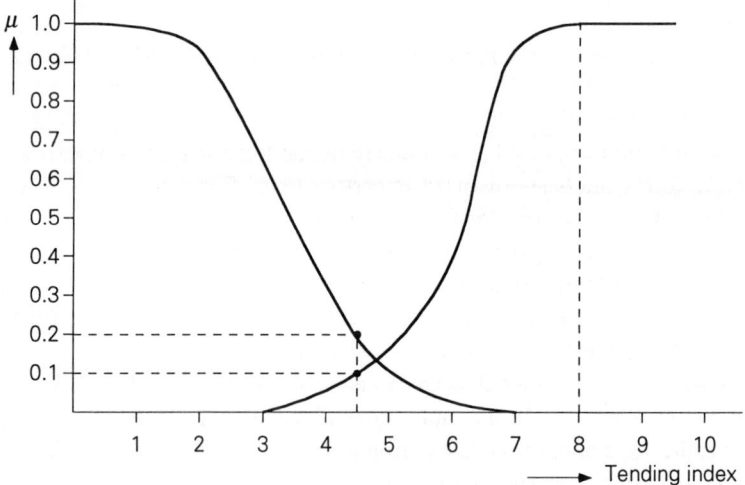

Figure 8.14: Fuzzy rule antecedents for adjustment of the number of clusters.

Now suppose that the tendency index yields a value of 4.5, then

$$\mu_{too\ weak}(4.5) = 0.2$$

$$\mu_{strong}(4.5) = 0.1$$

Both rules will become activated. As a result

$$\mu_{CHANGE\ THRESHOLD} = \max[\min(\mu_{too\ weak}(4.5))] = 0.2$$

and

$$\mu_{NO\ CHANGE} = \max[\min(\mu_{strong}(4.5))] = 0.1$$

(Note that for a single antecedent term, the max and min function are of no use.) As a result, we have two non-zero consequents, and we must decide on an action (defuzzification problem).

The *maximum method* will simply pick the rule with the largest membership grade. In this case, the *CHANGE THRESHOLD* action is chosen since its grade is 0.2 compared to 0.1 for the other rule.

A final remark should be made on *dispositional rules*. The term *disposition* means a proposition that is *usually* true:

$$USUALLY\ (X\ is\ R)$$

where *usually* is an implied fuzzy quantifier and *R* is a *constraining relation* acting on the variable *X* to limit the values it may take.

In fact, *common-sense knowledge* is basically a collection of dispositions about the real world. Take for example

$$USUALLY\ \text{spherical clusters are unimodal}$$

A disposition may be expressed as a heuristic rule like

IF *X* is spherical
THEN it is *likely* that *X* is unimodal

The inference rule is then of the form

$$USUALLY\ (X\ is\ F)$$
$$F \subset G$$
$$\overline{}$$
$$USUALLY\ (X\ is\ G).$$

Nearly all fuzzy expert systems use compositional rules of inference and allow dispositional reasoning.

8.8 Concluding remarks

In this chapter we have brought together some of the basic issues in expert systems; i.e. representation of knowledge, methods of inference, and reasoning under uncertainty. A special link has been made between the art of cluster analysis, on the one hand, and approximate reasoning, on the other.

It has been advocated that fuzzy rules (to represent fuzzy facts and knowledge about clustering data) and fuzzy logic (to combine and propagate fuzzy facts and fuzzy knowledge about a detected clustering) – together forming the basis for a fuzzy expert system – are well suited for success in reasoning about detected clustering in data (called cluster-oriented reasoning) and the validity of their presence in the data.

9 | Rule base developments for cluster analysis

While classification expert systems have been described as pure rule-based systems, they are really repositories of knowledge that summarize previous experiences. When knowledge turns up in unexpected forms, the expert system can become a scavenger that accepts knowledge in any form that proves useful. For expert systems, many useful constructs, such as intermediate hypotheses, have been introduced to simplify reasoning. One view of such constructs is that they are artificial features that are determined from the original features. Thus it would not be unreasonable for expert systems to diverge from absolute conformity with prespecified rule formats. In practice, it is quite common for a rule-based expert system to include composite features, such as a ratio of two feature values or a linear discriminant. An engineering expert system might readily incorporate physical and mathematical models. And if a neural net or some other new learning system can provide accurate information, it surely can be used in an expert system (Weiss and Kulikowski, 1991).

In the previous chapters, we have discussed the general concepts and methodology of cluster analysis and expert systems. This chapter presents general guidelines for developing rule bases to assist the cluster analyst in making valid diagnoses about the data under consideration. A simple language is described so that rule bases can be developed and can be used directly in EDAPLUS.

To a large extent, the development of a rule base will depend on the facts about the data which are either derived from algorithmic processing or supplied by the user. In view of the processing and modelling paradigm, as discussed in Chapter 7, the three general levels of cluster-oriented reasoning (facts, support and diagnosis) will be discussed in the sections that follow. Before starting, there should be a clear identification of implicit and explicit functional associations between pairs of data items, information and knowledge. Developing rules is related to the fact that such functional associations really exists.

Probably one of the most important questions to be answered is how rules can be learned from observations. In Chapter 6, we performed a series of experiments in order to establish informal links between facts about the data under consideration, and expert judgement about clustering tendency. While the observations expose all kinds of indeterminacy and uncertainty, the goal of *learning* from observations is still to transform implicit functional associations (case information) into explicit knowledge rules including the uncertainty associated with them. This learning problem of a rule base may be *the* major factor in development and may also be considered as a somewhat open-ended activity.

There are many learning methods available today Michalski and Stepp (1983a, 1983b), with advantages and disadvantages. Section 9.6 summarizes the characteristics of some of them. Linking together artificial neural nets (ANN) and expert systems will be discussed afterwards, Section 9.7. Hybrid solutions like these are proposed by Kandel and Langholz (1992), and still require further research.

9.1 Historical review

9.1.1 Knowledge-based clustering: the development of conceptual clustering

Generally speaking, the quality of an expert system depends principally on the quality of the knowledge base. Here, knowledge consists of abstractions and generalizations of voluminous material. The ability to generalize from examples and observations is well known as an essential capability of any learning system.

Generalization involves observing a set of training examples of some general concept, identifying the essential features common to these examples, and then formulating a *concept definition* based on these common features.

The principal form of classification of observed entities is performed by means of *numerical clustering*, whereas the principal form of learning from observations is called *conceptual clustering*, the goal of which is to construct classes of objects representing the concepts, and to discover *rules* which define relationships between objects in these classes. The term 'conceptual' indicates that, instead of an extensive enumeration of objects, the goal of clustering is primarily to represent sets of objects by concepts (Michalski, 1980; Michalski and Stepp, 1983a, 1983b; Michalski et al., 1986; Fu and Cheng, 1985; Ficher and Langley, 1986).

Conventional measures of *similarity* – as used in numerical clustering – are unable to capture the *context-sensitive* aspects of objects. To overcome this drawback, Michalski and Stepp (1983a) have proposed: if A and B are two objects under consideration, and E and C are the *environment* and a set of *predefined concepts*, respectively, then the similarity between A and B is a function of the descriptions of A and B, but also a function of the context-sensitive aspects of E and C.

In their paper, Shekar et al. (1987), proposed a semantic-based similarity measure for grouping objects into clusters, particularly suited for man-made objects as they are created from a *functional point of view* which is property-dependent (Shekar et al., 1987). This idea has been furthered by Srivastava and Murty (1990). They define finite feature sets that can be used to represent a pattern. These sets can be used along with *operational knowledge* to extract the object order of each object (data-dependent knowledge), and the classification order (based on the intended classification). While they identify conceptual knowledge, contextual knowledge, functional knowledge and background knowledge, the authors establish a unified framework for clustering in which the *object representation* (an instantiation of the weights of various features based on their importance in representing the object) is combined with the *classification intention* (an instantiation of the weights to various features based on their importance with respect of the intended classification).

Methods of conceptual clustering have proven to be efficient for constructing rules related to a knowledge base for expert systems. Research in this area has been presented by Ho et al. (1988). Their approach will be discussed in Section 9.6.

9.1.2 Knowledge-based clustering: the development of computer-assisted reasoning in cluster analysis

Parallel to the above developments in conceptual clustering, also in the field of numerical clustering, as an attempt to group pattern samples in a multi-dimensional space into meaningful categories without any a priori information, questions were posed about computer-assisted (knowledge-based) reasoning when several clustering methods are applied and the results have to be examined before attempting to draw conclusions. It was found to be worthwhile to explore the entire clustering toolbox to establish initial information (varied results on the same data under consideration) as much as possible, on the basis of which combining, supporting, and diagnostic reasoning could be performed using heuristic rules obtained from experts' experiences.

In search of a flexible method for integrating new knowledge incrementally into the existing store of knowledge, a *rule-based system* was found to provide the necessary infrastructure for all this. Since then, a variety of knowledge-based expert systems have been proposed which meet the basic needs of the clustering problem.

Backer and Eijlers (1986) proposed a system, CLUSAN 1, in which production rules were responsible for drawing (sub-)conclusions on intrinsic dimensionality, the existence and statistical influence of outliers, and the existence of irregular cluster shapes. Based on these findings, strategic planning for the statistical analysis could be derived and carried out. In 1988, the system was extended by incorporating reasoning about clustering tendency. In the above approach, numerical facts, derived from the data under consideration, include

– partition dominance (obtained after a large number of random initializations);
– the square error versus the number of clusters;
– a measure of fit of the MST edge-length distribution with the random MST edge-length distribution;
– the inconsistency of MST edges;
– an index for cluster separation.

Computed values are compared with a set of expert-supplied thresholds. Inferences are made about clustering tendency valuated by means of the legal values *nihil, small, significant, large, very large*.

Balasubramaniam et al. (1990) report on two types of knowledge-based systems: a knowledge-based system for the selection of clustering algorithms, and a knowledge-based system for identifying natural clusters. The development of the former is mainly motivated by the fact that understanding of the differences between various clustering algorithms with respect to their *strengths and weaknesses* (as we have encountered in Chapter 6), will lead to the system's ability automatically to choose the right and appropriate algorithm. The system is designed to have forward reasoning with backtracking in case of failure to achieve *stable clusters*.

The proposed system for identifying natural clusters is meant for a more general situation where the pattern space may not be uniformly structured. This system employs a hybrid approach which combines the *strong* points of both the partitional and hierarchical cluster methods. The initial compact clusters identified by a partitional method are merged on the basis of the pooled evidential strength of the proposal for merging two clusters. Here, the evidential intervals computed for different heuristics used in different hierarchical methods are properly combined by using the Dempster–Shafer rule of combination.

The development of EDAPLUS, the software package that comes with this book, has benefited from the above-mentioned ideas. The reader may implement all examples of rule bases as given in the following sections. Appendix B describes how to use EDAPLUS.

In order to present examples of rule bases consistenly and uniformly, we start our discussion with an easy-to-read/write language for rule bases.

9.2 A language for rule bases

It is important to note here that we certainly do not want to imply that knowledge representation in the form of rules, and stack-oriented inferencing by matching have to be considered as *the* ultimate solution. More advanced structures for data and inference models are likely to be more appropriate. The only advantage of the rule language here is comprehensibility. This is one of the goals here.

We start with the basic elements of the rule language.

The *syntax* of the rule language strongly resembles Pascal. Any white space only serves to separate tokens. At any place, comments can be included by using (* and *) and will be represented by only one space. There is no different interpretation of 'upper' and 'lower' case letters. Notice that whenever we use 'upper' case letters, it is only for easy reading purposes.

The language knows the following *tokens* (representing groups of characters that have special meaning for the language):
– keywords
– identifiers
– constants
– certainty factors
– operators.

Keywords
Keywords are part of the grammar. There are four of them defined:

> **IF**   ⟨precedes the condition of a rule⟩;

THEN   ⟨precedes the conclusion/action of a rule⟩;

BEGIN   ⟨precedes a group of statements⟩;

END   ⟨closes a group of statements⟩.

Thus, the general rule format is the following:

IF condition **THEN** (* a group of statements *)
BEGIN
      statement 1;
      statement 2;
      statement 3;
END;

Identifiers

Identifiers are strings of letters, numerals, underscores, beginning with a letter or an underscore. The maximum length of an identifier is 255 symbols.

Identifiers represent the names of functions, procedures, variables and operators.

Constants

The language recognizes four types of constant:

      numerical constants;

      fuzzy constants;

      logical constants;

      string constants.

Numerical constants are real numbers in scientific notation. *Fuzzy* constants are (sub)sets of the fuzzy domain of fuzzy values, like {*very low, low, medium, high, very high*}. The fuzzy domain (and thus any fuzzy constant) is an ordered set of fuzzy values. As such, the ordering is significant and cannot be changed. Examples of valid fuzzy constants are the following:

      {*low*};

      {*very low, very high*};

      {*medium, high*}.

Logical constants are **TRUE** and **FALSE**. *String* constants are strings such as 'this is a string'. The following example shows the use of a string constant.

query_yesno(string constant);

should read as

query_yesno('this is a string?');

which is a logical constant yielding **TRUE** or **FALSE**.

Certainty factors

Certainty factors are real decimal numbers in the interval [0,1]. They precede the keyword **IF**.

Operators

Operators identify the kind of manipulation of operands involved. We distinguish:

operands:	operators:
numerical	$-, +, *, /$
fuzzy	$-, +, *$
relational	$<, >, <=, >=, <>$, **in**
logical	**and, or** and **not**

The effect of fuzzy and relational operators will be discussed in the remaining part of this section.

As a result of the above, the general format of of any rule in the rule base is the following:

[cf]**IF** premise (conditional expression)

THEN statement;

(or **BEGIN** group of statements); **END**;

The premise results in a logical value **TRUE** or **FALSE**. A statement may include
– an *assignment*:   identifier := expression
– a *function*:   identifier(expression)
– a *procedure*:   identifier(expression) or just an identifier.

The condition part (premise) of any rule requires variables to be evaluated, whereas the statement part (action, conclusion) may include an assignment, a function call or a procedure call.

Variables (parameters) have to be declared in advance. They appear in the so-called *parameter file*.

In order to solve a problem, the program must have data and information with which it can reason. Information obtained from algorithmic processing is stored in the status register and is transmitted by so-called *status parameters* (numerical or symbolic algorithmic output). All status parameters have to appear in the parameter file, obeying the following syntax.

> **a** ⟨name (reference, used in code)⟩
> > **a** ⟨full name (used in displays)⟩
> > **??** ⟨textual description (used in Help)⟩
> **boolean** ⟨basic type (used for correct display format and value control)⟩
> ⟨empty line⟩

The following are all examples of valid parameters:

> **db**
> db_index
> Davies–Bouldin index
> real

> **V**
> validity
> final judgement for cluster validity
> fuzzy

> **w_db**
> weight
> instantiation of the weight of the db_index in the reasoning
> integer

Note that status parameters are uniquely related to their use in algorithmic processing: they carry information to be used in the reasoning process. If a rule uses a parameter which has not yet been evaluated, a message is generated to let the user know that the algorithm to which that parameter has been assigned has not yet been executed. Pieces of information stored in the status parameters are called facts and are collected in the *fact base*. These facts can be updated after a rule has been fired. In other words, facts representing information can be added and removed from the fact base.

To provide the basis for the examples troughout this chapter, the following set of status parameters is used (as defined in EDAPLUS).

– alpha Wilks's lambda statistic
(user data)
– m_d Number of different clusters detected in M trials
(user data)
– perc_d Cluster dominance (%) over M trials
(user data)

– *beta*	Wilks's lambda statistic
	(random data)
– *m_r*	Number of different clusters detected in *M* trials
	(random data)
– *perc_r*	Cluster dominance (%) over *M* trials
	(random data)
– *b0*	Branching factor
– *CPCC*	Cophenetic correlation coefficient
– *e_i*	Inconsistent edges MST
– *db*	Davies–Bouldin index
– *k*	Prespecified number of clusters
– *dim*	Dimensionality
	(user data)
– *num*	Pooled sample size
	(user data)

Notice that the fact base may contain values of other variables than status parameters. These *free variables* can be used throughout the knowledge program, though they are not related to any particular clustering algorithm. When used, the condition part of the rule requires the variables (like the status parameters) to be evaluated. Likewise, the statement part of the rule may include a value assignment to any of the these variables (like the status parameters).

Generally, rules are selected for firing – after matching the condition part – in the order of their appearance in the rule base, assuming that all certainty factors asigned to the rules are equal. If not, then the rule with the highest certainty factor is fired first, followed by the rule with the second highest certainty factor, and so forth. When rules have the same certainty factor, the order of firing is again determined by the order of appearance in the rule base.

Note that one should read 'the computed certainty factor' which is the computed combination of the rule certainty factor and the certainty factor of the condition part of the rule. The computation is determined by the operators of fuzzy logic, as discussed in Chapter 8.

The group of fuzzy operators on *ordered sets* of fuzzy values plays a dominant role in the language developed. Any *fuzzy value* is a subset of the fuzzy domain, here:

> {*very low, low, medium, high, very high*}

For example,

> {*low, medium*}
> {*medium, very high*}
> {*high*}
> { }

are valid fuzzy values, though { } is meaningless.

The following operators are instrumental in manipulating fuzzy values.

Ordered set operators

The ordered set operators include difference, union, and intersection.

$$C := A - B \Leftrightarrow \forall x \in A \wedge x \neg\in B[x \in C] \quad \text{(difference)}$$

$$C := A + B \Leftrightarrow \forall x \in A \vee x \in B[x \in C] \quad \text{(union)}$$

$$C := A \times B \Leftrightarrow \forall x \in A \wedge x \in B[x \in C] \quad \text{(intersection)}$$

Comparison of fuzzy values

This group of operators yields always a logical value **TRUE** or **FALSE**.

$$A < B \Leftrightarrow \forall x \in A, y \in B[x < y]$$

$$A \leq B \Leftrightarrow \forall x \in A, y \in B[x \leq y]$$

$$A = B \Leftrightarrow \forall x \in A, B[x \in A \wedge x \in B]$$

$$A \geq B \Leftrightarrow \forall x \in A, y \in B[x \geq y]$$

$$A > B \Leftrightarrow \forall x \in A, y \in B[x > y]$$

$$A <> B \Leftrightarrow \exists x[(x \in A \wedge x \neg\in B) \vee (x \in B \wedge x \neg\in A)]$$

$$A \text{ in } B \Leftrightarrow \forall x \in A[x \in B]$$

Functions of fuzzy values

We consider the following functions.

weaken(⟨name of fuzzy value⟩)
reinforce(⟨name of fuzzy value⟩)
min(⟨name of fuzzy value⟩)
max(⟨name of fuzzy value⟩)
average(⟨name of fuzzy value⟩)
complement(⟨name of fuzzy value⟩)

The functions *weaken* and *reinforce* result in a decrease or an increase of the value of the fuzzy identifier, respectively. For example,

weaken({*low, medium*})
reinforce({*low, medium*})

result in

> {*very low, low*}
> {*medium, high*},

respectively.

A rule including a *weaken* function should appear as follows:

> [0.7] **IF** (alpha > beta)
> **AND** (CT in {low, medium})
> **THEN** weaken(CT);

When this rule is fired, the net effect is CT := {*very low, low*}.

The *min*, *max* and *average* functions return a single value. Thus, for example,

> **min**({*medium, high, very high*})
> **max**({*medium, high, very high*})
> **average**({*medium, high, very high*})

return

> {*medium*}
> {*very high*}
> {*high*},

respectively. If the average value can not be represented by a single value, then the nearest higher value is returned.

Finally, *complement(A)* returns all fuzzy values of the fuzzy domain which are not included in *A*.

At this point, we observe that – as compared with Pascal – the *numerical functions round()*, *trunc()*, *frac()*, *sqr()* and *sqrt()* have the classical meaning and effect.

The final group of functions include query rules for user interaction, and some system functions. They appear as:

> **query_yesno**(⟨string⟩[,⟨message⟩])
> **query_int**(⟨string⟩)
> **query_real**(⟨string⟩)
> **query_fuzzy**(⟨string⟩)

corresponding to the data types *boolean, integer, real, and fuzzy*, respectively.

The system functions include

> **undefined**(⟨identifier⟩)
> **abort**
> **message**(⟨string⟩)

The undefined function returns **TRUE** or **FALSE** if variables involved have got a value or not.

The following rule sequence demonstrates a sample session of the use of queries and system functions.

IF a THEN abort

IF undefined(X) **THEN**
BEGIN
      message('the variable X is unknown');
      b := **query_yesno**('input X yourself?');
END;

IF b THEN
BEGIN
      X := **query_int**('the value for X is:');
END;

IF not b THEN
BEGIN
      message('execute algorithm_X');
      a := **query_yesno**('quit the session?');
END;

The above sample clearly shows that our language as developed is very easy to read.

9.3 Knowledge-based cluster analysis: a simple example

The rule language and the types of rule shown in Section 9.2 illustrate the easiness of constructing simple rule programs. This section describes the development of a simple rule base as proposed by Balasubramaniam, dealing with the *selection of clustering algorithms*. To keep complexity manageable, we limit ourselves to a reduced set of algorithms available, without loss of demonstration capability of the resulting rule base.

As stated in Section 9.1, the above proposed system was mainly motivated by the fact that understanding of the differences between various clustering algorithms with respect to their strengths and weaknesses, will lead to the system's ability automatically to choose the right and appropriate algorithm. Note that we follow the basic ideas of Balasubramaniam, not his exact implementation.

9.3.1 Summary of underlying ideas

(i) Selecting the clustering technique

In Part 1, we broadly classified clustering methods under three major groups, namely, the partitional, the hierarchical and the graph-theoretic methods. As the hierarchical methods are somewhat closely related to graph-theoretical methods (breaking the first few long edges of the MST is a divisive version of the single linkage algorithm), our main task is to choose between the partitional and hierarchical methods. We will consider here two radically different clustering algorithms, namely, Forgy's K-means algorithm (representing the partitional methods) and the single linkage algorithm (representing the hierarchical methods). It is hypothesized here that if we obtain two diametrically opposite clustering structures, one of them is near to the *natural grouping*.

As the K-means algorithm requires the value of K to be prespecified, it has been proposed to estimate this value roughly by viewing a 2D projection of the data, obtained by a structure-preserving transformation. Different transformations for 2D projection are described in Chapter 2, including linear and nonlinear transformations. Following Niemann (1979), it is hypothesized here that a linear projection method could be adequate for our purpose. Notice that in EDAPLUS all 2D projections are available in one view, on the basis of which the presence of particular structures can be concluded.

When the projected data show structures like *concentric rings*, we may immediately decide to go for an MST algorithm. But, if *long chain-like* structures are clearly visible in the projected data, we may go for a hierarchical method provided the *sample size* does not exceed 200. For large-sample problems, graph-theoretic methods (like MST) are known to be preferable to the hierarchical methods. Any rough idea we may obtain about the *separability* (and shape) of appearing data clusters could be used to guide our decision as to the choice of partitional or hierchical techniques.

If the appearing clusters are somewhat *spherical and well separated*, we may go for the partitional technique.

In all other cases, it may be good practice to compute the *Davies–Bouldin index* for clustering configurations in the vicinity of K^* (as the estimate of the *number of clusters*) for both, the K-means algorithm and the single linkage algorithm. For a natural cluster configuration, this index is known to give a minimum value. Hence, we decide on the clustering technique which yields the minimum value for the DB index.

(ii) Selecting the algorithm

Once our choice of the general clustering technique is made, the remaining thing to decide on is the choice of the right and appropriate clustering algorithm.

In cases where clusters contain *equal numbers of samples* from multivariate normal distributions, we may use the complete linkage algorithm instead of a partitional algorithm if the sample size is low, and if the separability of the clusters is low. However, if clusters exhibit wide disparity in sample size, the average linkage algorithm may be prefered among all hierarchical algorithms.

As a general guide, we never use hierarchical algorithms if the sample size exceeds 200 pattern samples.

If a partitional technique is chosen, we automatically imply that Forgy's K-means algorithm is the appropriate choice. It should be noticed that a variety of alternatives could be considered, like McQueen's *K*-means algorithm, the minimum squared error criterion algorithm, Wishart's convergent *K*-means algorithm, and so forth, as has been considered by Balasubramaniam. The same holds for the family of hierarchical clustering algorithms.

Other appropriate choices could be Ward's algorithm, the centroid algorithm, the median algorithm, McQuitty's algorithm, and so forth. These all have their own strengths and weaknesses which can be taken into account in deciding on which algorithm is the most appropriate one for a given data clustering problem.

The foregoing considerations lead to a sequence of decisions which are brought together in Figure 9.1.

We are now able to construct the knowledge base, keeping Figure 9.1 in mind.

9.3.2 The knowledge base

We present here the rules which are used in the proposed system. They are related to the concepts presented above. Since they are self-explanatory no further explanation is provided.

```
IF a THEN (* global abort flag *)
BEGIN
      a := not a; (* reset a *)
      abort;
END;
```

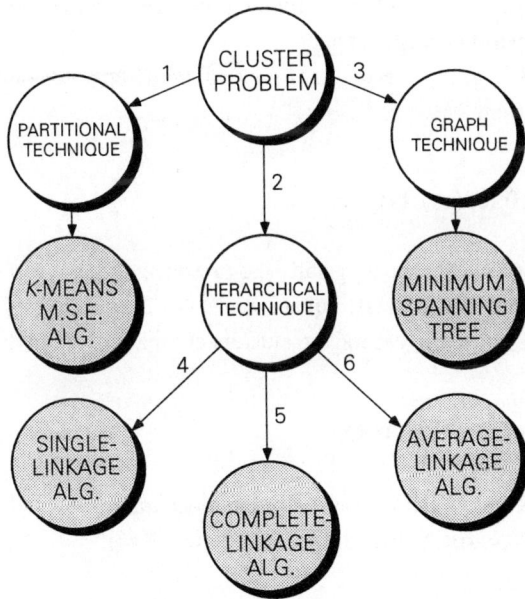

Figure 9.1: Decision tree for the selection of the clustering algorithms.

```
IF undefined(VIEW) THEN
BEGIN
    message('execute a 2D-projection algorithm');
    VIEW := 1;
END;

IF undefined(v1) THEN
BEGIN
    message('inspect visually the presence of some sort of concentric
    cluster shapes');
    v1 := query_yesno('are concentric cluster shapes present?');
END;

IF undefined(v2) THEN
BEGIN
    message('inspect visually the presence of some sort of long chain-like
    cluster shapes');
    v2 := query_yesno('are long chain-like cluster shapes present?');
END;

IF undefined(v3) THEN
BEGIN
    message('inspect visually the presence of some sort of spherical well-
    separated cluster shapes');
    v3 := query_yesno('are spherical well-separated cluster shapes
    present?');
END;

IF undefined(v4) THEN
BEGIN
    message('inspect visually the presence of some sort of compact
    clusters with nearly equal sizes');
    v4 := query_yesno('are clusters compact with nearly equal sizes?');
END;

IF undefined(K*) THEN
BEGIN
    message('estimate visually the most appropriate value for k');
    K* := query_int('the best estimate K* =');
END;
```

```
IF undefined(b) THEN
BEGIN
      message('execute single linkage algorithm');
      b := TRUE;
END;

IF undefined(c) THEN
BEGIN
      message('execute Forgy's k-means algorithm');
      c := TRUE;
END;

IF b AND c THEN
BEGIN
      db_SL := db_SL(K*);
      db_PART := db_PART(K*);
      db := min(db_SL,db_PART);
END;

(* we now start to choose the appropriate clustering algorithm *)

IF v1 THEN
BEGIN
      message('choose the MST algorithm');
      a := query_yesno('execute the MST algorithm?');
END; (* the program will abort here, and the MST will be executed *)

IF (num <= 200) THEN
BEGIN
      message('choose in principle a hierarchical clustering technique');
      v5 := TRUE;
END;

IF v2 AND not v5 THEN
BEGIN
      message('choose the MST algorithm');
      a := query_yesno('execute the MST algorithm?');
END; (* the program will abort here, and the MST will be executed *)

IF v3 AND not v5 THEN
BEGIN
      message('choose the partitional clustering technique');
```

a := **query_yesno**('execute Forgy's K-means algorithm?');
END; (* the program will abort here, and Forgy's K-means algorithm will be executed for K* *)

IF v2 **AND** v5 **THEN**
BEGIN

  message('choose the single-linkage algorithm');

  a := **query_yesno**('execute the single-linkage algorithm?');
END; (* the program will abort here, and the SL algorithm will be executed *)

IF v4 **AND** v5 **AND** (num > 16) **THEN**
BEGIN

  message('choose the complete-linkage algorithm');

  a := **query_yesno**('execute the complete-linkage algorithm?');
END; (* the program will abort here, and the CL algorithm will be executed *)

IF v5 **AND** not v4 **THEN**
BEGIN

  message('choose the average-linkage algorithm');

  a := **query_yesno**('execute the average-linkage algorithm?');
END; (* the program will abort here, and the AL algorithm will be executed *)

IF (not(v1 **OR** v2 **OR** v3 **OR** v5)) **AND** (db=db_SL) **THEN**
BEGIN

  message('choose the single-linkage algorithm');

  a := **query_yesno**('execute the single-linkage algorithm or MST?');
END; (* the program will abort here, and the SL algorithm or MST will be executed *)

IF (not(v1 **OR** v2 **OR** v3 **OR** v5)) **AND** (db=db_PART) **THEN**
BEGIN

  message('choose the partitional clustering technique');

  a := **query_yesno**('execute Forgy's K-means algorithm?');
END; (* the program will abort here, and Forgy's K-means algorithm will be executed for K* *)

IF VIEW=1 **AND** not a **THEN**
BEGIN

  message('the proposed algorithm has deliberately not been executed;

       the program ends here');
       a := **TRUE**;
  END;

The *performance* of the above proposed system for the selection of a suitable algorithm is not studied here in detail. The results affirm what we have encountered in the experiments of Chapter 6.

Moreover, the use of a single algorithm imposes a uniform structure on the entire data, rather than infering (sub) structures from the data. The data analyst, being aware of this phenomenon, generally will apply a number of algorithms of varied nature, aiming to obtain ideas about the true (sub)structure(s) of the data.

For example, we may have a situation where elongated and spherical clusters are interspersed in the pattern space. Then, partitional algorithms may successfully detect the spherical clusters, but at the same time will also group samples into spherical clusters, which should have been grouped into a single elongated cluster. On the other hand, a single-linkage algorithm will identify an elongated structure successfully but may fail in correctly identify spherical clusters.

Thus, it may be difficult to find an algorithm which is consistently better than any other algorithm throughout the whole pattern space.

In what follows, we will concentrate on so-called *hybrid approaches* in which we seek the utilization of several algorithms by proposing different heuristics for mutual support and evidential combination.

9.4 Cluster-oriented reasoning

In Chapter 7, we identified cluster-oriented reasoning as the resultant of the processing paradigm and the modelling paradigm. These paradigms were characterized by multi-level architectures: (algorithms, experiments, inferences) and (data, information, knowledge), respectively. The levels were – within the context of cluster-oriented reasoning – referred to as the *initiation level, support level* and *diagnostic level*. Modelling, processing and reasoning are interrelated as shown in Table 9.1. The main objective of each of the levels is as follows.

Table 9.1: Reasoning as the resultant of modelling and processing paradigm.

	modelling	processing	*reasoning*
I	data	algorithms	*initiation*
II	information	experiments	*support*
III	knowledge	inferences	*diagnosis*

Initiation level. Contains the algorithmical core of clustering methodology and related techniques. When data to be considered, is supplied to a variety of algorithms, the results can be regarded as *facts* about the data. (*Those facts include the Davies–Bouldin statistic, the Wilks lambda statistic for some values of k, the cluster dominance for some values of k, the CPCC value for different linkage procedures, the edge-length distribution of the MST, the number of patterns, the dimensionality, and estimated intrinsic dimensionality.*)

Support level. Contains an interface which makes it feasible for (naive) users to explore a variety of different experiments supporting the evaluation and validation of the facts obtained from the user data. (*Supporting information includes data-dependent baseline distributions obtained from random experimentation for major statistics, cluster dominance, CPCC values, edge-length distributions.*)

Diagnostic level. Contains an inference engine to reason on the basis of the facts obtained at level I and supporting information provided by level II, and cluster analyst's expertise. (*The diagnostic level includes clustering expertise to interpret the facts, to weight supporting evidence by random experimentation, to combine different results, and to evaluate the belief in the ultimate findings.*)

Each of the levels are controlled by strategic and/or procedural and/or interpretational rules. Level I includes the initiation rules on the basis of which supporting experimentation has to be carried out. Level II includes support rules on the basis of which random experimentation has to be used to validate facts and to draw intermediate conclusions. Finally, at level III evidence combining rules are used to yield a final diagnostic conclusion about the data under consideration, as well as the final belief in the analysis as a whole.

Figure 9.2 shows the general rule hierarchy in which facts and experimental results are combined to yield a final diagnosis. As such, we obtain a decision network implemented as a rule-based system as already discussed.

Now, it is important to note that the transition of facts to evidential (supporting) information, as well as the transition of facts to pertinent knowledge, can be represented by *conceptual associations* between *structural variables* (as discussed in Chapter 7) aiming at constructing rules to be used in the reasoning at the appropriate levels. Commonly, these rules (as generalized descriptions) are to be constructed from large sets of *observations*. We call this *learning*.

9.4.1 Constructing rules from observations

First, let us recall the notion of implicit and explicit *conceptual association*.

In a relation, the concept domain C is said to be conceptually associated with (dependent on) the set of domains $\{A\}$ if to each tuple of values in the set of domains $\{A\}$ there corresponds precisely one value in domain C, at any given time.

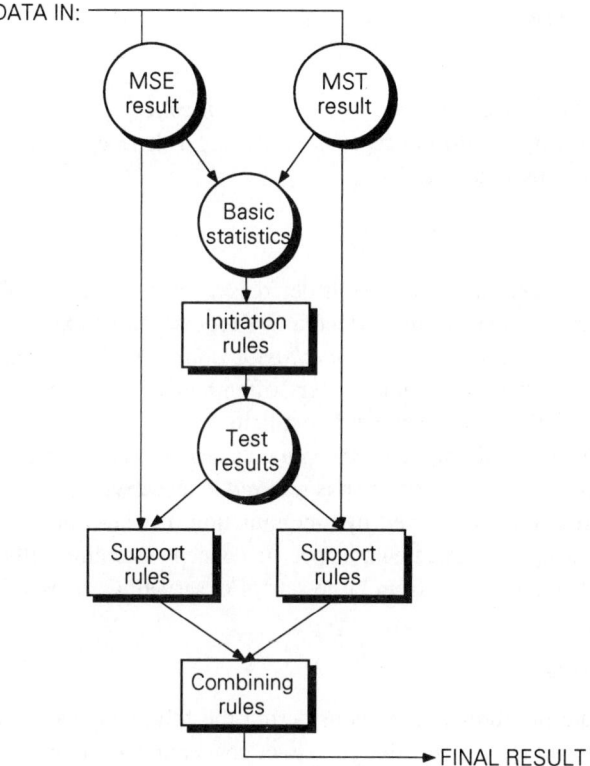

Figure 9.2: General rule hierarchy in EDAPLUS.

For example, consider k different concepts C_1, C_2, ..., C_k (the domain of C) and assume that the set W of observations is related to these concepts: each observation is pertinent to one and only one concept. Each observation $\mathbf{w}$ of W can be described by a set of p attributes (facts, structural variables) $A = \{a_1, a_2, ..., a_p\}$. Then, a conceptual association (following Chapter 7) can be stated as

$$R(a_i(\mathbf{w}),C_j) \leftarrow (a_i < v), \mathbf{w} \in C_j$$

or

IF $a_i(\mathbf{w}) < v$ THEN $\mathbf{w}$ is a member of C_j.

Note that R is a *generalization* of the observations

$$\mathbf{w}_1: \ a_i(\mathbf{w}_1) = v_1, \text{ AND } \mathbf{w}_1 \in C_j$$

$$\mathbf{w}_2: \ a_i(\mathbf{w}_2) = v_2, \text{ AND } \mathbf{w}_2 \in C_j$$

$$\vdots$$

$$\mathbf{w}_N: a_i(\mathbf{w}_N) = v_N, \text{ AND } \mathbf{w}_N \in C_j,$$

where $v_n < v$, $(n = 1, ..., N)$.

If R always holds, then R is said to be an *explicit conceptual association* between the structural variable a_i and the concept C_j. However, if there exists a $\mathbf{w}_k$ ($k \in \{1,N\}$) for which it has been observed that

$$R(a_i(\mathbf{w}_k) = v_k, \text{ AND } \mathbf{w}_k \in C_h$$

with $v_k > v$, this observation is to be considered as a counter-example. Thus, all $\mathbf{w}$, except $\mathbf{w}_k$, are instances of a general rule, whereas $\mathbf{w}_k$ represents the exception to this general rule and is said to be an *implicit conceptual association*. The distinction between implicit and explicit conceptual associations is very often hard to draw. In practice, conceptual associations are partly implicit and partly explicit.

As already mentioned, implicit conceptual association is referred to as *information*, whereas explicit conceptual association is referred to as *knowledge*.

If we consider R to be defined by a conjunction of some subset of attribute–value couples conceptually associated with a specific concept, learning methods are needed to contruct rules R automatically from large sets of observations. This will be discussed in Section 9.6.

9.4.2 An example

We will illustrate the foregoing by considering the following set of observations. The question arises whether it is possible to extract conceptual associations between structural variables and concepts. Note that the concepts represent the expert's judgement about the clustering tendency of the observed data.

Suppose that we have 19 different data sets $\mathbf{w}_1, \mathbf{w}_2, ..., \mathbf{w}_{19}$ to be analysed through clustering. For each data set we have established five different facts by using a subset of the available algorithms. These facts constitute the structural variables $a_1, a_2, ..., a_5$.

We have

a_1: alpha (Wilks's lambda statistic/user data)

a_2: beta (Wilks's lambda statistic/random data)

a_3: %dom (cluster dominancy/user data)

a_4: CPCC (cophenetic correlation coefficient)

a_5: db (Davies–Boulden index).

The *concepts* involved are the expert's judgements about clustering tendency, $C_1, C_2, ..., C_5$: *very high* (VH), *high* (H), *medium* (M), *low* (L), and *very low* (VL), respectively.

Table 9.2 shows the set of observations from which we wish to derive conceptual associations, on the basis of which we wish to construct a knowledge base to infer a conclusion automatically about the clustering tendency of any data set supplied to the system. Note that each data set is supposed to have two natural clusters: $k = 2$. Thus, a_1, a_2, a_3 are computed for the case of $k = 2$.

Table 9.2: Set of observations used for constructing conceptual associations.

w	a_1	a_2	a_3	a_4	a_5	C
1	0.591	0.620	32	0.53	0.529	VL
2	0.608	0.598	18	0.51	0.476	VL
3	0.632	0.578	10	0.50	0.476	VL
4	0.631	0.632	16	0.46	0.392	VL
5	0.516	0.541	30	0.43	0.591	L
6	0.603	0.602	36	0.52	0.344	L
7	0.495	0.606	72	0.80	0.391	M
8	0.313	0.613	98	0.83	0.200	M
9	0.179	0.641	100	0.90	0.215	M
10	0.339	0.619	100	0.71		H
11	0.383	0.609	94	0.80	0.549	H
12	0.359	0.593	100	0.82	0.227	H
13	0.349	0.624	96	0.86	0.455	H
14	0.340	0.521	96	0.86	0.530	H
15	0.416	0.627	84	VH	0.527	VH
16	0.366	0.599	94	VH	0.227	VH
17	0.350	0.519	96	VH	0.455	VH
18	0.382	0.609	88	0.86	0.467	VH
19	0.037	0.625	100	1.00	0.529	VH

After inspecting the contents of the set of observations, we notice that a_1, a_3, and a_4 exhibit roughly the same kind of association with respect to the domain of concepts. a_1 tends to be large (0.5 to 0.6) for very low to low clustering tendency, and small (0.3 to 0.4) for high to very high clustering tendency, thus giving rise to conceptual associations like:

$$R_1(a_1(\mathbf{w}),C) \leftarrow (a_1 < 0.42), \mathbf{w} \text{ in } \{H,VH\}$$

$$R_2(a_1(\mathbf{w}),C) \leftarrow (0.42 < a_1 < 0.59), \mathbf{w} \text{ in } \{M\}$$

$$R_3(a_1(\mathbf{w}),C) \leftarrow (0.59 < a_1), \mathbf{w} \text{ in } \{VL,L\}$$

$$R_4(0.516,C) \leftarrow \mathbf{w} \text{ in } \{L\}$$

$$R_5(0.313,C) \leftarrow \mathbf{w} \text{ in } \{M\}$$

$$R_6(0.179,C) \leftarrow \mathbf{w} \text{ in } \{M\}$$

Clearly, the first three conceptual associations R_1, R_2 and R_3, are general rules (knowledge rather than information) while the remaining associations represent the exceptions to these general rules (information rather than knowledge).

Likewise, we may come up with

$$R_7(a_3(\mathbf{w}),C) \leftarrow (a_3 < 70), \mathbf{w} \text{ in } \{\text{VL,L}\}$$

$$R_8(a_3(\mathbf{w}),C) \leftarrow (a_3 > 70), \mathbf{w} \text{ in } \{\text{M,H,VH}\}$$

$$R_9(a_4(\mathbf{w}),C) \leftarrow (a_4 < 0.80), \mathbf{w} \text{ in } \{\text{VL,L}\}$$

$$R_{10}(a_4(\mathbf{w}),C) \leftarrow (a_4 > 0.80), \mathbf{w} \text{ in } \{\text{M,H,VH}\}$$

$$R_{11}(0.71,C) \leftarrow \mathbf{w} \text{ in } \{\text{H}\}$$

The above associations could be constructed at first glance from the observation table, neither guaranteeing optimality, nor completeness. We will not try to optimize these associations here. We will return to this problem in Section 9.6.

It may appear, on the basis of exploration of the above conceptual associations, that the design of a knowledge base is straightforward. We will now show that this is far from trivial.

As we have discussed at the beginning of this section, the fundamental architecture of cluster-oriented reasoning includes fact-driven initialization (which we have established in the foregoing), combined and experimental support, and finally, drawing diagnostic conclusions by taking all facts and support into account. In view of this, two questions may arise: how we *evaluate* each of the conceptual associations, and how we *combine* the evaluated results of each of them.

Recall the conceptual association

$$R_1(a_1(\mathbf{w}),C) \leftarrow (a_1 < 0.42), \mathbf{w} \text{ in } \{\text{H,VH}\}.$$

If we scale the concepts VH, H, M, L, VL by assigning a numerical value 5, 4, 3, 2 and 1, respectively, R_1 may be rewritten

$$R_1(a_1(\mathbf{w}),C) \leftarrow (a_1 < 0.42), \text{CT}(a_1(\mathbf{w})) = (4 + 5)/2 = 4.5,$$

where CT denotes clustering tendency, being the concept domain of interest.

Likewise, for example

$$R_8(a_3(\mathbf{w}),C) \leftarrow (a_3 > 70), \mathbf{w} \text{ in } \{\text{M,H,VH}\}$$

could be rewritten as

$$R_8(a_3(\mathbf{w}),C) \leftarrow (a_3 > 70), \text{CT}(a_3(\mathbf{w})) = (3 + 4 + 5)/3 = 4,$$

and

$$R_8(a_3(\mathbf{w}),C) \leftarrow (a_3 < 70), \text{CT}(a_3(\mathbf{w})) = (1 + 2)/2 = 1.5.$$

As a result, and since explicit conceptual associations are mutually exclusive, we obtain $CT(a_1(\mathbf{w}))$, $CT(a_3(\mathbf{w}))$ and $CT(a_4(\mathbf{w}))$ for each observed data set $\mathbf{w}$. From that table, we can derive the conceptual associations

$$R_{12}(CT(a_1(\mathbf{w})),CT(a_3(\mathbf{w})),CT(a_4(\mathbf{w})),C) \leftarrow$$
$$(CT(a_1,a_3,a_4(\mathbf{w})) < 1.5), \mathbf{w} \text{ in } \{VL\},$$

$$R_{13}(CT(a_1(\mathbf{w})),CT(a_3(\mathbf{w})),CT(a_4(\mathbf{w})),C) \leftarrow$$
$$(1.5 < CT(a_1,a_3,a_4(\mathbf{w})) < 2.5), \mathbf{w} \text{ in } \{L\}$$

$$R_{14}(CT(a_1(\mathbf{w})),CT(a_3(\mathbf{w})),CT(a_4(\mathbf{w})),C) \leftarrow$$
$$(2.5 < CT(a_1,a_3,a_4(\mathbf{w})) < 3.5), \mathbf{w} \text{ in } \{M\}$$

$$R_{15}(CT(a_1(\mathbf{w})),CT(a_3(\mathbf{w})),CT(a_4(\mathbf{w})),C) \leftarrow$$
$$(3.5 < CT(a_1,a_3,a_4(\mathbf{w})) < 4.5), \mathbf{w} \text{ in } \{H\}$$

$$R_{16}(CT(a_1(\mathbf{w})),CT(a_3(\mathbf{w})),CT(a_4(\mathbf{w})),C) \leftarrow$$
$$(4.5 < CT(a_1,a_3,a_4(\mathbf{w})) < 5), \mathbf{w} \text{ in } \{VH\},$$

where

$$CT(a_1,a_3,a_4(\mathbf{w})) =$$
$$p_1 \times CT(a_1(\mathbf{w})) + p_3 \times CT(a_3(\mathbf{w})) + p_4 \times CT(a_4(\mathbf{w})).$$

The coefficients p_i represent the weights of the various subconclusions derived from the facts. They have to be supplied by the expert or have to be determined by some learning method.

In conclusion, we see that considerable effort is needed to extract all kinds of conceptual associations which are necessary for designing a rule base.

However, once we have obtained an appropriate set $\{R\}$, the implementation of the rule base using our rule language turns out to be very straightforward. The crucial part of it will be as follows:

```
.(* sample rule base *)

.
fired1 := FALSE;
fired2 := FALSE;
fired3 := FALSE;
fired4 := FALSE;
fired5 := FALSE;
fired6 := FALSE;

IF (not fired1) AND (k <> 2) THEN
BEGIN
```

```
        fired1 := TRUE;
        fired2 := TRUE;
END; (* diagnosis for K = 2 only *)
```

(* the following seven rules implement the associations R1, R2, R3, R7, R8, R9, R10 *)

```
IF (not fired2) AND (0.59 < alpha) THEN
BEGIN
        fired2 := TRUE;
        CT_alpha := 1.5;
END;
```

```
IF (not fired2) AND (0.42 < alpha < 0.59) THEN
BEGIN
        fired2 := TRUE;
        CT_alpha := 3;
END;
```

```
IF (not fired2) AND (alpha < 0.42) THEN
BEGIN
        fired2 := TRUE;
        CT_alpha := 4.5;
END;
```

```
IF (not fired3) AND (%dom > 70) THEN
BEGIN
        fired3 := TRUE;
        CT_%dom := 1.5;
END;
```

```
IF (not fired3) AND (%dom < 70) THEN
BEGIN
        fired3 := TRUE;
        CT_%dom := 4;
END;
```

```
IF (not fired4) AND (CPCC < 0.80) THEN
BEGIN
        fired4 := TRUE;
        CT_CPCC := 1.5;
END;
```

```
IF (not fired4) AND (CPCC > 0.80) THEN
BEGIN
      fired4 := TRUE;
      CT_CPCC := 4;
END;

(* the following six rules implement R12, R13, R14, R15, and R16 *)
IF (not fired5) THEN
BEGIN
      fired5 := TRUE;
      CT_end :=  p_alpha * CT_alpha +
                 p_%dom * CT_%dom +
                 p_CPCC * CT_CPCC;
      phase:= 1;
END;

IF (not fired6) AND (CT_end < 1.5) THEN
BEGIN
      fired6 := TRUE;
      V := {very low};
END;

IF (not fired6) AND (1.5 < CT_end < 2.5) THEN
BEGIN
      fired6 := TRUE;
      V := {low};
END;

IF (not fired6) AND (2.5 < CT_end < 3.5) THEN
BEGIN
      fired6 := TRUE;
      V := {medium};
END;

IF (not fired6) AND (3.5 < CT_end < 4.5) THEN
BEGIN
      fired6 := TRUE;
      V := {high};
END;

IF (not fired6) AND (4.5 < CT_end < 5) THEN
BEGIN
```

```
        fired6 := TRUE;
        V := {very high};
END;
```

(* the final rule is conclusive *)
IF phase = 1 **THEN**
BEGIN
    CT := V;
END;
(* the program ends here; CT is the diagnostic conclusion about clustering
tendency *)

The example ends here with the remark that major domains such as the Davies–Bouldin
index, the dimensionality, and the sample size, though not dealt with in the sample
program, are included in the rule base RULEBASE1.EDA as provided on the
accompanying diskette.

We will now discuss the construction of support rules which could be derived from
additional experimentation with random data. At the same time, it will be clear that rules
in general should be independent of K.

9.4.3 K-independent support rules

As we have seen, strong and weak clustering tendency could be revealed by the domain
variables *alpha* (the Wilks lambda statistic) and %dom (the clustering dominance),
among others. Also, the use of the same statistic for random data, *beta*, using an equal
number of random patterns of the same dimensionality as the user data, within the same
sampling window, has been found to be of significance.

From the previous example (Table 9.2), we may conclude

$$R(a_1(\mathbf{w}),a_2(\mathbf{w})) \leftarrow a_1(\mathbf{w}) < a_2(\mathbf{w})$$

or

$$R(\text{alpha}(\mathbf{w}),\text{beta}(r)) \leftarrow \text{alpha}(\mathbf{w}) < \text{beta}(r)$$

where alpha($\mathbf{w}$) represents Wilks's lambda statistic for the user data ($\mathbf{w}$) and beta(r)
represents the same statistic for random data (r). Thus, this association could be represented by

$$R(\text{alpha}(\mathbf{w}),\text{beta}(r)) \leftarrow \text{alpha}(\mathbf{w}) < \text{beta}(r)$$

$$R(0.608,0.598)$$

$$R(0.603,0.602)$$

The first line represents the general rule, while the two other lines represent the two
exceptions. It is easy to see that the general rule holds if the user data is well clustered.

Keep in mind that the above association is valid for $K = 2$ only. Table 9.3 shows a set of observations for $K = 3$ and $K = 4$.

As is known, alpha and beta depend on the number of clusters K, which also can be observed in Table 9.3. Therefore, the association should read

$$R(\text{alpha}(\mathbf{w},K),\text{beta}(r,K)) \leftarrow \text{alpha}(\mathbf{w},K) < \text{beta}(r,K)$$

From Table 9.3, we learn that data sample $\mathbf{w} = 20, 21, 35$ and 36 represent exceptions.

If we recall the definition of *cluster dominance*, being the maximum number of times that a same cluster is detected over the N trials that the K-means algorithm runs with different starting configurations randomly chosen from the data set, another association can be established:

$$R(\%\text{dom}(\mathbf{w}),\%\text{dom}(r)) \leftarrow \%\text{dom}(\mathbf{w}) > \%\text{dom}(r)$$

Obviously, also this association is k-dependent, thus:

$$R(\%\text{dom}(\mathbf{w},k),\%\text{dom}(r,k)) \leftarrow \%\text{dom}(\mathbf{w},k) > \%\text{dom}(r,k)$$

with the exceptions $\mathbf{w} = 20, 35$ and 36.

On the basis of large scale experimentation (of which Table 9.2 and 9.3 are sample observations), we observe that generally (except in some cases of very low clustering tendency)

$$\text{alpha}(\mathbf{w},k) < 1/k,$$
$$\text{beta } (r,k) > 1/k,$$
$$\%\text{dom}(\mathbf{w},k) > 100/k,$$
$$\%\text{dom}(r,k) < 100/k.$$

From this, we can derive four conceptual associations.

$$R(\text{alpha}(\mathbf{w},k),k) \leftarrow \text{alpha}(\mathbf{w},k) < 1/k$$
$$R(\text{beta}(r,k),k) \leftarrow \text{beta}(r,k) > 1/k$$
$$R(\%\text{dom}(\mathbf{w},k),k) \leftarrow \%\text{dom}(\mathbf{w},k) > 100/k$$
$$R(\%\text{dom}(r,k),k) \leftarrow \%\text{dom}(r,k) < 100/k.$$

Figure 9.3 illustrates the above associations in the (alpha($\mathbf{w}$)/beta(r), %dom($\mathbf{w}$)/%dom(r))-plane. Figure 9.3a shows the case of a *strong* clustering tendency for the user data, whereas Figure 9.3b shows the case in which the user data tend to have a *weak* clustering tendency.

Table 9.3: Observations for k = 3 and k =4.

($k = 3$)

w	alpha(w)	beta(r)	%dom(w)	%dom(r)
20	0.361	0.333	30	52
21	0.362	0.351	40	16
22	0.337	0.381	34	22
23	0.259	0.339	88	14
24	0.289	0.349	58	24
25	0.278	0.364	88	28
26	0.176	0.305	90	60
27	0.229	0.389	74	24
28	0.182	0.336	92	38
29	0.117	0.353	88	26
30	0.109	0.391	92	26
31	0.121	0.369	88	28
32	0.105	0.387	88	28
33	0.101	0.368	86	28
34	0.128	0.408	80	12

($k = 4$)

w	alpha(w)	beta(r)	%dom(w)	%dom(r)
35	0.262	0.221	26	80
36	0.267	0.248	22	28
37	0.219	0.239	46	28
38	0.245	0.301	54	20
39	0.116	0.250	82	26
40	0.139	0.254	84	14
41	0.146	0.226	82	42
42	0.158	0.252	70	22
43	0.132	0.254	80	34
44	0.099	0.255	84	44

If the user data are strongly clustered then the N trials for the user data will generally yield high values for the cluster dominance and small values for alpha. On the other hand, the N trials for random data will systematically yield higher values for beta and low values for the cluster dominance. Evidently, if the user data are weakly clustered, the N pairs of values of alpha and %dom for the user data shift towards the N pairs of values of beta and %dom for the random data.

Figure 9.4 shows two extremes: 'circles' ($\circ$), representing the values for alpha(w)/

beta(*r*) and %dom(**w**)/%dom(*r*), for a number of trials of artificial data with very high cluster tendency; and 'squares' (∎), representing the values for alpha(**w**)/beta(*r*) and %dom(**w**)/%dom(*r*), for a number of trials of real data with presumably very low cluster tendency. If we are aiming to draw conclusions about the cluster tendency of the user data to be analyzed, we seek evidential information in favour of a preconclusion about cluster tendency, followed by some kind of supporting experimentation, which may lead to *reinforcement* or *weakening* of such a preconclusion.

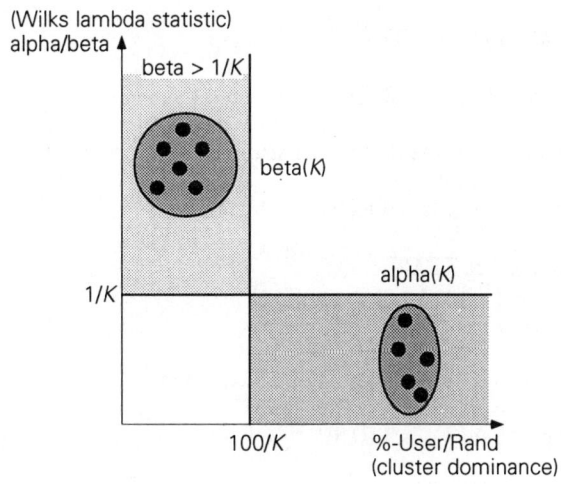

(a) strong clustering

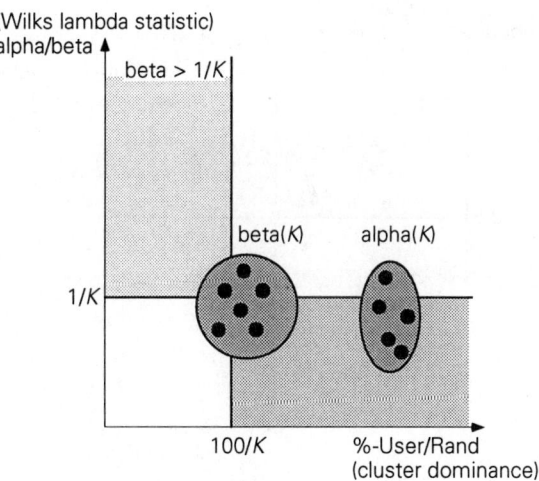

(b) weak clustering

Figure 9.3: Strong and weak cluster tendency.

If we consider the facts from the user data (here alpha(**w**,*k*) and %dom(**w**,*k*)), then the strongest piece of evidential information about the presence of clustering tendency is represented by

R(alpha(**w**,*k*),*k*) is **TRUE**

and R(%dom(**w**,*k*),*k*) is **TRUE.**

Clearly, less strong evidential information might be obtained when

R(alpha(**w**,*k*),*k*) is **TRUE**

or R(%dom(**w**,*k*),*k*) is **TRUE.**

Then, after supporting experimentation, the *strongest* support is obtained when

R(alpha(*k*),beta(*k*)) is **TRUE**

and   [R(beta(*k*),*k*) is **TRUE**

and R(%dom(*r*,*k*),*k*) is **TRUE**].

On the other hand, the *weakest* support is obtained when

R(alpha(*k*),beta(*k*)) is **TRUE**

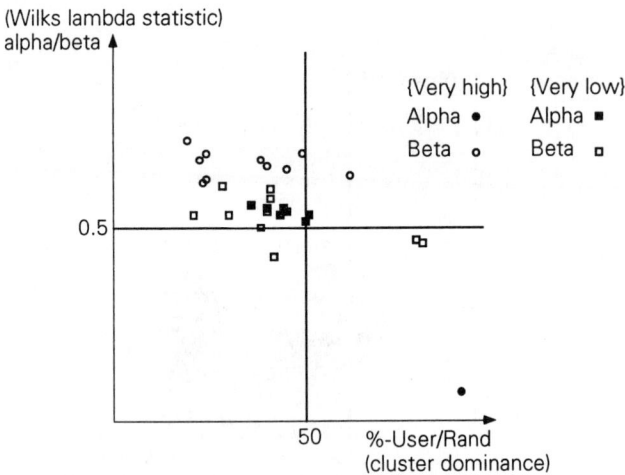

Figure 9.4: Two (α,β)-plots of a number of trials with very high and very low clustering tendency, respectively.

and   [R(beta(k),k) is not **TRUE**

and R(%dom(r,k),k) **TRUE**]

or   [R(beta(k),k) is **TRUE**

and R(%dom(r,k),k) is not **TRUE**].

If R(alpha(k),beta(k)) is not **TRUE**, no support or even negative support can be considered. From the above, it is clear that negative support (*weakening*), no support (*neutral*), and positive support (*reinforcement*) are to be considered as 'operators' on the intiative pre-conclusion about cluster tendency based upon the facts from the data. This was dealt with in Section 9.2.

As an example, if the pre-conclusion is that the clustering tendency CT can be represented by

$$CT := \{medium, high, very\ high\},$$

a weakening leads to

$$CT := \{low, medium, high\},$$

no support will not change CT, and a reinforcement will lead to

$$CT := \{high, very\ high\}.$$

Other structural variables may lead to different pieces of evidential information, and thus to different pre-conclusions. Also here, we may obtain a supported conclusion, say

$$CT(2) := \{medium, high\};$$

then, at the diagnostic level, we should combine these two conclusions. So, assume our first set of structural variables yielded

$$CT(1) := \{high, very\ high\};$$

then, the diagnostic combination may be obtained by some 'termset' operator like *min, max, medean* (as discussed in Section 9.2) used for the intersection or union of CT(1) and CT(2).

For example, if we take min(CT(1),CT(2)), the final conclusion will read

$$CT := \{high\},$$

assuming that both conclusions have equal weight. Very often, the weights will be different due to a priori known pecularities of the structural variables considered, and their sensitivity to the number of patterns used or the dimensionality of the patterns. This is the kind of knowledge which typically belongs to the diagnostic level.

9.4.4 Diagnostic rules

In Section 9.4, the diagnostic level was indicated as the level that includes clustering expertise to interpret facts (1), to weight supporting evidence by random experimentation (2), to combine different results (3), and to evaluate the belief in the ultimate findings (4).

Let us consider two pre-conclusions, derived from facts, CT(1) and CT(2):

$$CT(1) = \{high, very\ high\}$$

and $$CT(2) = \{medium, high\}$$

Clustering expertise may be such that if a specific condition is TRUE then CT(1) is optimistically biased, (1).

Experimentation (support level) shows positive support for CT(2) and no support (neutral) for CT(1), (2).

Generally, CT(1) and CT(2) are combined as

$$CT := CT(1) \times CT(2)\ (intersection),\ (3).$$

Now, the following diagnostic implementation is possible.

        IF specific condition is TRUE
            THEN CT(1) := weaken(CT(1))

Thus $CT(1) := \{medium, high\}$.

Due to negative support we get

$$CT(2) := reinforce(CT(2)),\ and\ consequently,\ we\ obtain$$

$$CT(2) := \{high, very\ high\}$$

Then, a rule for combining CT(1) and CT(2) yields

$$CT := CT(1) \times CT(2) := \{high\}$$

Or in words: 'The two pre-conclusions, derived from the facts, supported by additional testing, and weighted on the basis of some clustering expertise, were combined by taking the intersection of the two ordered sets of fuzzy values, yielding a diagnostic conclusion that the clustering tendency of the data to be analysed is found to be *high*'.

The set of rules which are of the above format are called the diagnostic rule set. Most importantly, clustering expertise, domain knowledge, etc. are really accounted for at the diagnostic level.

As will be clear from Chapter 10, this is also the place where reference classifications, expert intuition and the like can be tested and can influence the ultimate diagnostic conclusion.

9.5 A design example

The following example has been successfully used in advanced cluster analysis courses. The objective is to produce the *detailed design* of a knowledge base for computer-assisted reasoning in cluster analysis. The example is meant to serve as an educational project for students who take part in academic courses at postgraduate level.

We will follow the linear model of *expert system development life cycle*, as described in Bochsler (1988) and summarized by Giarratano and Riley (1989). The life cycle consists of a number of stages (planning, knowledge definition, knowledge design, actual implementation, verification and evaluation) and describes the development of a system to some point at which its functional capabilities will be evaluated.

9.5.1 Planning

The purpose of the planning stage is to produce a *work plan* for the system development. Major factors here include feasibility, management, phasing, scheduling, preliminary functional layout and high-level requirements. Table 9.4 summarizes the planning stage.

9.5.2 Knowledge definition

The object of the knowledge definition stage is to define the knowledge requirements for the system development. This stage consists of two main tasks:
– knowledge source *identification* and selection;
– knowledge *acquisition*, analysis and extraction.

These tasks are summarized in Table 9.5 and Table 9.6.

Table 9.4: Planning stage.

task	objective
feasibility	the value of the development is educational;
management	course supervisor, course hours, EDA-PLUS, PC requirements;
phasing	embedded in the course program;
functional layout (preliminary)	the purpose of the system is to master the basic issues in computer-assisted reasoning in cluster analysis;
high-level	EDAPLUS and the on-board editor should
requirements	allow incremental development of the knowledge base.

Table 9.5: Source identification.

task	objective
identification	cluster analysis methodology, cluster analysis experts, large sets of observations;
importance (priority)	the above issues are equally important;
availability	books, course material, single human expert, data;
selection	the above issues are all required.

Table 9.6: Knowledge acquisition.

task	objective
strategy	attending the course, reading books, interviewing the course supervisor;
identification	clustering algorithms and large-scale experimentation;
functional	the system's capability should be
layout (detailed)	(semi)automatic assessment of the clustering tendency of the data at hand, and the belief in the analysis (validity);
execution flow	problem knowledge should be decomposed into groups of rules: initiation rules (facts), support rules and diagnostic rules; the execution flow should correspond to these logical groups;
preliminary report	a system's description from the user's viewpoint is an essential part of system's development (right from the beginning);
requirements	the system is expected to perform a dialog with the user about the problem, and to pursue the above described reasoning steps;
specification	

9.5.3 Knowledge design

The next stage, knowledge design, has the objective of producing a detailed design of the *knowledge program*. The main factors are summarized in Table 9.7.

Table 9.7: Knowledge problem.

task	objective
knowledge representation	EDAPLUS only supports rule representation;
control structure	the most important structure is the meta-level control structure for executing rules;
fact structure	determined by EDAPLUS;
user interface	determined by EDAPLUS; dialog messages are to be developed by the system designer;
test plan	testing of the code; test data are made available;
implementation	the program has to be written in the rule language supported by EDAPLUS;
user interface	determined by EDAPLUS;
report	document of the design;
test plan	we want to have specified here a series of controlled experiments with test data.

9.5.4 The program code

This stage deals with the actual *implementation* of the code. Major factors are summarized in Table 9.8.

Table 9.8: Implementation.

task	objective
coding	actual coding;
tests	we want to have a prespecified test procedure;
listings	commented coding;
user's manual	a systems description from the user's viewpoint;
installation guide	how to install the system;
documentation	document functionality, limitations, etc.

9.5.5 Verification and evaluation

This stage has the objective of determing the *correctness*, *completeness*, and *consistency* of the system. The main factors are summarized in Table 9.9.

Table 9.9: Correctness, completeness and consistency.

task	objective
test procedures	perform tests;
evaluations	analyze the test results;
recommendations	are changes to the system to be considered;
validation	is the system in agreement with user needs and the requirements;
final report	if the system is complete, then issue a final report; if not, issue an interim report.

Note that the system verification must be performed in conjunction with all the knowledge incorporated in the system.

A software engineering approach like this does not only guarantee a reliable program, but at the same time offers a manageable backbone for presenting the basic issues.

9.5.6 An operational example

Within the course of the educational project, the following simple knowledge rule base for diagnosing clustering tendency is likely to be achieved.

To a reasonably degree, previously discussed ideas could have been taken into consideration.

The rule base
(* this rule base sample is also available on the accompanying diskette *)

```
IF a THEN (* global abort flag *)
BEGIN
    a := not a; (* reset a *)
    abort;
END;

(* here the initial dialog with the user begins *)
IF undefined(count) THEN
BEGIN
    message('Welcome to EDAPLUS decision network');
    count := 1;
END;

IF undefined(b) AND undefined(K) THEN
b := query_yesno('number of clusters (K) is unknown','input k yourself?');

IF b AND undefined(K) THEN
```

k := **query_int**('the number of clusters (K) is:');

IF not b **AND** undefined(K) **AND** count<>0 **THEN**
BEGIN
    message('run <u>HIERARCHICAL/NR OF CLUST</u> first!', use F4 (show
    partition) to define K');
    a := **query_yesno**('leaving the decision network?');
    count := 0;
END;

IF count = 0 **AND undefined**(K) **THEN**
BEGIN
    message('you decided to go on! <u>WARNING</u>: K is still unknown');
    count := 1;
END;

IF undefined(c) **AND undefined**(alpha) **THEN**
c := **query_yesno**('input alpha yourself?','the Wilks lambda statistic (alpha)
is unknown');

IF c **AND undefined**(alpha) **THEN**
alpha := query_real('alpha is:');

IF undefined(alpha) **THEN**
BEGIN
    message('run <u>PARTITIONAL/k-MEANS/USER DATA</u>','use # of
    passes = 100');
    a := **query_yesno**('leaving the decision network?');
END;

(* we now enter the first decision level: initialization *)
IF count = 1 **THEN**
BEGIN
    message('we now start the *INITIAL* decision level');
    count := 2;
END;

IF undefined(v1) **AND** alpha >= 1/K **THEN**
v1 := {*very low, low*};

IF undefined(v2) **AND undefined**(%dom_user) **THEN**
v2 := {*very low*};

IF undefined(v2) **AND** %dom_user <= 100/K **THEN**
v2 := {*very low, low*};

```
IF undefined(v2) AND %dom_user > 100/K THEN
v2 := {medium, high, very high};

IF undefined(beta) THEN
BEGIN
      message('for further analysis we need a random data experiment',
      'run PARTITIONAL/k-MEANS/RANDOM DATA',
      'this decision session will end here!');
      abort;
END;

(* we enter the second decision level: support *)
IF count = 2 AND beta <= 1/K AND %dom_random >= 100/K THEN
BEGIN
      weaken(v1);
      count := 3;
END;

IF count = 2 AND beta >1/K AND %dom_random < 100/K THEN
BEGIN
      reinforce(v1);
      count := 3;
END;

IF count = 2 THEN count := 3;
(* none of the above was valid *)

IF count = 3 AND alpha <= 1/K AND %dom_user <= 100/K THEN
BEGIN
      weaken(v1);
      count := 4;
END;

IF count = 3 AND alpha > 1/k AND %dom_user > 100/K THEN
BEGIN
      reinforce(v1);
      count := 4;
END;

IF count = 3 OR count = 4 THEN
BEGIN
      message('we now start the support decision level');
      count := 5;
END;
```

```
IF count = 5 AND (dim > 3 OR num < 50) THEN
BEGIN
      weaken(v1);
      weaken(v2);
      count := 6;
END;

IF count = 5 THEN count := 6;

IF undefined(CPCC) THEN
BEGIN
      message('we need to run PARTITIONAL/MST analysis','this session
      ends here!');
      abort;
END;

IF count = 6 AND CPCC in {high, very high} THEN
BEGIN
      message('we now come to the final diagnosis');
      reinforce(v1);
      reinforce(v2);
      count := 7;
END;

IF count = 6 THEN
BEGIN
      message('we now come to the final diagnosis');
      count := 7;
END;

IF count = 7 AND edges_i in {high, very high} THEN
BEGIN
      reinforce(v2);
      count := 8;
END;

IF (count = 7 OR count = 8) AND num < 100 THEN
BEGIN
      weaken(v2);
      count := 9;
END;

IF (count = 6 OR count = 7 OR count = 8) AND num >= 100 THEN count
:= 9;
```

```
IF count = 9 THEN
BEGIN
      CT := v1;
      VAL := v2;
END;
```
(* the program ends here by assigning a value to the cluster tendency (CT)
and the validity of the analysis (VAL) *)

Even if we are not able to perform large-scale testing to determine the correctness,
completeness and consistency of the above program, it is worth mentioning some of the
test results. They show the potential capabilities of such a knowledge program. They also
show some of the aspects of *evaluation* and *validation*.

Some test results

The first test aims to verify the correctness of the knowledge program with respect to
user needs and requirements, i.e. assigning a value to the clustering tendency (CT) and
the validity of the analysis (VAL) for varied data sets.

From a series of 100 data experiments with fixed sample size (100) and varying
number of clusters, dimensionality, and clustering tendency, out of which the
observations in Table 9.2 and Table 9.3 were selected, we observe that in all cases the
expert's judgement {C} was **in** {*the system's multiple assignment*}. The numerical facts
are summarized in Table 9.10.

Table 9.10: Summary of the results of the series of experiments A and B.

	case A	case B
expert's judgement	{L}	{VH}
alpha	0.604	0.377
m_user	20	6
%dom_user	55	98
beta	0.624	0.609
m_rand	36	36
%dom_rand	30	20
CPCC	*low*	*high*
edges_i	*very low*	*very high*
k	2	2
db_index	0.519	0.418
k_db	4	4
dim	2	2
num	100	100
system's diagnosis	{L,M}	{H,VH}

The overall impression is that the system tends to be *optimistically* biased when clustering tendency is *very low*; on the other hand, the system tends to be *pessimistically* biased when clustering tendency is *very high*.

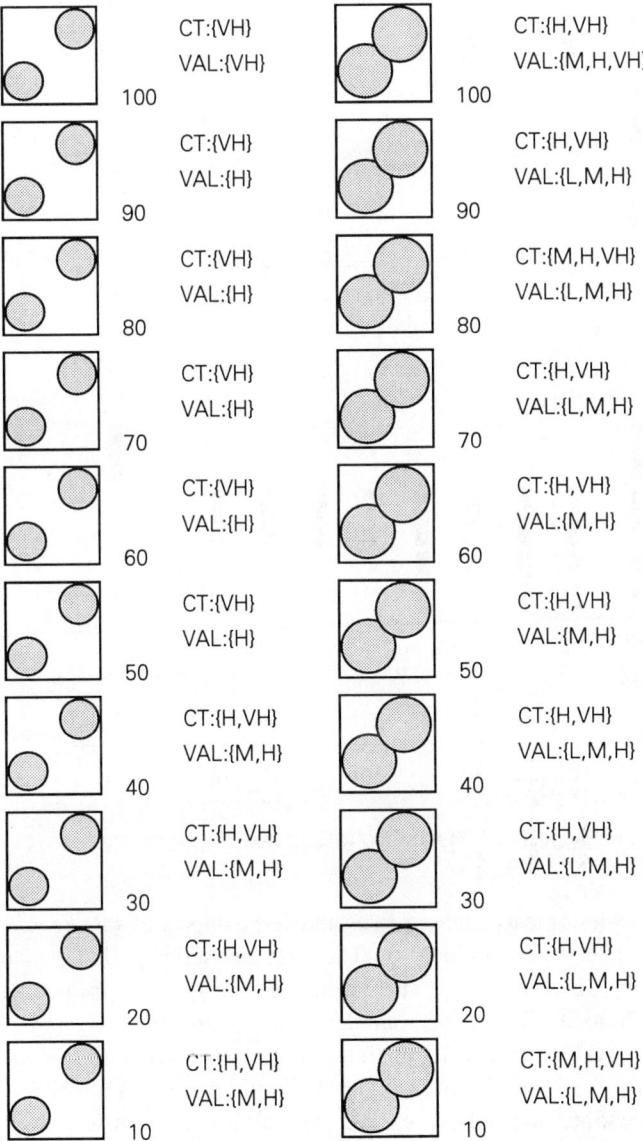

Figure 9.5: The impact of sample size on the system's diagnosis.

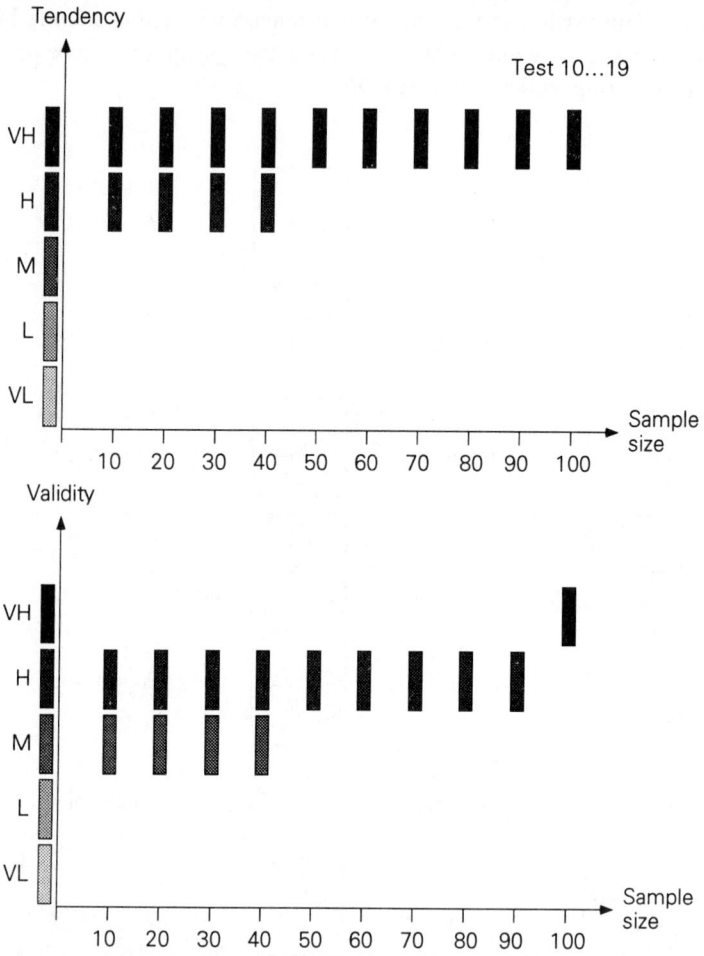

Figure 9.6: Suggested tendency and validity versus sample size (test 10...19).

The second series of tests attempts to establish the impact of *sample size* on the clustering tendency and clustering validity. The first test series includes extremely well-separated clusters (TEST 10 to 19), whereas the second test series include well-separated clusters (TEST 20 to 29). These series are depicted in Figure 9.5.

The system's performance as a function of the sample size with respect to tendency (CT) and validity (VAL) is shown in Figure 9.6 and Figure 9.7. We observe, when the sample size drops, an intended decrease in the validity, in both series. Generally, the tendency is less affected by decreasing sample size. The expert's opinion was that the above behaviour might be appropriate in the above series of tests, but it was found to be too premature to form any judgment concerning the impact of the sample size on varied cluster shapes. This deserves more elaborated experimentation.

The third and last experiment refers to real data (URL01) as used in Chapter 6. In this experiment, we aim to illustrate the impact of the *number of clusters* (K) on the clustering tendency and clustering validity of a given data set. Figure 9.8 shows the result of the computed db_index as a function of K. From that, we certainly may conclude that $K = 3$ is the appropriate choice. This also appears (though less convincingly) in the alpha/beta-dominance-plane, Figure 9.9, where $K = 3$ seems to be favourable with respect to a low value for alpha and high cluster dominance.

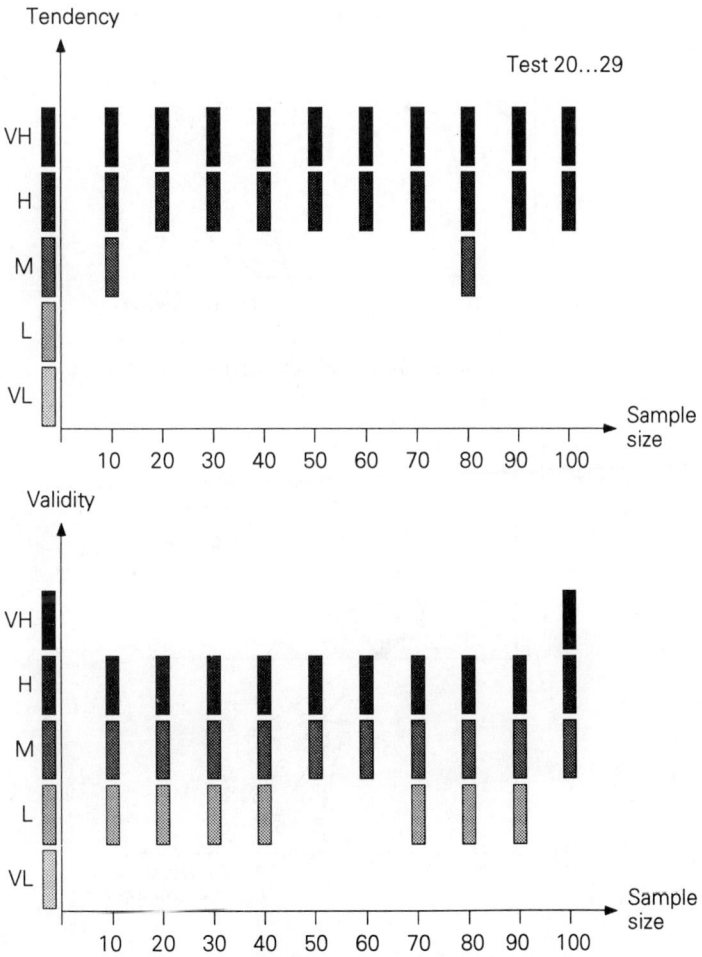

Figure 9.7: Suggested tendendcy and validity versus sample size (20...29).

As the validity is largely determined by dimensionality and sample size, it is not affected by the number of clusters considered. It is questionable whether this behaviour is desirable. Further analysis appears to be necessary. On the other hand, the impact of the number of clusters on the clustering tendency is almost what one could desire. For K = 3, the system's response is as clear as possible for the given data. Then it drops for increasing values of K, see Figure 9.10.

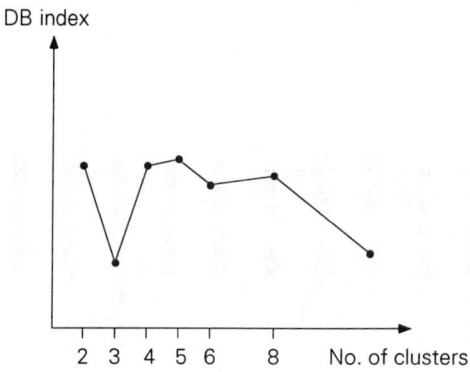

Figure 9.8: Computed DB index as a function of k (URL data).

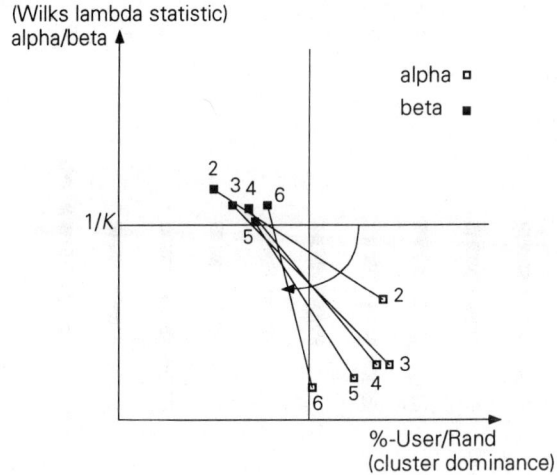

Figure 9.9: (α,β)-plot for the URL data.

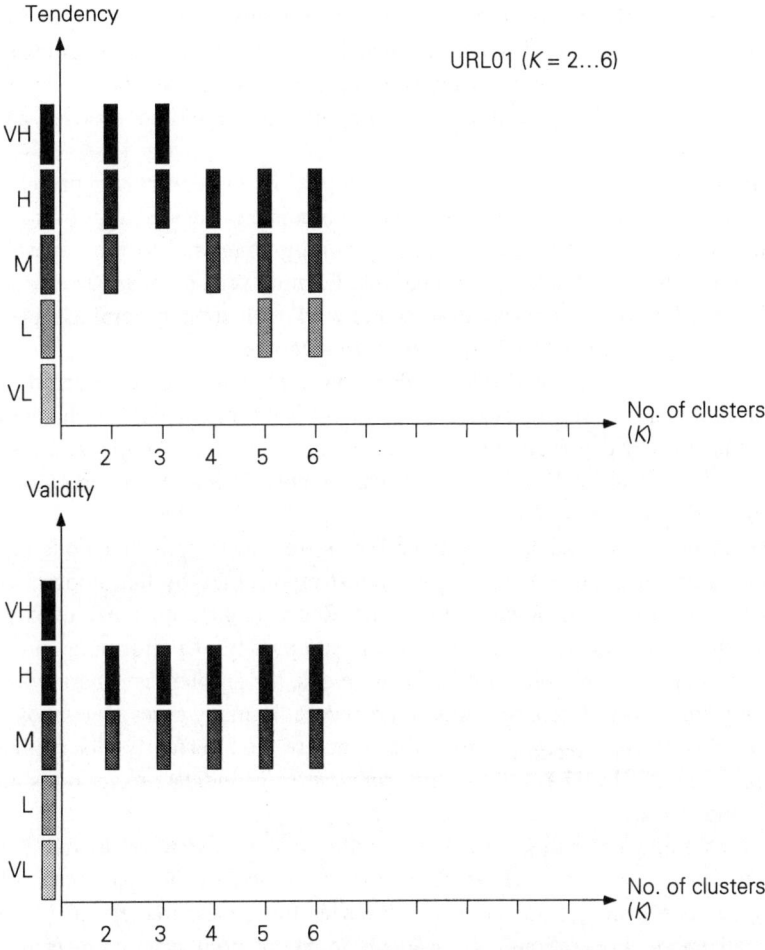

Figure 9.10. Suggested tendency and validity versus number of clusters.

In conclusion, by consistenly following a life-cycle approach, it is possible to build a reliable knowledge program. However, due to testing, analyzing test results and incorporating recommendations, many iterations are likely to be necessary to obtain a quality product.

9.6 Learning

In Section 7.4, we discussed the essential role of conceptual associations as a means of relating informational facts with concepts (classes, clusters), and how to extract them from large sets of observations (Section 9.4.1). Roughly speaking, we have identified such a process as *empirical learning* (machine learning, concept learning) with the fundamental goal of extracting rules from sample data, which will be applicable to new samples of data sets, though without specifying how this could be achieved algorithmically.

In this section, we will first concentrate on partitioning observed representations of sample data sets into a set of concept-covering (decision) rules. Here, optimal rules are those that produce the fewest errors in partitioning compared to the expert-specified conceptual partitioning. Such a process of rule learning is also known as *rule induction*. Typical rule induction systems are designed to work with some general model, such as a *decision tree*, a discriminant function or a *neural network*.

In Section 8.4.1, we stated that all decision trees could be represented as sets of production rules. However, rules as used in the decision tree model are characterized by the fact that they are principally designed to be *mutually exclusive*. Relaxing the restrictions on mutual exclusivity of rules can potentially lead to coverage of concepts more efficiently and compactly.

In the decision tree model, the main problems are how to split each node into groups by distinct attribute (feature) values for the test represented by that node, and how to evaluate the discriminative power of that node. Relaxing the mutual exclusivity requirement will turn out to be very complex. As the complexity of performing an exhaustive search through all possible sets of rules is enormous, the problem is generally tackled by considering some sort of *heuristic search* procedure. In many cases, one tends to accept 'reasonable' solutions by allowing rules that are not perfectly discriminatory or sufficiently general. In what follows, *discriminant rules* indicate exact rules as well as approximative rules.

The remaining part of this section deals with the *neural network* model. On the one hand, neural networks can be regarded as just another type of nonparametric classifier; on the other hand, the neural network classifier format (characterized by relatively simple mathematical operations) is considered to be potentially more powerful than any other general model, because the classifier can have arbitrary complexity. The training procedures appear to be surprisingly simple, but their capabilities for making correct inferences must be empirically tested.

Generally, learning systems that use sample data can be contrasted with rule-based expert systems that attempt explicitly to capture the knowledge of human experts in a computer program. The relationship between the machine learning methods and expert systems is considered, as are the relative benefits of each of the two alternative, yet potentially complementary, approaches.

Conventionally, the learning system has been applied independently of the knowledge-based system. However, within the context of our discussion about cluster analysis, the development of a *hybrid system* of neural network learning and rule-based reasoning is worth thinking about. Development of hybrid systems in general is at the forefront of current research (Kandel and Langholz, 1992).

9.6.1 Rule induction by concept learning

Given our discussion of conceptual associations (Section 9.4.1), and following Ho et al. (1988), we first summarize the problem as follows.

Consider the task of determining some sort of *descriptive generalization* (by a set of rules) of k concepts, C_1, C_2,...,C_k from a set W of data set observations related to these concepts. Assume further the set $\{C_1, C_2,..., C_k\}$ of k concepts to be associated (by an expert or by any other means of classification) to a partition in k classes of the set W, $P = \{P_1, P_2,..., P_k\}$. Each sample data set $\mathbf{w} \in W$ is described by a set of p attributes (structural variables) $A = \{a_1, a_2,..., a_p\}$, each of which presents a certain number of modalities, denoted by $\{v_{j1}, v_{j2},..., v_{jm}\}$, where m possible modalities of the attribute a_j are considered.

Then, a rule R is defined as the conjunction of attribute-value couples $\{a_j,v_j^*\}$, where v_j^* is the modality required by R for attribute a_j. These attributes are said to be *tied* to R; the others are said to be *free* from R. In other words, a sample data set $\mathbf{w} \in W$ can be recognized by the rule R if the value of $\mathbf{w}$ along each attribute a_j tied to R is v_j^*. An object recognized by R is a member of the concept C_k if R is representative for C_k.

Thus, this representative rule can be expressed as a production rule, such as

$$\textbf{IF} \quad a_i(\mathbf{w}) = v_i^*$$
$$\textbf{AND}$$
$$a_j(\mathbf{w}) = v_j^*$$
$$\textbf{AND}$$
$$\vdots$$
$$\textbf{AND}$$
$$a_n(\mathbf{w}) = v_n^*$$
$$\textbf{THEN}$$
$$\mathbf{w} \text{ is a member of the concept } C_k;$$

where $a_i, a_j,..., a_n$ are tied attributes.

The set of q representative rules $\{R_{k1}, R_{k2},..., R_{kq}\}$ is denoted by $\mathbf{R}_k$. The set of objects recognized by the rule R_{kh} is denoted as $P_k(R_{kh})$. Or, in other words, R_{kh} covers $P_k(R_{kh})$.

In finding representative rules, two constraints have to be satisfied.

- *Covering*: Each observed data set (object) of the learning set has to be recognized by a representative rule of $\mathbf{R} = \cup \mathbf{R}_k$.
- *Discriminating*: The representative rules, $\mathbf{R}_k$ of the concept C_k, are not supposed to recognize members of the other concepts.

It is important to note that W is assumed to be *sufficient* to cover most of the elements in the domain: *the hypothesis of completeness*. As a consequence, the number of observations will be very large.

The approach of Ho et al. (1988)

In their approach, the goal is to detect the *sub-structures* representing the *common properties* of the objects belonging to subclasses of a concept, so that they may be described by *discriminant representative* rules.

Each rule of the above type (*type 1*) defines a *sufficient condition* for an object to belong to a concept. This leads to a clustering problem in itself.
– Find a clustering for each class P_k satisfying the discriminant constraint.
– The number of subclasses of each clustering is not known a priori.

The discriminant constraint implies detecting the minimum number of rules R_{kh} that satisfies $P_{k'}(R_{kh}) = 0,\ V_{k'} \neq k$.

Minimizing the number of misrecognized objects and *maximizing* the covering capacity of the rules leads to the following optimization criterion:

minimize $J_P(\mathbf{R})$

$$J_P(\mathbf{R}) = \frac{\displaystyle\sum_{k_h=1} \mathrm{card}(P_k)\left(1 + \sum_{k' \neq k} \mathrm{card}(P_{k'}(R_{kh}))\right)}{\displaystyle\sum_{k=1,K} \mathrm{card}(P_k(R_{kh}))}$$

The *quality* of a representative rule R_{kh} of the class P_k can be computed as follows:

$$Q(R_{kh}) = \frac{\mathrm{card}(P_k(R_{kh}))}{\mathrm{card}(P_k(R_{kh})) + \displaystyle\sum_{k' \neq k} \mathrm{card}(P_{k'}(R_{kh}))}$$

Note that the quality of an exact rule is equal to 1 because

$$\sum_{k' \neq k} (\mathrm{card}(P_{k'}(R_{kh}))) = 0.$$

Search of rules of *type 1*

The search begins by determining the representative rule of the concept C_k generated by an object $\mathbf{w} \in P_k$. In order to determine the 'discrimination' of rules, it is necessary to give an acceptable threshold q which is a measure of the quality requested for the rules. To sort out the discriminant rules which are generated from an object $\mathbf{w} \in P_k$, the following definitions are necessary.

Definition 9.1

The rule R_1 is *more general* than the rule R_2 if the object set of P_k recognized by R_1 is larger than the object set of P_k recognized by R_2.

Clearly, the less general a rule, the more discriminant it is. For example, in the trivial case where all the values $a_j(\mathbf{w})$ are taken, the rule is always discriminant but recognizes only the object $\mathbf{w}$ itself. This rule is of no interest.

Definition 9.2

R is a *representative rule* generated by an object **w** means that R is one of the most general rules generated by the conjunction of the discriminant attribute values of **w**.

The basic idea of the algorithm is the following.

1. Start with an '*empty*' rule (this is the most general one as all its attributes are free);

2. Tie the attribute $a_j(\mathbf{w})$ to this rule; accept attribute a_j if $Q(R(a_j(\mathbf{w})) \geq q$.

3. Repeat step 2 for all j.

4. Choose the rule that covers the maximum number of objects in P_k.

5. Repeat steps 1 to 4 for all **w**.

6. Select for each concept the most general rule and add this rule to the rule set.

7. Repeat step 6 until all objects in P_k are covered.

The above method has proven to be efficient for constructing rules related to a knowledge base for expert systems. But the quality of the results depends on the given data.

9.6.2 Learning (fuzzy) rules from examples

After discussing a rule induction method in the previous section showing a straight-forward approach to generating general rules from specific examples, we re-emphasize in this section the real problem of acquiring operational rules for decision-making or diagnosis, as well as the problem of allowing explicit representation of fuzziness in the rule and a degree of belief the rule as a whole.

The organization of the represented knowledge is generally found to be a serious bottleneck because it usually involves in-depth analysis of interviews with an expert and the final formulation of relations encountered. In the absence of an explicit criterion for evaluating the relevant relations, their validity is very often not guaranteed. Instead of asking the expert to formalize his or her knowledge in terms of decision rules, a better approach – whenever possible – might be to ask the expert to provide examples of the decision-making process.

As such, *inductive learning* from examples (Ho et al., 1988), as discussed in the previous section, is a prime example of such an attempt. In essence, the method has also been used in the specific solution for incremental hierarchical knowledge organization, as proposed in Appendix A.

A program that learns from examples must produce general rules from specific training instances (examples). The space of all possible training instances is called the *instance space*. The space of all possible rules is called the *rule space*. The problem is to construct a program that chooses the rule that fits the given instances in an optimal way when the predicates that are involved in the decision-making are known. Furthermore, the program must choose in an optimal way the various parameters that are included in the predicates. This is exactly what has been established in Section 9.6.1. Other interesting examples of learning programs are described in Michalski et al. (1986), Quinlan (1983) and Pitas and Venetsanopoulos (1992), to name but a few.

In what follows, we take the method of Wang and Mendel (1992) as a general method to generate *fuzzy rules* from numerical data. Their method consists of five steps. Step 1 divides the input and output spaces of the given numerical data into fuzzy regions (see also Appendix A). Step 2 generates fuzzy rules from the given data. Step 3 assigns a degree to each of the generated rules in order to resolve conflicts among the generated rules. Step 4 creates a combined fuzzy rule base based on both the generated rules and linguistic rules of human experts. Step 5 determines a mapping from input space to output space based on the combined fuzzy rule base using a defuzzifying procedure. We consider a simple two-input, one-output case.

Let y (output) be the numerical assessment of the reliability of the data match between a CL-recovered clustering and the true categories. Let the CL-recovered clustering be characterized by the CPCC value (x_1) and the data match itself by the value of the Rand coefficient (x_2) measuring the degree to which the CL-recovered clustering and the true categories match.

Suppose we are given a set of desired input–output data pairs

$$(x_1^{(1)}, x_2^{(1)}; y^{(1)}), (x_1^{(2)}, x_2^{(2)}; y^{(2)}), \ldots$$

where x_1 and x_2 are inputs, and y is the output. The task here is to generate a set of fuzzy rules from the desired input-output pairs, and use these fuzzy rules to determine a mapping

$$f : (x_1, x_2) \rightarrow y$$

Then, the five steps to carry out are as follows.

Step 1 – Divide the input and output spaces into fuzzy regions.
Assume that the domain intervals of x_1, x_2 and y are $[x_1^-, x_1^+]$, $[x_2^-, x_2^+]$ and $[y^-, y^+]$, each of them divided into $2N + 1$ regions, denoted by

small N, small N – 1, ..., centre, ..., big N – 1, big N

For reasons of simplicity we assume x_1 and x_2 operating on the same domain interval [0.5,1] and divided into five fuzzy regions ($N = 2$) which will be denoted in our terminology

very low, low, medium, high, very high

We simply adopt triangular-shaped membership functions but, of course, other divisions of the domain regions and other shapes of membership functions are possible, if not

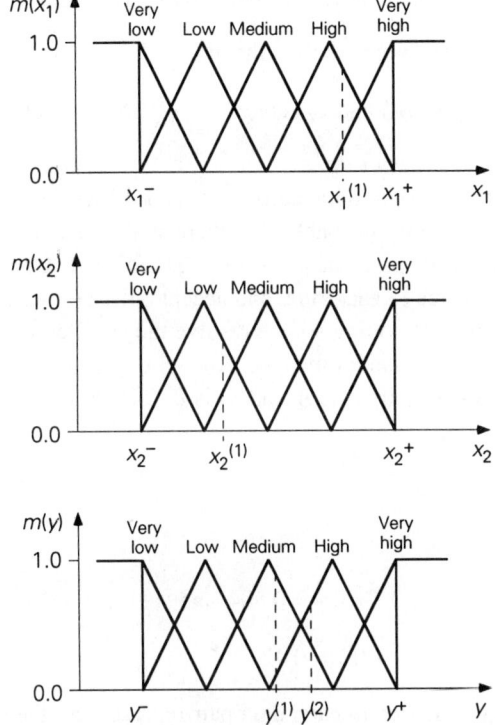

Figure 9.11: Divisions of the input and output space into fuzzy regions and the corresponding membership functions.

likely.

Figure 9.11 shows the domain division and membership functions for x_1 and x_2. Likewise, we expect y to be divided into five regions.

Step 2 – Generate fuzzy rules from given data pairs.
First, determine the degrees of given $x_1^{(i)}$, $x_2^{(i)}$ and $y^{(i)}$ in different regions. For example, $x_1^{(1)}$ in Figure 9.11 has degree 0.8 in *high*, degree 0.2 in *very high*, and zero degree in all other regions. Similarly, $x_2^{(2)}$ has degree 1 in *Medium*, and zero degree in all other regions. Second, assign a given $x_1^{(i)}$, $x_2^{(i)}$ or $y^{(i)}$ to the region with maximum degree.

As a result, we obtain one rule from one pair of desired input–output data, e.g.,

$$(x_1^{(1)}, x_2^{(1)}; y^{(1)}) \rightarrow$$
$$[x_1^{(1)} \ (0.8 \text{ in } high, \text{ max}), x_2^{(1)} \ (0.7 \text{ in } low, \text{ max});$$
$$y^{(1)} \ (0.9 \text{ in } medium, \text{ max})] \rightarrow$$

Rule 1: IF x_1 is *high* and x_2 is *low*, THEN y is *medium*;

Likewise, the second input–output pair results in:

Rule 2: IF x_1 is *high* and x_2 is *medium*, THEN y is *high*.

Step 3 – Assign a degree to each rule
Since there are usually lots of data pairs to take into account, and each data pair generates one rule, it is highly probable that there will be some conflicting rules, i.e., rules that have the same IF part but a different THEN part. One way to resolve this conflict is to assign a degree to each rule, and accept only the rule from a conflict group that has maximum degree. By doing so, we resolve the conflict problem, but reduce the number of rules greatly at the same time.

For a rule which says that IF x_1 is A and x_2 is B, THEN y is C, the degree of such a rule is defined as

$$D(\text{rule}) = m_A(x_1)m_B(x_2)m_C(y).$$

Consequently,

$$D(\text{Rule 1}) = 0.8 \times 0.7 \times 0.9 = 0.504, \text{ and}$$

$$D(\text{Rule 2}) = 0.6 \times 1 \times 0.7 = 0.42.$$

We may further assign a degree to each data pair representing the expert's belief in its usefulness.

Suppose the data pair $(x_1^{(1)}, x_2^{(1)}; y^{(1)})$ has degree $m^{(1)}$, then we redefine the degree of Rule 1 as

$$D(\text{Rule}) = m_{high}(x_1)m_{low}(x_2)m_{medium}(y)m^{(1)}.$$

Step 4 – Create a combined fuzzy rule base
The form of a fuzzy rule base is illustrated in Figure 9.12. We fill the boxes of the base with fuzzy rules according the following strategy.

(1) The rule base combines generated rules from numerical data and domain rules from an expert.
(2) If there is more than one rule in one box of the fuzzy rule base, use the rule that has maximum degree.
(3) If a rule is an AND rule, it fills only one box of the rule base; but, if the rule is an OR rule, it fills all the boxes in the rows or columns corresponding to the regions of the IF part.

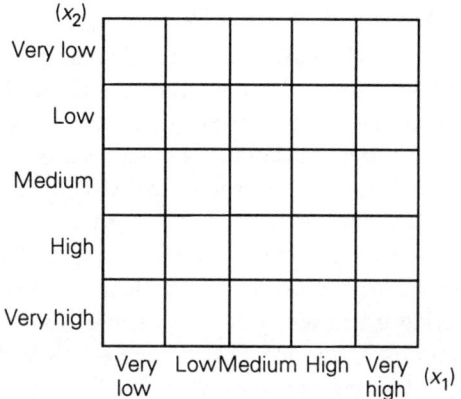

Figure 9.12: The form of a fuzzy rule base.

Step 5 – Determine a mapping based on the combined fuzzy rule base

The following defuzzification is used to determine the output value y for given inputs (x_1, x_2): first, for given inputs (x_1, x_2), we combine the antecedents of the ith fuzzy rule using product operation to determine the degree m_{Oi}, of the y corresponding to (x_1, x_2), i.e.

$$m_{Oi} = m_{I_1^i}(x_1) m_{I_2^i}(x_2)$$

where O^i denotes the output region of Rule i, and I_j^i denotes the input region of Rule i for the jth component.

Then, the centroid defuzzification formula reads as follows:

$$y = \frac{\sum_{i=1}^{k} m_{Oi}^i \overline{y}^i}{\sum_{i=1}^{k} m_{Oi}^i}$$

where $\overline{y}^i$ denotes the centre value of region O^i (the centre of a fuzzy region is defined as the point that has the smallest absolute value among all points at which the membership function for this region has membership value equal to unity), and k is the number of fuzzy rules in the combined fuzzy rule base.

Thus, in conclusion, the above method 'learns' from the 'examples' and generates rules that constitute the 'generalization property' such that when new inputs are presented the mapping continues to give desired or successful outputs y.

9.6.3 Artificial neural network (ANN) learning[1]

Artificial neural networks (ANNs) have been used in numerous applications, and they have shown great potential for acurate learning from examples. Moreover, ANNs have a definite learning function even when the reasoning logic cannot be determined explicitly. However, after an ANN has been trained, it has often been used as a black box; it has been hard to understand the knowledge learned by the ANN through training. The reason for this is that the weights and links in an ANN do not generally make intuitive sense. On the other hand, fuzzy logic models and fuzzy rules are especially attractive because of their similarity to the reasoning process used by humans. As a result, modern research has focused on deciphering the meaning of the structure and parameters of ANNs by converting them into fuzzy rules. This would also make ANNs useful for the inference of fuzzy rules dirrectly from training examples when the direct acquisition of the rules is very difficult or expensive, and it would loosen the knowledge acquisition bottleneck. To this end, significantly different approaches have been experimented with to generate fuzzy rules from ANNs.

Artificial neural networks are computational models designed to generate performance similar to that of the human brain in fields such as speech and image recognition (Pao, 1989). They mimic the brain and other biological neural networks in that they encompass a massively parallel architecture, and they derive their power from the sheer number of neurons, rather than the complexity of each single neuron. These systems lend themselves to fast implementations in parallel, especially when done in hardware. They differ from conventional systems in that a conventional computer processes instructions one at a time, sequentially, while a neural network processes many competing hypotheses at the same time. Also, since there are many nodes, neural networks are more noise tolerant than conventional computer systems; if a few nodes are erroneous, there are still many others and the overall performance of the net might be unaffected.

The multi-layer perceptron (Rumelhart and McClelland, 1986)
One type of neural network – the *multi-layer perceptron* – is composed of many simple computational elements, or nodes, that form layers. These layers are linked by weights, which are adapted in a supervised learning process so that the neural network will correctly classify a user-supplied pattern. This can be done via backpropagation, an algorithm that 'trains' a network on sample data so that the network will respond correctly. Figure 9.13 shows the architecture of a multi-layer perceptron.

The input to the first layer is the pattern to be classified. A node's input is the scalar product of the output vector from the previous level and the weight vector between that level and the node:

$$\text{net}_j = \sum_i \mathbf{w}_{ji} O_i$$

---

[1]   Credit is due to Tina Majchrzak, Minesh Patel and Amine Bensaid, University of South Florida.

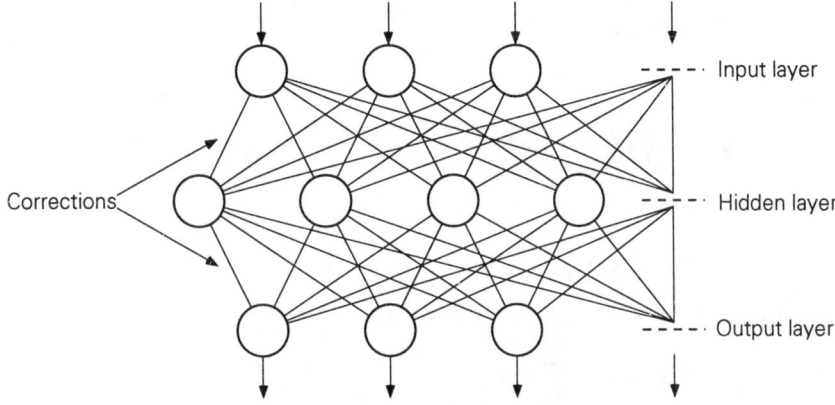

Figure 9.13: The architecture of a multi-layer perception.

where net_j is the input to layer j, $\mathbf{w}_{ji}$ is the weight matrix between layer j and layer i, and O_i is the output of layer i. A node's output is this value passed through the sigmoidal nonlinearity

$$O_j = f(\text{net}_j) = \frac{1}{1 + e^{-(\text{net}_j + \theta_j)/\theta_0}}$$

where θ_j is the threshold or bias. For inputs much larger than the bias, the output is close to 1. For small inputs, it is close to zero. These correspond to 'on' and 'off' states, respectively. For values close to the threshold, the node emits a number between 0 and 1, corresponding to an undecided state. The threshold is treated as a link from a virtual node whose output value is always 1. This way, the value also can be learned. θ_0 is a constant that determines the shape of the sigmoid. A small θ_0 results in a sigmoid that is like a threshold logic unit, with an abrupt increase from 0 to 1. A large θ_0 creates a more gentle graduation between 0 and 1.

Generally, a pattern p will have an output O_{pk} that is different from the desired output t_{pk} (k is an output layer). A learning procedure, such as backpropagation, helps minimize this.

Backpropagation

The *backpropagation algorithm* seeks to minimize the normalized system error

$$E = \frac{1}{2p} \sum_p \sum_k (t_{pk} - O_{pk})^2$$

with respect to the weights in the neural network. It does this by using an iterative, steepest-descent approach, making incremental changes in the weight space in proportion to the error's rate of change with respect to the weight:

$$\Delta \mathbf{w}_{ji} = -\eta \frac{\partial E}{\partial \mathbf{w}_{ji}}$$

This simplifies to

$$\Delta \mathbf{w}_{ji} = \eta \delta_{pj} O_{pi}$$

where for output layers

$$\delta_{pk} = (t_{pk} - O_{pk}) O_{pk} (1 - O_{pk})$$

and for input and hidden layers

$$\delta_{pj} = O_{pj} (1 - O_{pj}) \sum_k \delta_{pk} \mathbf{w}_{kj}$$

In these equations, $O_j(1 - O_j)$ is at its maximum when $O_j = 0.5$, and its minimum when $O_j = 0.1$. This makes sense in light of the fact that a node whose output is 0.5 is considered 'undecided', whereas outputs of 0 or 1 indicate nodes set 'on' or 'off'.

Fuzzy rule generation: a survey

In Keller and Tahani (1992a), a standard backpropagation NN is used to implement fuzzy logic functions. Each variable in the antecedent of a fuzzy rule is presented to the NN as 11 input nodes which are (11) samples from the variable's possibility (or membership) function. Each variable in the consequence of the rule is similarly represented by 11 output nodes in the NN. One hidden layer is used to internalize a representation of the relationship between the antecedent and the consequence. The network is tested with rules containing single and multiple antecedent clauses and a variety of uncertain input data, and it is concluded that the simple NN used can learn and extrapolate complex relationships between the possibility distributions representing the antecedent and consequence clauses.

In Keller and Tahani (1992b), this NN is tested for insensitivity to noisy input distributions and for the ability to internalize multiple conjunctive and disjunctive clause rules. The results show that the neural network function performs well.

The emphasis in Horikawa et al. (1992) is on developing *fuzzy neural networks*, trained by the backpropagation algorithm, can not only acquire fuzzy inference rules and tune membership functions simultaneously, but which also have a structure that makes the identified rules very easy to understand. The performance of the NNs is illustrated by inferring fuzzy rules for some specific functions. If the fuzzy NNs proposed in Horikawa et al. (1992) always provide rules that are as readable as those obtained by their model, these NN will prove to be quite useful for fuzzy modelling. However, the rules obtained in Horikawa et al. (1992) may not yield an optimal covering. Some may be redundant, since the total number of rules is based solely on the product of the membership functions of each of the inputs. For example, a proposed NN structure with two inputs, x and y, where x has two membership functions and y has three, would result in exactly six fuzzy rules, regardless of how many are actually required to model the relationships between inputs and outputs in the target system.

In Tagaki and Hayashi (1992), on the other hand, the goal is not to work on the structure of the NN used to implement fuzzy rules, nor is it to implement a set of rules already known. Instead, a conventional backpropagation network is used at two different stages to achieve what would, otherwise, have been achieved by a set of rules. In fact, the process starts by dividing up the set of possible antecedent input vectors into training and testing sets. Then the training set is clustered in order to discover antecedents that are similar enough to imply that the same rule should be used for them; an optimal number of clusters (C) is sought to determine the number of rules that the system will have. The next step is to teach a (first) NN to what extent (measured by a membership grade) a given rule is applicable to a given input vector. Then a second NN is used to learn the outputs (consequences of the rules) corresponding to the input patterns (antecedents of the rules). We note that the term 'fuzzy rule' was not used in Tagaki and Hayashi (1992) to mean a rule that necessarily deals with fuzzy variables; rather, 'fuzzy rule' referred to a rule that is applicable to a (crisp) input to a certain degree only. We also wish to highlight the fact that the performance of this system is critically dependent on the quality of the clustering performed early in the process and on the (implicit) assumption that in a backpropagation neural network, an output that is not exactly 0 or 1 conveys some information in the fuzzy logic sense.

Romaniuk and Hall (1993) present an NN which, given only domain examples, is capable of generating a fuzzy partitioning of attributes that will later represent the premises for rules. The NN is an incremental learner with a dynamic architecture; a recruitment of cells algorithm (RCA) is used, every time a training example is presented to the NN, to decide if a new node should be added to the net and how the NN parameters should be updated. After training, the network may be pruned using a global attribute covering (GAC) algorithm which yields fewer and more general rules. The rules are obtained directly from the network's structure by following the connections from the inputs to the outputs. The authors claim that the resulting rules perform well compared to certain instance-based learners.

Fuzzy rules, acquired from the domain experts, and a membership function represent the inputs in Kawamura et al. (1992). First, a fuzzy model is constructed with the following elements: a membership function delineating the relationship between the variables in the system under study (this function can be converted to a set of rules); the antecedent membership functions; a definition of the fuzzy logic operation(s) to be used when combining antecedents; and the consequence membership functions. A (single) NN is then constructed to depict the information about each input, and design data are used to train the NN (using backpropagation); this yields a membership function that better approximates the relationship between the system variables than the initial fuzzy model. Finally, the weight changes caused by training the NN are interpreted as modifications in the elements of the initial fuzzy model; this offers an intuitive explanation of the changes undertaken by the NN during training. However, all the concepts in Kawamura et al. (1992) are only discussed for a specific example, and no general procedure is proposed for constructing the NN or performing the final interpretation.

9.7 *Intelligent hybrid systems (IHS)* (Kandel and Langholz, 1992)

In this section, we discuss the *intelligent hybrid system* architecture that provides domain interaction between two computational paradigms that we have dealt with: rule-based and neural-based. It surely appears to be logical – and most promising – to seek synergism of the domain of expert systems (logical, cognitive, though static) driven by explicit knowledge, and the domain of neural networks (numeric, associative, adaptive and dynamic) representing implicit knowledge.

Both domain tools complement each other remarkably well in that the expert system's inability to adjust dynamically is compensated by the neural network's ability to update dynamically. The proposed integration is expected to be much more effective and flexible in its power successfully to handle a wider scope of application variations.

The expert system component in the IHS architecture is a *fuzzy expert system* (Kandel and Langholz, 1992) which reflects the fact human expertise embedded in the knowledge base is imprecise, incomplete, or not totally reliable. Fuzzy logic thus provides a systematic framework for dealing with fuzzy quantifiers. As such, a fuzzy expert system offers a knowledge-based technique with *approximate reasoning* to emulate human decision-making better.

On the other hand – as we have already seen – *artificial neural networks* provide a greater degree of robustness, or fault tolerance, than conventional sequential machines. In addition, neural networks also possess the ability to handle inconsistencies or conflicts in the data.

As human expertise embedded in the knowledge base is thought to be explicit knowledge, the drawbacks of an expert system approach to solving problems lie in the fact that the knowledge acquisition still is a bottleneck, and in the system's inability to synthesize new knowledge or to adjust resident knowledge dynamically. Even if integrating the expert system with a neural network may compensate or resolve those drawbacks to some extent, the neural network also suffers from a number of serious pitfalls like the way it may converge – if it does – and the *ad hoc* way of updating. Moreover, its architecture is most often unknown at the beginning. Nevertheless, the ability of the neural network to *learn* in an imprecise environment enables the fuzzy expert system to modify and enrich its knowledge structure autonomously.

Organization and knowledge transfer

The IHS architecture is supposed to facilitate, to improve and to support application (domain-) oriented algorithmic processing like signal processing, image processing, or – in the context of our subject matter – algorithmic processing in exploratory data analysis (cluster analysis).

A real IHS architecture should accomodate two modes of operation. First, if we assume that the knowledge is pre-organized but has to be modified or updated, then the neural network is expected to take the knowledge from the fuzzy expert system and to modify it through learning. However, if we assume that the knowledge is only partly pre-organized and has to be extended, modified and updated incrementally, then the neural network is expected to take the data from the algorithmic processing (signal processing, image processing and so on) and has to learn (fuzzy) rules from the data through association and feed the learned rules to the expert system. To some extent, the latter mode is pursued in Appendix A.

So, the *integrated intelligent processing* environment demands an organization in which knowledge-based, neural-based, model-based and algorithmic-based processing may interact with each other: a platform for implementing a multi-domain intelligent processing environment. Clearly, such a structure is still in its infancy, though will dominate future research.

More on the development of hybrid architecture for intelligent systems, as well as its applications, can be found in Kandel and Langholz (1992).

9.8 Concluding remarks

This chapter has dealt with the development of rule bases for cluster analysis. The major part was examplified using a simple language for a fuzzy knowledge base, such that sample rules could be tested directly in EDAPLUS. The suggested levels of cluster-oriented reasoning (facts, support, diagnosis) – as proposed as a logical framework for reasoning – were merely meant to guide the reader through the structure and organization of knowledge in general as it implicitly represents a strategy to reason, rather than to imply that such a structure is the only possible organization of knowledge involved. We note that a structure like the suggested levels of reasoning, in fact is to be considered as part of the domain knowledge.

It is not surprising that the most important aspect of this chapter is the problem of constructing rules from observations. This has been called learning. Much attention has been given to the problem of learning, as will be clear from the literature; however, the problem still is too complex to solve completely and to understand the characteristics of the specific method to hand. Some aspects, as well as an overview, have been presented, particularly concerning fuzzy rules and artificial neural networks. Though further research is needed, the intelligent hybrid system architecture has been launched as the promising dual-domain approach that combines fuzzy expert systems and neural networks.

10 | Case study: analysis of delphinid sonar sound signals

Anyone who is prepared to learn quite a deal of matrix algebra, some classical mathematical statistics, some advanced geometry, a little set theory, perhaps a little information theory, and some computer technique, and who has access to a good computer and enjoys mathematics ... will probably find the development of new taximetric methods much more rewarding, more up-to-date, more 'general', and hence more prestigious than merely classifying plants or animals or working out their phylogenies (Johnson, 1968).

In this chapter we describe a research application which may illustrate how to frame and validate the usage of cluster analysis. More specifically, we illustrate how hierarchical trees and partitions produced by cluster analysis may become information. That is, we show the steps that link the gathering of data and the generation of information needed to draw conclusions.

Chapter 4 discussed the problem of how to frame the analysis and belief, how to choose the options, and how to make the decisions. The steps which we have to go through were represented in the diagram of Figure 4.3. Keeping this diagram in mind, we can summarize the underlying ideas best by quoting from Romesburg (1984):

Beginning on the left, the researcher frames the problem, deciding what multivariate data need to be collected and how to tailor the cluster analysis. This requires that a vantage point be chosen that can frame the pattern that will eventually be seen. Next, data are collected, the cluster analysis is made, and a pattern of similarities is found in the tree. The researcher studies the tree, and from this further frames the pattern, again deliberately selecting and abstracting its essential properties. Then, using this twice-framed pattern, the researcher mentally passes it through an information function to decide what should believed. In this chain, raw data become reduced data, reduced data are interpreted as patterns, and the interpreted patterns become information – a serial metamorphosis of data into thought. From this chain, it is also clear that there are two points in the analysis where the pattern of data is framed. The first is when the researcher chooses objects, attributes, scales of measurement, a method of standardization, a resemblance coefficient, and a clustering method all in accordance with the dictates of the research goal. The second point is when the researcher examines the tree and abstracts the relevant features from the irrelevant ones, consciously or unconsciously.

Note that a research application might include a series of different research goals for which different 'data to thought' paths apply. This will be the case in our research application on delphinid sonar sound signals, as will be clear in the following.

10.1 Introduction

We examine here the research of Dudok van Heel (1962; 1966; 1981) and Kamminga et al. (1981; 1987; 1990), on dolphin sonar sound production for navigation in relation to the mass stranding phenomenon. The performance and properties of dolphin sonar systems have been the subject of several studies and investigations during the past twenty years. As a result, there is now a fairly detailed description of the echolocation signals of dolphins belonging to quite different families and subspiecies.

Based upon in-depth studies on echolocation carried out on both captive animals and free-ranging wild animals, of different families of Odontocetes over many years, it has recently been postulated (Kamminga and Beitsma 1990) that – from an ecological point of view – a different sonar system performance is to be found for *pelagic* dolphins than for *littoral* animals in response to different navigational demands imposed on them.

Evolution

It is tempting to believe that evolution has influenced the sonar system to resolve different navigational problems. Ranging in open ocean requires only a simple navigational signal in the absence of echoes and reverberations of nearby objects, rocks, sandbanks and underwater flora. On the other hand, we expect to find a more coastal sonar behaviour, for the littoral and coastal species, due to a different ecological habitat.

Adaptation

As a result of preliminary experimental work, it has also been hypothesized that dolphins may be capable of modifying their echolocational signal parameters to create 'optimal' ecological adaptation.

Two-component sonar

It has been found by Kamminga and Wiersma (1981) that some species reveal a high-frequency component in their sonar signals for foodfinding in the nearby environment while simultaneously emitting a high-energy, low-frequency component well suited for navigation at long range.

The two-component sonar hypothesis have served to shed new light on the significance of the failing of sonar as a cause of mass stranding in pelagic species and the reason why some inshore species fail to show the same phenomenon.

As more and more data became available, it was speculated that the two-component sonar sound production, as suggested by Kamminga, could be considered as the most 'appropriate' working hypothesis at the moment. The hypothesis is supported by dolphins' skull morphology, and is in agreement with established ecological classifications (Dudok van Heel) and existing taxonomic classifications.

Moreover, it was pointed out by Dudok van Heel (1966) that '*the pelagic Odontocete when feeding in coastal waters might be so intent on listening to the information to be gleaned from the nearby environment (echoes from pursued prey) that information from the surroundings, if returning at all from these pelagic type signals, was not registered.*

In this way a pelagic animal might find itself unexpectedly in such confined quarters that panic could lead to a mass stranding'.
This in contrast with some *pelagic* species who seem to have a tool for survival, acoustically provided by the two-component sonar.

We now are prepared to identify the research problem and to state the research goals. After that, several framings have to be defined, each of which is subject to its appropriate primary and secondary validation.

10.2 The research problem

From the introduction above, it is clear that the *research hypothesis*, Hr, should read:

> (Hr): 'A two-component sonar prevents some species of *Odontocetes* from erroneous echolocation and thus from stranding'.

We now construct a series of research goals in order to confirm Hr.

Suppose we were to test this hypothesis by using cluster analysis; to do so requires that we assess the relation between two variables: (1) a reliable, numerical data representation of the true sonar signals produced by the species of interest, and (2) an ecological classification of the same species of which the stranding behaviour is known a priori.

Variable (2) can be observed and is well documented, with a universally accepted objective definition for it. But variable (1) – being a quantitative variable – can only be dealt with if we have a 'sound' signal model at hand which maps the true sonar signals onto a quantitative feature space, and, at the same time, preserves existing classifications, including the ecological classification. As will become clear, the Gabor signal model appears to be a justified parametric description of the sonar click wave forms.

So, our *framing hypothesis*, Hf, may read as follows.

> (Hf): 'Gabor sonar sound signal modelling preserves all existing classifications related to sonar sound production, including the ecological classification'.

Now, the above reasoning might not be valid if and only if the following *research question*, (Qr), had not been raised in the past.

> (Qr): 'Is sonar sound production for echolocation related to ecological habitat?'

As it has generally been accepted that sonar sound production is related to the soft anatomy in the head and to dolphins' skull morphology, a cluster analysis of the physical skull measurements suggested that the resulting tree pattern bore a remarkable resemblance to existing ecological tree patterns.

From what could be observed, it was logical to raise the above research question because it opened ways to further insight into the significance of failing of sonar as a cause of mass stranding.

The above research question (as a shot in the dark), the necessary framing hypothesis (mapping true signals on multivariate data), and the ultimate research hypothesis (discovering a law of nature) have been brought together in the research reasoning diagram shown in Figure 10.1. Clearly, ?(1) – being Qr – is the shot in the dark. Next, confirmation of ?(2), the framing hypothesis Hf, justifies applying Gabor sonar sound signal modelling. Finally, ?(3) includes the ultimate, challenging research hypothesis.

Note that if the research hypothesis, Hr, produces vital knowledge and understanding, when tested, then our research question, Qr, which led to the creation of Hr, must in retrospect be judged as good.

As the *research planning* has now been established, the actual framing of inputs and outputs that will condition the analysis of the successive research goals will take place in the following subsections.

Generally speaking – as pointed out in Chapter 4 – the researcher must make several (often subjective) decisions that will condition the analysis. The actual choices are called framing decisions. They include the choice of

– objects
– attributes
– scales of measurements
– standardization (if so, and how)

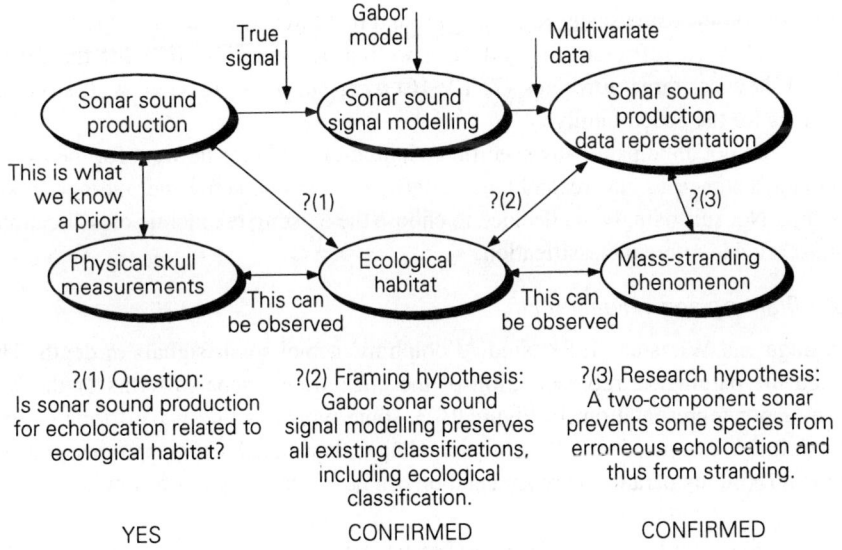

Figure 10.1: Outline of the research planning; the research reasoning diagram.

- resemblance coefficient
- clustering method.

After implementing these decisions, a wide range of numerical results, visual interpretation, internal indices, and measures of confidence may generate interesting and useful conclusions. With respect to the specific research goal at hand, very often there are certain additional features that we would like the cluster analysis to have. They are
- well-structured clusters
- agreement with existing classifications
- agreement with expert intuition
- demonstration of stability and robustness.

So, in view of the above, it is clear that a cluster analysis requires *subjective decision making*, generally depending on the research goal and its context. Thus, a cluster analysis that aims to generate an interesting scientific question will be framed and validated differently than one that aims to verify an existing hypothesis.

10.2.1 Framing and validating Qr

The intention of Kamminga's research was to unravel the structure of the mechanism of underwater sound production for a variety of dolphin species. For that purpose, as there was good reason to believe that physical skull dimensions were related to the mechanism of underwater sonar production for echolocation, he examined 26 skulls of *Sotalia fluviatilis*, 40 skulls of *Sousa sinensis*, and 14 skulls of *Sousa teuszii*.

On these he measured 16 attributes, as indicated in Figure 10.2. Clearly, they were all quantitative on the same scale. Kamminga found that three key attributes (7, 10, 14; see Figure 10.2) were sufficient to produce a well-structured clustering for the *Sotalia* family. Likewise, three attributes (7, 10, 16) were sufficient to yield well-structured clusterings for the *Sousa* family.

As he did not anticipated any specific weighing of attributes he went for the average Euclidean distance as the resemblance coefficient, and selected the average-linkage algorithm. Not surprisingly, he decided to choose the existing taxonomic classification as the most reliable referent classification.

10.2.2 Framing and validating Hf

Kamminga and Wiersma (1981) studied dolphins' actual sonar signals in depth. They justified the parametric representation of the click wave shape in terms of the well-known Gabor representation. In Figure 10.3, some typical actual wave forms of sonar clicks (left) are compared with the corresponding Gabor model representations (right). In analytical form, the parametric representation can be written down as follows.:

$$f(t) = \exp\left[-\frac{\pi^2}{\Delta t^2}(t - t_0)^2\right] \cdot \cos(2\pi f_0(t - t_0) + \varphi),$$

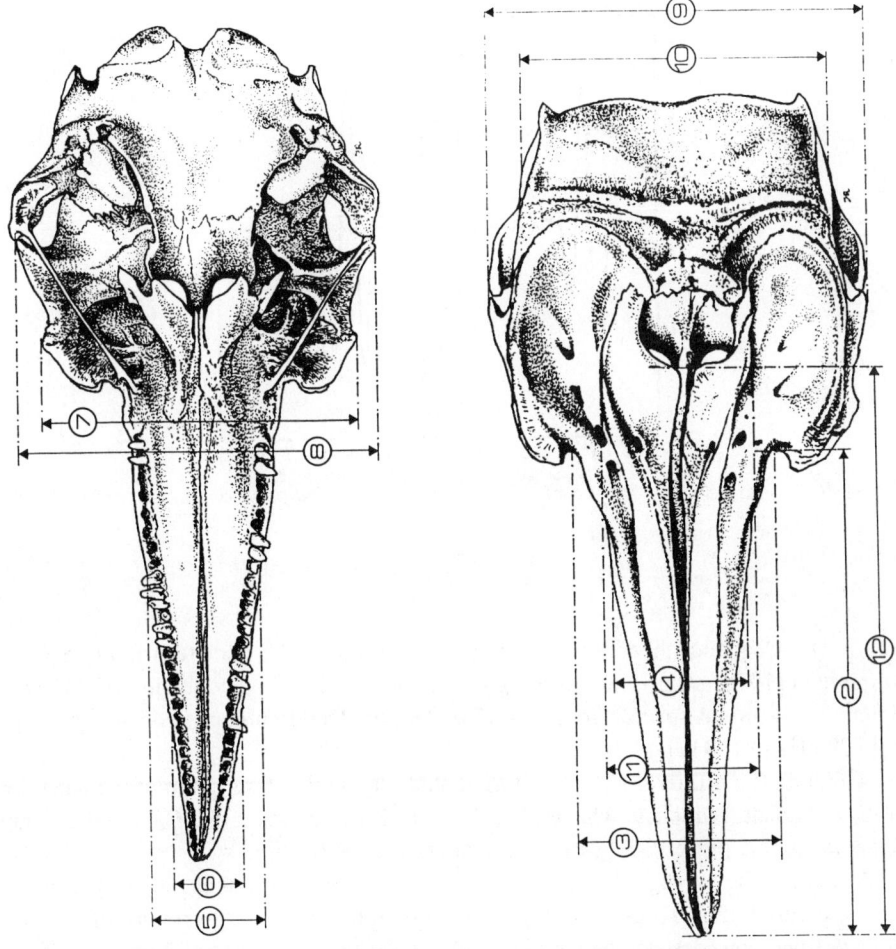

Figure 10.2: Physical skull measurements; skull top view (left), skull bottom view (right). Courtesy of Dr P.J.H. van Bree, Amsterdam University.

where

$f_0 =$ dominant frequency,

$\Delta t =$ time duration, related to the standard deviation of the envelope of f_0,

$\varphi =$ phase of f_0 within the envelope,

$t_0 =$ mid epoch of the time function.

As a result, each actual sonar click can be represented by a four-dimensional feature vector

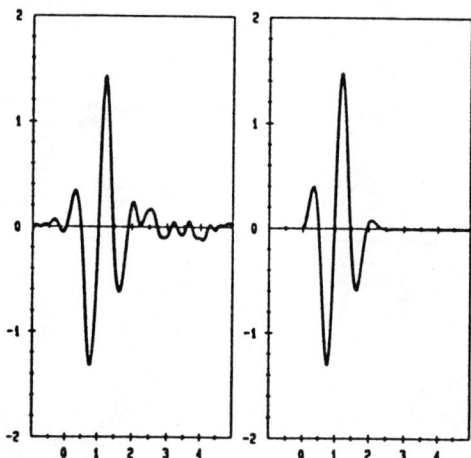

Figure 10.3: Actual dolphin sonar clicks (left), corresponding Gabor model representations (right).

$$\mathbf{x} = (f_0, \Delta t, \Delta f, \varphi)^{\mathrm{T}},$$

where Δf =frequency bandwidth (standard deviation of the spectrum relative to f_0). Kamminga (1979) report on clustering results obtained from a collection of 343 sonar signals for different species of the *Tursiops* family. The dominant frequency f_0 ranged from 28 kHz to 130 kHz.

Although the primary objective of their study lay in the description of the behaviour of the dominant frequency, what emerged was a clear indication that Gabor sonar sound signal modelling preserved the existing ecological classification.

As before, the average Euclidean distance and the average-linkage algorithm yielded well-structured clusterings, and the clustering into the six specimens of *Tursiops truncatus* was considered to be the logical choice for the referent classification.

10.2.3 Framing and validating Hr

It was found by Kamminga and Beitsma (1990) that (for example) *Sotalia fluviatilis guianensis* possesses two-component sonar. This is can be seen in Figure 10.4, where the actual sonar click is decomposed into a high-frequency and a low-frequency component. This mechanism indicates non-stranding behaviour. The research hypothesis Hr allows a prediction of the tree pattern in the clustering such that non-stranding two-component sonar species 'close' to mass stranding species (low-frequency region) can also be found 'close' to non-stranding species (high-frequency region). The assumed truth of the hypothesis must follow from the fact that the observed tree obtained by a cluster analysis resembles the predicted tree pattern as discussed above. Thus, the test includes the predicted classification as the referent classification.

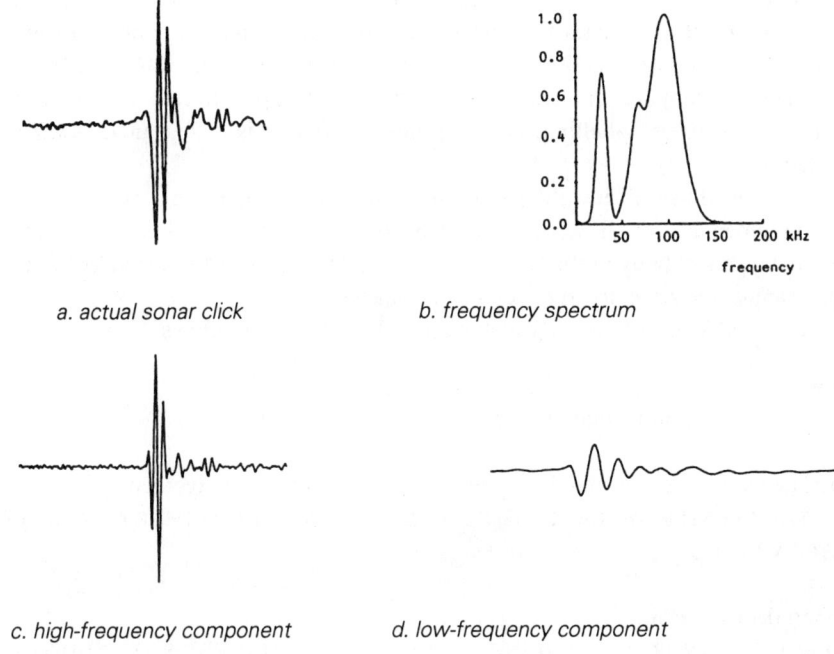

a. actual sonar click   *b. frequency spectrum*

c. high-frequency component   *d. low-frequency component*

Figure 10.4: Two-component sonar click.

10.2.4 Implementing primary and secondary validation in the knowledge base

We have seen (Chapter 4, and this chapter) that the usefulness and reliability of conclusions drawn from a cluster analysis is a measure of primary validity. Features like well-structuredness, agreement with existing classifications, and agreement with expert intuition are measures of secondary validity.

Typical examples of referent classifications for secondary validation, as identified in the foregoing, are the following:

research goal	referent classification
Qr	taxonomic classification
Hf	specimen classification
Hr	predicted classification

Any pattern matrix (see Section 2.1) permits optional labels (from any referent classification).

What has to be determined is the degree of agreement between the obtained cluster result and one of the appropriate referent classifications, provided by the labelling assigned to the pattern matrix.

In Section 6.1, we encountered the *Rand coefficient* as an 'objective' means to assess the degree to which two classifications of the same data match. In contrast to *internal indices*, used to validate the usefulness of a single cluster result, the Rand coefficient, *R*, is designed to compare the cluster result with any appropriately chosen referent classification or expert labelling. As such, this coefficient is called an *external index* (secondary validation).

We will now illustrate how the knowledge base can accomodate external indices. The actual implementation will be outlined in Section 10.5. As we intend to expand the domain-independent body of the knowledge base with some domain-dependent rules, we here re-examine the structure as developed in Chapter 9.

The essential levels of decision-making have been listed as follows:

rule base
(* global abort flag; initial dialogue *);
//
(* initial decision level sensing the numerical results and visual judgements *)
[HERE, THE CLUSTER TENDENCY IS DENOTED V1, AND THE PRIMARY VALIDITY IS DENOTED V2];
//
(* support decision level *)
[WEAKENING AND/OR REINFORCING V1 AND V2 ON THE BASIS OF ADDITIONAL TESTING];
//
(* final diagnosis *)
[WEAKENING AND/OR REINFORCING V1 AND V2 ON THE BASIS OF COMBINED EVIDENCE].
//

The program ends by assigning a value to the cluster tendency (CT) and the validity of the analysis (VAL) by the terminating rule

```
IF ⟨condition⟩ THEN
BEGIN
CT := v1;
VAL := v2;
END;
```

Now, let v3 be the fuzzy constant representing the secondary validity. Then, for example, v3 is designed to assess the structuredness of the tree pattern and the degree of agreement between the inferred classification and the referent classification.

So, at the initial decision level, we expect to encounter

```
[i]        IF undefined(v3) THEN
           BEGIN
```

v3 := {*medium, high, very high*};
S := {*very low, low, medium, high, very high*};
message('the rule base may contain domain-specific rules!');
END;

At the support level, we may find

[ii]   IF undefined(R) THEN
             BEGIN

             message('make sure that the appropriate domain-specific labelling is
             assigned to the pattern matrix considered',
             'run the RAND statistic';
             a := query_yesno('leaving the decision network?');
             END;

             followed by the domain-specific rule

[iii]   IF ⟨count⟩ AND R < 0.7 THEN
             weaken(weaken(v3));

Finally, at the diagnosis level, we expect to have domain-specific rules like

[iv]   IF ⟨count⟩ AND v1 in {*very low, low, medium*} AND R in S THEN
             BEGIN
             weaken(v3);
             END;
             [v] IF ⟨count⟩ AND v1 in {*very high*} AND R in S THEN
             BEGIN
             reinforce(v3);
             END;

The program will terminate with the rule

             [vi] IF ⟨count⟩ AND undefined(R) THEN
             v3 := v2;

[vii]   IF ⟨count⟩ THEN
             BEGIN
             CT := v1;
             VAL := min(v2,v3);
             END;
             (* end of the program *)

Rules [i]–[vii] show how concepts of information, tolerances and norms illuminate subjective decision-making: the researcher must decide about conclusions – what to believe.

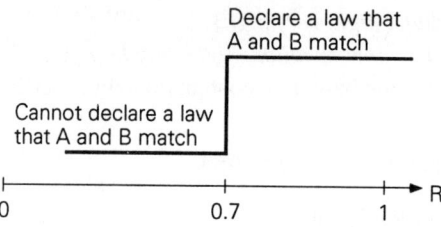

Figure 10.5: A hypothetical information function for deciding whether R inductively infers a law.

Information function

Rule [iii] is an attempt to map a specific interval of R onto the fuzzy constant for secondary validity, v3. This is called an *information function*. It relates one or more variables to information. The information function here relates R to two classes of information: I1, "A and B are match", and I2, "A and B do not match", where A is a classification under consideration, and B some referent classification (see Figure 10.5).

The role of R is to 'turn on' the proper belief. The researcher must decide whether the value of R is high enough to favour confirmation of the matching. This is done by setting a *normative threshold* (here R = 0.7). This is done by trial and error based on past experience. However, different fields of science draw the line for the normative threshold at different values.

The effect of firing rule [iii] will be

v3 := {*very low, low, medium*}.

The net result of firing rule [iv] or [v] is

v3 := {*very low, low*}

or

v3 := {*high, very high*},

respectively. In effect, the rules have turned some of the observed facts into beliefs.

From a practical point of view, we notice that it seems quite straightforward to interweave the domain-independent body of the knowledge base with some domain-specific rules, without ruining the intentional functionality if it were not decided to determine R.

10.3 The analysis

So far, we have presented the research planning, the distinct research goals, and the corresponding framings and validity criteria. We now have reached the stage of interpreting the results and putting them into perspective.

The research work of Dudok van Heel and Kamminga aims to bring together biological classifications (taxonomic and ecological) and non-biological classifications like morphological and sonar sound signal classifications. They have found Gabor sonar sound signal modelling to be the 'most natural' mapping interface.

In their research, the species listed in Table 10.1 were considered. Most of them were first recorded by Kamminga, partly as captive animals, partly as free-ranging wild animals. The main ecological classes are *pelagic*, *littoral*, and *riverine*; *littoral* habitat is further subdivided into *offshore*, *inshore*, and *estuarine*. From the signal recordings it could be gleaned that the species possess sonar clicks with dominant frequencies ranging from 20 kHz to 130 kHz, as listed in Table 10.2. Note that the sonar signal of SFG (*Sotalia fluviatilis guianensis*) appeared to have two major frequency components. Later, Kamminga related his observation of two-component sonar signals with echolocation and stranding behaviour.

Table 10.1: Distribution of species, clustered into species and clustered into ecological habitat.

	odontocetes	ecological habitat
PC	PSEUDORCA	PELAGIC, OFFSHORE
CC	OEPHALORHYNCHUS	INSHORE, ESTUARINE
PP	PHOCOENA PHOCOENA	INSHORE, ESTUARINE
NP	NEOPHOCAENA	INSHORE, ESTUARINE, RIVERINE
TT	TURSIOPS TRUNCATUS	OFFSHORE, INSHORE, ESTUARINE
SFG	SOTALIA	INSHORE
SFF	SOUSA	ESTUARINE, RIVERINE

Table 10.2: Distribution of species with respect to their sonar click dominant frequency.

specimen	dominant frequency (kHz)
PC	28
CC	125
PP	120
NP	130
TT	35
SFG	95/29

As already mentioned, the biological classifications form the backbone of the analysis of sonar sound production. This is mainly because they direct the path of reasoning and provide the researcher with logical choices for referent classifications. In other words, non-biological classifications like Kamminga's sonar sound signal classification are expected to match these referent biological classifications.

In what follows, we will address
– the morphological classification (Qr)
– the sonar sound classification (Hf, Hr)

We will then discuss and comment on the successive findings.

10.3.1 Analysis of Qr

The 26 *Sotalia* skulls (made available by the Taxonomic Institute of the University of Amsterdam) were measured as discussed before (see Figure 10.2). Some attribute values were missing and were estimated by the Dixon 3 method (see Section 2.1). Some typical values for the major *Sotalia* skull features (7, 10, 14) are presented in Table 10.3.

The average-linkage algorithm produced a tree pattern (dendrogram), as shown in Figure 10.6, which allows us to believe that there are two distinct groups, I and II, of which some of the skull sample measurements are listed in Table 10.3. The tree pattern is well structured and matches the ecological classification (*litteral*, *riverine*) almost perfectly. The only exception is object 7 which may have suffered from correcting missing data.

A second experiment included 41 *Sousa* skulls and yielded also a well-structured tree-pattern (see Figure 10.7). More importantly, the taxonomic classification into *Sousa chinensis* and *Sousa teuszii* was retrieved with only one sample (28) misclassified.

Table 10.3: Typical values for the major Sotalia skull features.

	7	10	14
I	129	133	308
	137	123	312
	143	123	306
	136	119	306
	.	.	.
	.	.	.
	.	.	.
II	124	108	291
	114	112	265
	120	110	276
	.	.	.
	.	.	.

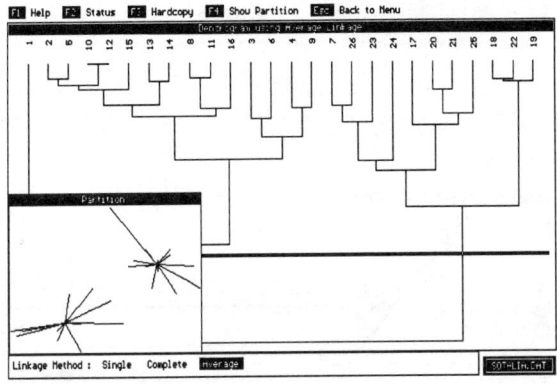

Figure 10.6: The Sotalia tree pattern for physical skull measurements.

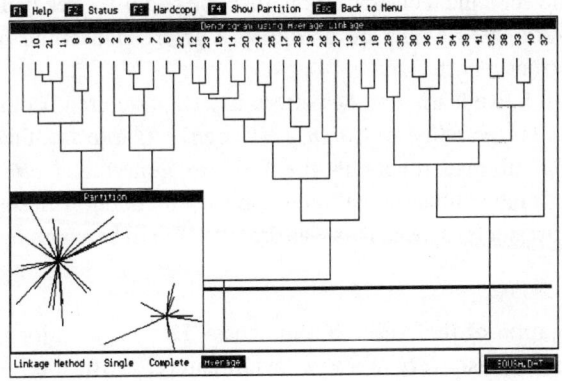

Figure 10.7: The Sousa tree pattern for physical skull measurements.

These experiments, unintentionally, showed groupings which appear to have such a remarkable agreement with ecological habitat that the question Qr sooner or later had to be raised. As revealed by the above analysis, we now tend to believe that sonar sound production for echolocation indeed plays a major role in ecological habitat.

10.3.2 Analysis of Hf

A set of 100 true signals, originating from the different species listed in Table 10.1, were modelled as points in a four-dimensional parameter space (f_0, Δt, Δf, φ). Kamminga and Wiersma showed that over 90% of the variance could be preserved by taking only the two-dimensional representation (f_0, Δf).

If Hf were true, that is, if Gabor modelling should preserve all existing classifica-

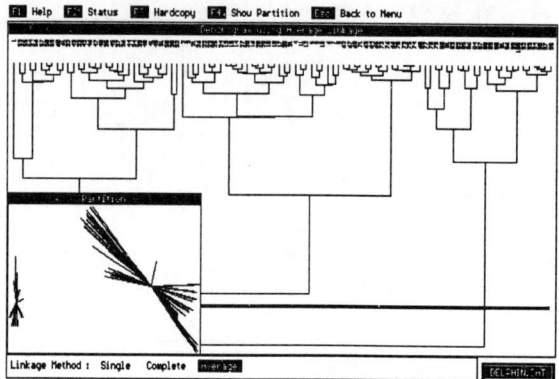

Figure 10.8: Tree pattern for 100 true signals represented in a two-dimensional Gabor space (f₀, df).

tions, including the ecological classification, then the resulting tree pattern from a cluster analysis (see Figure 10.8) would be predicted a priori, and could be compared with referent classifications a posteriori.

From Figure 10.8, we learn that the pattern is well structured. Both the primary validity and the secondary validity are {*high, very high*}. As a result, there will be enough agreement among all researchers in the field to generate a *consensus* about the appropriateness of Gabor sonar signal modelling for mapping true sonar signals onto a low-dimensional parameter space. This was the very aim of research hypothesis Hf.

10.3.3 Analysis of Hr

Although confirmation of the research hypothesis, Hr, has a major impact on science because its implies the discovery of a new law of nature, verification of Hr itself is to be considered as straightforward.

As mentioned earlier, it was found that *Sotalia fluviatilis guianensis* possesses a two-component sonar signal. It was then hypothesized that a two-component sonar should indicate non-stranding behaviour.

Following the ecological classification, pelagic animals are typical mass-stranders, whereas inshore, estuarine, and riverine dolphins certainly belong to the class of non-stranders. Most of the offshore specimens possess two-component sonar, and can be encountered in open sea, as well as in inshore situations. What we want to establish is that true signals from the same specimen (*Sotalia*) can be found 'close' to the mass-stranding Pelagic samples, though not showing any stranding behaviour, as well as 'close' to non-stranding inshore samples. Again, all true signals are modelled as points in a two-dimensional parameter space (the Gabor space). The true sonar signals of *Sotalia* are then decomposed into their constituting sonar components and offered to the clustering algorithm, together with all the samples of species with only one frequency

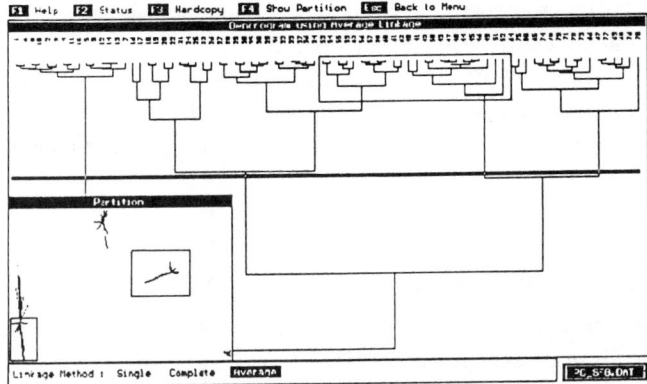

Figure 10.9: Tree pattern for the Sotalia sonar clicks (shaded areas) represented in a two-dimensional Gabor space (f_0, df).

component. All *Sotalia* samples are within the shaded areas, in both the tree pattern and the corresponding partition, as shown in Figure 10.9. Also, this experiment yields primary and secondary validity of degree {*high, very high*}. Figure 10.10 illustrates the significance of the CPCC and the inconsistent edges in the tree pattern, ruling out any possibility of being just a random event.

Without limiting the importance of the above findings, it may be too early to declare a new law of nature. In looking over the above analysis, it is apparent that many gaps have to be filled in, especially for the pelagic dolphins because of the difficulties in recording echolocation signals in open sea (Kamminga, 1994). At the time of writing we still have too few and too sparse descriptions of clicks at sea to answer fundamental questions. In spite of this subject matter can be considered as a prime example of how

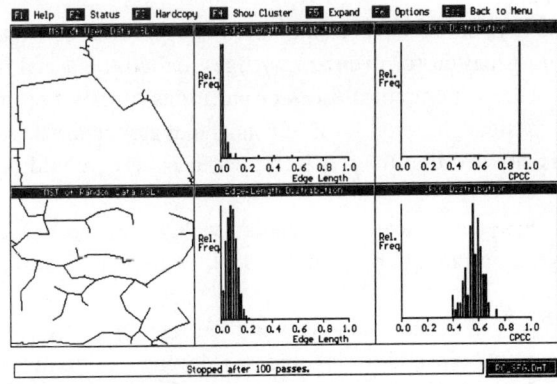

Figure 10.10: Significance of the CPCC value and the inconsistent edges of the resulting tree pattern.

cluster analysis fits into complex research and reasoning. The underlying concepts of framing and validating have been the major objectives.

As can be expected, explorative research like this has to be redone over and over before ultimate framing and validation can be considered as final. However, meanwhile, the research application has shown how concepts of "from data-to-understanding" bear upon cluster analysis.

10.4 Computer-assisted analysis

Let us conclude this chapter by discussing the impact of computer-assisted reasoning in cluster analysis for research types as described.

First, we recall the notion of conceptual association (Chapters 7 and 9), representing the transition from facts to evidential information, as well as the transition from facts to pertinent knowledge. Conceptual associations aim to construct rules, to be used in the reasoning, from large sets of observations (learning).

Next, we examine from the case study, as has been described, what kind of observations might be available and useful, particularly for generating rules for secondary validity.

Then, we will see how expert intuition can be captured and expressed as rules.

Finally, we illustrate the approach by drafting a knowledge base for hypothesis testing on both sides (confirmation and falsification).

10.4.1 Conceptual associations

Recalling the notion of conceptual association (Chapters 7 and 9), we re-emphasize the fact that in a relation, a concept domain, C, is said to be conceptually associated with a set of domain facts, $\{A\}$, if to each tuple of values in $\{A\}$, say a tuple of observed parameter values or measurements, there corresponds precisely one value in the domain C, at any given time.

In the foregoing analysis, we have encountered the process of transition from facts (observations) to information (or even knowledge); called the serial metamorphosis of data into thought, or how hierarchical trees or partitions of data become information or knowledge. This is basically a serial form of conceptual associations.

So, the analysis showed that the research hypothesis (Hr) could be modelled as the seriation of

(1) $R(\mathbf{w},A) \rightarrow R(\mathbf{w},C)$; conceptual association between observed facts and some concept domain (grouping in a tree, a cluster label, confidence level, or any conceptual classification); and

(2) $R(\mathbf{w},C) \rightarrow R(\mathbf{w},C')$; conceptual association between the observed hierarchical tree or partition (conceptual classification) and a reference domain (here taxonomical classification or ecological classification, or any kind of expert intuition that will justify the intermediate conceptual outcome).

Note that $R(\mathbf{w},A)$ has represented the observations regarding the skull morphology in the first place, which has produced a conceptual classification that has been found not to conflict with the taxonomic and ecological (reference) classifications. Note that both the taxonomic and ecological classifications are well documented. Secondly, $R(\mathbf{w},A)$ represented the observations regarding Gabor modelling. Also in this case, a conceptual classification (different from the above) was produced that was found not to conflict with the existing reference classification.

When associations (1) and (2) are – within the domain of research – found to be true for all specimens to hand, then the observations really make the transition from fact to information or even knowledge. Then, also, the rules can easily be tested in a computer-assisted environment.

More of interest is the case when observations give rise to fuzzy rules. In that sense, we recall the simple two-input, one-output example of Section 9.6.2. Such an example may account for the fact that we want to represent the seriation of the two conceptual associations in one (fuzzy) rule. Then the numerical assessment of the reliability of (or confidence in) the first association is given by the CPCC value, whereas the Rand coefficient measures the degree to which the induced conceptual classification and the reference classification (taxonomy or ecological classification) match. In Section 9.6.2, we saw how well given input–output pairs can yield a fuzzy rule base with a resolution of fuzzy regions that may be appropriate.

Clearly, in the case of more complex seriations of conceptual associations, and in the case of multiple paths from observation to conceptual information or knowledge, we are always able to construct (fuzzy) rules that include secondary validation. In Section 4.3, we listed a complex of factors to be taken into account when validating the ultimate induced association. Such a complex of factors reads as follows:

1. Is the resulting tree pattern well structured (this may involve a variety of tree indices; CPCC is one of them)?
2. Is the result in agreement with existing classifications (like taxonomic and ecological classifications; the Rand coefficient accounts for the degree of match)?
3. Is the result in agreement with other data analysis methods (goodness of fit with respect to population descriptors; regression analysis and the like)?
4. Are the results valid for different data from the same research objective (how representative were the data to hand)?
5. Is the result in agreement with existing expert intuition?

Factor 5 in particular should be handled with care. This will be the subject of the next subsection.

10.4.2 Capturing expert intuition

Most of the work has been defined in terms of acquiring knowledge from the expert through training instances of observed variables and conceptual output. Certainly, the expert's intuitive knowledge has been explored to the extent of interpretation of facts related to desired output. No problem with that: on the contrary, it has been stressed that such an approach could circumvent the bottleneck of full knowledge acquisition by any kind of interviewing. As a result, the approach has been deliberately data-driven/case-based.

However, including the expert's intuition about what he or she is expecting (and probably wants) to see from the induced tree patterns may steer the result towards a confirmation of the research hypothesis too much. So, the question is not how the expert's intuition can be captured and expressed as rules (this is not very different from the above secondary validation); the real question is whether it is desirable or not to mix a rule base made up of rules that represent processing knowledge (about algorithms), support knowledge (what to test and how to interpret the results), and diagnostic knowledge (how to combine different findings and how to validate (primary and secondary) intermediate and ultimate findings), all related to domain knowledge, and representations of the expert's intuition about what to observe and what to conclude. If one wants to include the expert's intuition, then the representing rules should be used separately from the formal computer-assisted analysis and interpretation.

10.4.3 A knowledge-based test for confirmation and falsification

The final paragraphs again deal with the so-called information function (as mentioned in Section 10.2.4) for deciding whether a specific subset of the observation space (or input space) implies a law; this is done by setting normative thresholds on all observable (or input) variables that are relevant for a (sub)conclusion. Thresholds are set by trial and error based on past experience.

The value set of all thresholds determines when the (sub)conclusion is confirmed as a law; likewise there exists a value set of all thresholds which determines when the (sub)conclusion is falsified as a law. Figure 10.11 shows a simple example for a two-input, one-output case of thresholding a confirmation or a falsification. Whenever necessarry, one can incorporate such confirmations and falsifications into the computer-assisted analysis by generating those tests when needed.

10.5 Concluding remarks

In this chapter we have described a research application which illustrates how to frame and validate the usage of cluster analysis. More specifically, we have gone through the process of how hierarchical trees and partitions become information or knowledge: a seriation of conceptual associations, ultimately confirming or falsifying a given hypothesis.

This was clearly one example to show that cluster analysis is not an end in itself but is used to generate initial, supporting and diagnostic observable facts on the basis of which goal-oriented reasoning and judgement could take place.

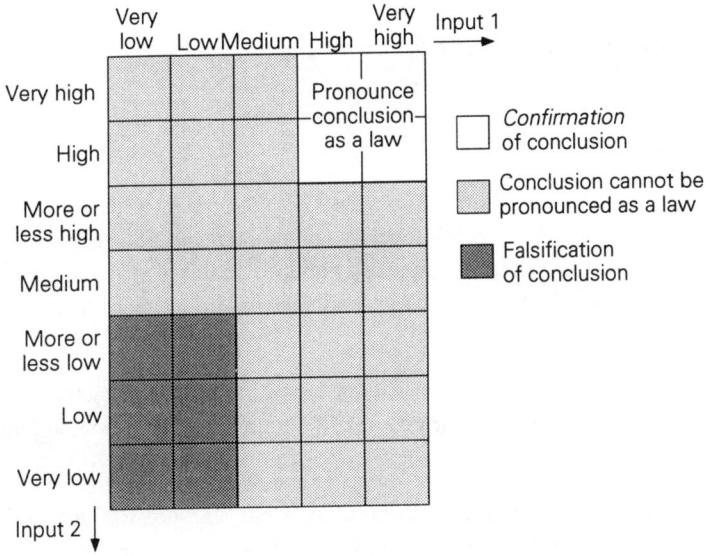

Figure 10.11: Information function for confirmation or falsification of a conclusion.

The example also clarifies the fact that a true analysis is the resultant of the processing paradigm (the tools), the modelling paradigm (data–information–knowledge association), and the research framing.

Practice shows that (fuzzy) approximate reasoning is the rule rather than the exception. It can also be concluded that computer-assisted analysis (generating algorithmic facts, supporting evidence and diagnostic interpretation) brings considerable consistency to the art of performing cluster analysis.

Glossary of key terms[1]

abduction *Artificial intelligence.* Inference in which the result of a causal relationship is taken to imply the cause. While not logically valid, this kind of reasoning is involved in areas such as medical diagnosis, in which a disease is inferred from the symptoms it causes.

approximate *Science.* **1.** Describing a value or result that is not exactly correct, but is close enough for a predetermined purpose. **2.** To obtain a value or result approaching the actual or desired one.

artificial intelligence *Computer science.* A field of study concerned with the development and use of computer systems that have some resemblance to human intelligence, including such operations as natural language recognition and use, problem-solving, selection from alternatives, pattern recognition, generalization based on experience, and analysis of novel situations.

association theory *Psychology.* A theory that concepts are learned by a simple, reinforced connection between a stimulus and a desired response.

association *Behaviour.* The forming of a learned connection between a stimulus and a response, or between one stimulus and another.

attribute *Computer programming.* **1.** Any characteristic of a data variable, such as length or format. **2.** A characteristic or data field of a record that may take on a variety of values, such as the address or telephone number in a customer file record. *Statistics.* A qualitative random variable.

classification The process of dividing things into classes, or the divisions arrived at by such a process; specific uses include: *Engineering.* the process of grading and sorting of particles by size, shape, or density. *Statistics.* the assignment of observations to groups; used in prediction, as in medical diagnosis, and description, as in botany.

cluster A group of similar items considered as a unit; specific uses include: *Psychology.* any group of objects, words or events that an individual perceives as belonging together. *Computer programming.* **1.** A group of items having similar content identifiers. **2.** A group of related data types and function definitions.

---

[1] Academic Press Dictionary of Science and Technology, 1992.

cluster analysis *Statistics.* In multivariate analysis, the study of the relations among characters or individuals, based on grouping units that show similar patterns.

clustering algorithm *Computer programming.* In pattern recognition, a statistical method of establishing any sets of samples, among a given group of unlabelled samples, that are more similar to each other than to samples outside the set.

cophenetic correlation coefficient *Systematics.* a statistic that measures the degree of fit between a phenogram and a table of phenetic distances.

data The plural form of **datum**. *Science.* **1.** Two or more individual facts or pieces of information. **2.** A body of facts, information, or knowledge, particularly when derived from scientific observation or experimentation. *Computer programming.* The representation of facts, numbers or concepts that can be communicated, stored and processed by computer to form information.

decision rule *Statistics.* Any sample-based criterion for choosing among alternatives; for example, in hypothesis testing, a statement about the sample statistic that determines whether the null hypothesis is to be accepted or rejected.

deduction *Science.* A formal reasoning process in which the conclusion is reached as a result of a finite sequence of logical steps, each of which is an axiom, a given statement, or an immediate consequence of earlier statements. If the specific premises given are true, then it follows that the conclusion must also be true.

dendrogram: *Biology.* A branching diagram used to show relationships between members of a group; a family tree with the oldest common ancestor at the base, and branches for various divisions of lineage.

ecology *Biology.* The branch of the biological sciences that deals with the relationship between organisms and their environment, including their relationship with other organisms.

expert system *Artificial intelligence.* An application in which problems are solved by means of an information base containing rules and data from which inferences are drawn on the basis of human experience and previously encountered problems.

fuzzy set *Mathematics.* A generalization of the concept of set; in particular, a subset F of a given set S for which the characteristic function $S_F: S \to [0,1]$ has the entire unit interval as its range, rather than just the two values $\{0,1\}$.

fuzzy-set theory *Artificial intelligence.* An application of the concept of fuzzy set in expert systems for estimating the degree of certainty of conclusions.

heuristics *Science*. Any of various problem-solving techniques that involve the use of subjective knowledge, hunches, trial and error, rules of thumb, and other such informal but generally accurate methods.

hierarchy *Behaviour*. A social system within a group of animals of the same species, in which individuals exercise dominance over those ranking lower in status and are in turn dominated by those ranking higher.

induction *Science*. A process of reasoning in which a general conclusion is reached from specific data, especially when the conclusion does not necessarily or directly follow from these data.

inference *Science*. A conclusion that follows logically from the available evidence but that is not the direct and incontrovertible result of that evidence. *Statistics*. A statement about a population based on a sample drawn from it.

information *Science*. A general term for any data that have been recorded, classified, organized related, or interpreted within a certain context so that meaning is apparent.

knowledge *Artificial intelligence*. **1**. A general term for the aggregation of facts, principles and other information that is characteristic of human intelligence. **2**. A level of capability or sophistication in a machine system that is regarded as analogous to human knowledge.

knowledge-based system *Artificial intelligence*. A computer system that stores and uses a very large amount of information about an application and serves as an intelligent assistant, rather than as an expert.

logic *Science*. The use of correct reasoning, or the principles involved in correct reasoning. *Mathematics*. The branch of mathematics that formulates and studies principles of reasoning. *Computer programming*. The algorithm or decision procedures used by a program.

machine learning *Computer programming*. The ability of a machine to improve its behavior based on previous trials and past performance.

Monte Carlo method *Statistics*. A method of approximating the solution of a mathematical problem by investigating the properties of a random process or system, often using computer-generated random numbers.

neural network *Artificial intelligence*. A computational network, often for pattern recognition, composed of mathematically defined elements that are thought to approximate the working of biological neurons; often composed of a layer that receives and organizes inputs, a hidden layer, and an output layer in which individual neurons identify particular patterns. Networks can be trained by backpropagation.

validation *Computer programming*. The procedures involved in checking data or programs for correctness, compliance with standards, and conformance with the requirement specifications.

phenetic classification *Systematics*. A classification based on general or overall similarity.

phenogram *Systematics*. A branching diagram depicting phenetic relationships among taxa.

reasoning under uncertainty *Artificial intelligence*. Reasoning about situations, e.g., in medical diagnosis, in which good or complete data are not available, and for which decisions must be made based on available data and knowledge of likelihoods of the various possibilities.

sonar *Engineering*. Any system that uses underwater sound waves to determine the location of objects or for communication. *Acoustics*. Specifically, a system that uses transmitted acoustic signals and echo returns, as well as acoustic signals originating from other sources, for navigating and determining position and bearing. (An acronym for *so*und *n*avigation *and r*anging.)

taxon, *plural* **taxa**. *Systematics*. Any group of organisms that has been scientifically designated as belonging to a specific taxon, and has been given a position within the taxonomic hierarchy.

taxonomy *Systematics*. The theories and techniques of describing, naming and classifying organisms.

References

Anderberg, M.R., 1973, *Cluster analysis for applications*. Academic Press, New York.

Andrews, D.F., 1972, 'Plots of high dimensional data', *Biometrics*, vol. 28, pp. 125–136.

Backer, E., 1978, *Cluster analysis by optimal decomposition of induced fuzzy sets*. Delft University Press, Delft.

Backer, E., 1979, 'Cluster analysis formalized as a process of fuzzy identification based on fuzzy relations', *Revue HF*, vol. 11, pp. 1–68,.

Backer, E., 1988a, 'Developments towards a knowledge base for cluster analysis', in H.H. Bock, Ed., *Classification and related methods of data analysis*, North Holland, Amsterdam, pp. 661–672.

Backer, E., 1988b, 'CLUSAN: A knowledge base for approximate reasoning in exploratory data analysis', in E.S. Gelsema and L. Kanal, Eds., *Pattern recognition and artificial intelligence*, Elsevier Science Publishers, pp. 395–411.

Backer, E., and Jain, A.K., 1981, 'A clustering performance measure based on fuzzy set decomposition', *IEEE Transactions on Pattern Analysis and Machine Intelligence*, vol. 3, pp. 66–75.

Backer, E., and Eijlers, E.J., 1986, 'CLUSAN: A knowledge base for cluster analysis', *Proc. Seventh Symposium Information Theory in the Benelux*, pp. 113–120.

Backer, E., Haas, H.P.A. and Getreuer, R., 1983, 'On the stability of shared near neighbor clustering', *IEEE Transactions on Pattern Analysis and Machine Intelligence*, vol. 5, pp. 220–224.

Backer, E., Gerbrands, J.J., Reiber, J.H.C., Reijs, A.E.M.W., Krijgsman, W. and van den Herik, H.J., 1988, 'Modelling uncertainty in ESATS by classification inference', *Pattern Recognition Letters*, vol. 8, pp. 103–112.

Balasubramaniam, A., Parthasarathy, G. and Chatterji, B.N., 1990, 'Knowledge based approach to cluster algorithm selection', *Pattern Recognition Letters*, vol. 11, pp. 651–661.

Ball, G.H. and Hall, D.J., 1964, *ISODATA, a noval method of data analysis and pattern classification*. Stanford Research Institute, AD–699616.

Barnes, G.R., 1976, 'Fuzzy sets and cluster analysis', *Proc. Third Int. Joint Conf. on Pattern Recognition*, San Diego, pp. 371–375.

Barr, A. and Feigenbaum, E.A., Eds., 1981, *The handbook of artificial intelligence, vol. 1*. William Kauffman, Inc., San Mateo, CA.

Barr, A. and Feigenbaum, E.A., Eds., 1982, *The handbook of artificial intelligence, vol. 2*. Pitman, London.

Bayne, C.K., Beauchamp, J.J., Begovich, C.L. and Kane, V.E., 1980, 'Monte Carlo comparisons of selected clustering procedures', *Pattern Recognition*, vol. 12, pp. 51–62.

Bellman, R.E., Kalabe, R. and Zadeh, L.A., 1966, 'Abstraction and pattern classification', *Journal Math. Anal. Appl.*, vol. 13, pp. 1–7.

Bezdek, J.C., 1974, 'Cluster validity with fuzzy sets', *Journal of Cybernetics*, vol. 3, pp. 58–73.

Bezdek, J.C., 1981, *Pattern recognition with fuzzy objective function algorithms*. Plenum Press, New York.

Bezdek, J.C. and Harris, J.D., 1978, 'Fuzzy relations and partitions: an axiomatic basis for clustering', *Journal for Fuzzy Sets and Systems*, vol. 1, pp. 111–127.

Bezdek, J.C. and Pal, S.K., 1992 *Fuzzy models for pattern recognition*. IEEE Press, New York, NY.

Bochsler, D.C., 1988, 'A project management approach to expert system applications', *ISA 88*, pp. 1458–1466.

Bonissone, P.P., 1987, 'Summarizing and propagating uncertain information with triangular norms', *Int. Journal of Approximate Reasoning*, vol. 1, pp. 71–101.

Brailovsky, V.L., 1991, 'A probabilistic approach to clustering', *Pattern Recognition Letters*, vol. 12, pp. 193–198.

Buchanan, B.G. and Shortliffe, E.H., 1984, *Rule-based expert systems*. Addison-Wesley, Reading, MA.

Chang, C.L. and Lee, R.C.T., 1973, 'A heuristic relaxation method for non-linear mapping in cluster analysis', *IEEE Transactions on Systems, Man, and Cybernetics*, vol. 3, pp. 197–200.

Cormack, R.M., 1971, 'A review of classification', *Journal of the Royal Statistical Society*, Series A, vol. 134, pp. 321–367.

Dave, R.N., 1990, 'Fuzzy shell-clustering and applications to circle detection in digital images', *Int. Journal of General Systems*, vol. 16, pp. 343–355.

Davies, D.L. and Bouldin, D.W., 1979, 'A cluster separation measure', *IEEE Transactions on Pattern Analysis and Machine Intelligence*, vol. 1, pp. 224–227.

Day, W.H.E., 1990, 'Graphs as structural models: The application of graphs and multigraphs in cluster analysis (book review)', *Journal of Classification*, 7, p. 141.

Debenham, J.K., 1989, *Knowledge systems design*. Prentice Hall, Sydney.

Dempster, A.P., 1967 'Upper and lower probabilities induced by multivalued mappings', *Annals of Mathematical Statistics*, vol. 38, pp. 325–329.

Devijver, P.A. and Kittler, J., 1982, *Pattern recognition: a statistical approach*. Prentice Hall, Englewood Cliffs, NJ.

Diday, E., 1973, 'The dynamic cluster method in non-hierarchical clustering', *Int. Journal of Computer and Information Sciences*, vol. 2, pp. 61–88.

Diday, E. and Simon, J.C., 1976 'Clustering analysis', in K.S. Fu, Ed., *Digital pattern recognition*, Springer-Verlag, New York, pp. 47–94.

Dixon, J.K., 1979, 'Pattern recognition with partly missing data', *IEEE Transactions on Systems, Man, and Cybernetics*, vol. 9, pp. 617–621.

Dubes, R.C. 1987, 'How many clusters are best? - An experiment', *Pattern Recognition*, vol. 20, pp. 645–663.

Dubes, R.C. and Jain, A.K., 1976, 'Clustering techniques: the user's dilemma', *Pattern Recognition*, vol. 8, pp. 247–260.

Dubes, R.C. and Jain, A.K., 1979, 'Validity studies in clustering methodologies', *Pattern Recognition*, vol. 11, pp. 235–254.

Duda, R.O. and Hart, P.E., 1973, *Pattern classification and scene analysis*. John Wiley, New York.

Duda, R.O., Gaschnig, H. and Hart, P.E., 1979, 'Model design in the PROSPECTOR Consultant System for Mineral Exploration', in D. Michie, Ed., *Expert Systems in the Micro-electronic Age*, Edinburgh University Press, Edinburgh, pp. 153–167.

Dudok van Heel, W.H., 1962, 'Sound and Cetacea', *Neth. Journal for Sea Research*, vol. 1, pp. 407–508.

Dudok van Heel, W.H., 1966, 'Navigation in Cetacea', in K.S. Norris, Ed., *Whales, Dolphins and Porpoises*, University of California Press, Berkeley and Los Angeles, 27, pp. 597–602.

Dudok van Heel, W.H., 1981, 'Investigations on Cetacean Sonar III: a proposal for an ecological classification of odontocetes in relation with sonar', *Aquatic Mammals*, vol. 8, pp. 65–68.

Duin, R.P.W. and Backer, E., 1988, 'Discriminant analysis in a non-probabilistic context based on fuzzy labels', in E.S. Gelsema and L. Kanal, Eds., *Pattern recognition in practice and artificial intelligence*, North Holland, Amsterdam, pp. 229–235.

Dunn, J.C., 1973, 'A fuzzy relative of the ISODATA process and its use in detecting compact, well-separated clusters', *Journal of Cybernetics*, vol. 3, pp. 32–57.

Dunn, J.C., 1974, 'Well-separated clusters and optimal fuzzy partitions', *Journal of Cybernetics*, vol. 4., pp. 95–104.

Everitt, B.S., 1974, *Cluster analysis*. Heinemann, London.

Everitt, B.S., 1979, 'Unresolved problems in cluster analysis', *Biometrics*, vol. 35, pp. 169–181.

Fisher, D. and Langley, P., 1986, 'Conceptual clustering and its relation to numerical taxonomy', in W.A. Gale, Ed., *Artificial intelligence and statistics*, Addison-Wesley, Reading, MA, pp. 77–116.

Forgy, E., 1965, 'Cluster analysis of multivariate data: efficiency versus interpretability of classifications', *Biometrics*, vol. 21, 768 (abstract).

Fu, K.S., 1982, *Syntactic pattern recognition and applications*. Prentice Hall, Englewood Cliffs, NJ.

Fu, K.S. and Cheng, Y., 1985, 'Conceptual clustering in knowledge organization', *IEEE Transactions on Pattern Analysis and Machine Intelligence*, vol. 7, pp. 592–598.

Fukunaga K. and Koontz, W.L.G., 1970, 'A criterion and an algorithm for grouping data', *IEEE Transactions on Computers*, vol. 19, pp. 917–923.

Gale, W.A., 1986, *Artificial Intelligence and Statistics*. Addisson-Wesley, Reading, MA.

Garvey, T.D., Lowrance, J.D. and Fischler, M.A., 1981, 'An inference technique for integrating knowledge from disparate sources', *Proc. Seventh Int. Joint Conference on Artificial Intelligence*, pp. 319–325.

Gerbrands, J.J., Backer, E. and Cheng, X.S., 1986, 'Multiresolutional cluster segmentation using spatial context', in P.A. Devijver and J. Kittler, Eds., *Pattern recognition theory and applications*, Springer-Verlag, Berlin and Heidelberg, pp. 133–140.

Giarratano, J. and Riley, G., 1989, *Expert systems: principles and programming*. PWS-Kent, Boston, MA.

Gitman, I. and Levine, M., 1970, 'An algorithm for detecting unimodal fuzzy sets and its application as a clustering technique', *IEEE Transactions on Computers*, vol. 19, pp. 583–593.

A.D. Gordon, A.D. and Henderson, J.T., 1977, 'Algorithm for Euclidean sum of squares classification', *Biometrics*, vol. 33, pp. 355–362.

Gowda, K. and Krishna, G., 1978, 'Disaggregative clustering using the concept of mutual nearest neighborhood', *IEEE Transactions on Systems, Man and Cybernetics*, vol. 8, pp. 888–894.

Hand, D.J., 1985, 'Statistical expert systems: mecessary attributes', *Journal of Applied Statistics*, vol. 12.

Hartigan, J.A., 1975, *Clustering algorithms*. John Wiley, New York.

Hayes-Roth, F., 1985, 'Rule-based systems', *Communications of the ACM*, vol. 29, pp. 921–932.

Ho, T.B., Diday, E. and Gettler-Summa, M., 1988, 'Generating rules for expert systems from observations', *Pattern Recognition Letters*, vol. 7, pp. 265–271.

Horikawa, S.I., Furuhashi, T. and Uchikawa, Y., 1992, 'On fuzzy modelling using fuzzy neural networks with the back-propagation algorithm', *IEEE Transactions on Neural Networks*, vol. 3, pp. 801–806.

Jain, A.K., Smith, S.P. and Backer, E., 1980, 'Segmentation of muscle-cell pictures', *IEEE Transactions on Pattern Analysis and Machine Intelligence*, vol. 2, pp. 232–242.

Jain, A.K., 1986, 'Cluster analysis', in T.Y. Young and K.S. Fu, Eds., *Handbook of pattern recognition and image processing*, Academic Press, New York, pp. 35–57.

Jain, A.K. and Dubes, R.C., 1988, *Algorithms for clustering data*. Prentice Hall, Englewood Cliffs, NJ.

Jardine, N. and Sibson, R., 1971, *Matthematical Taxonomy*. John Wiley, London.

Jarvis, R.A. and Patrick, E.A., 1973, 'Clustering using a similarity measure based on shared near neighbors', *IEEE Transactions on Computers*, vol. 22, pp. 1025–1034.

Johnson, R.A. and Wichern, D.W., 1982, *Applied multivariate statistical analysis*. Prentice Hall, Englewood Cliffs, NJ.

Johnson, L.A.S., 1968, 'Rainbow's end: the quest for an optimal taxonomy'. *Proc. Linn. Soc. N.S.W.*, 93, pp. 8–45.

Johnson, S.C., 1967, 'Hierarchical clustering schemes', *Psychometrika*, vol. 32, pp. 241–254.

Kamminga, C., 1979, 'Remarks on dominant frequencies of cetacean sonar', *Aquatic Mammals*, vol. 7, pp. 93–100.

Kamminga, C., 1994, *Research on dolphin sounds*. Delft University of Technology, PhD thesis.

Kamminga, C. and Beitsma, G.R., 1990, 'Investigations on cetacean sonar IX: remarks on dominant sonar frequencies from *Tursiops truncatus*', *Aquatic Mammals*, vol. 16, pp. 14–20.

Kamminga, C. and J.G. van Velden, 1987, 'Investigations on cetacean sonar VII: sonar signals of *Pseudorca crassidens* in comparison with *Tursiops truncatus*', *Aquatic Mammals*, vol. 13, pp. 43–49.

Kamminga, C. and Wiersma, H., 1981, 'Investigations on cetacean sonar II: acoustical similarities and differences in odontocete sonar signals', *Aguatic Mammals*, vol. 8, pp. 41–62.

Kandel, A., Ed., 1991, *Fuzzy expert systems*. CRC Press, Boca Raton, FL.

Kandel, A. and Langholz, G., Eds., 1992, *Hybrid architectures for intelligent systems*. CRC Press, Boca Raton, FL.

Kaufman, L. and Rousseeuw, P.J., 1990, *Finding groups in data, an introduction to cluster analysis*. John Wiley, New York.

Kawamura, A., Watanabe, N., Okada, H. and Asakawa, K., 1992, 'A prototype of neuro-fuzzy cooperation system', *IEEE Int. Conf. on Fuzzy Systems*, pp. 1275–1282.

Keller, J.M. and H. Tahani, H., 1992a, 'Backprop neural networks for fuzzy logic', *Information Sciences*, vol. 62, pp. 205–221.

Keller, J.M. and H. Tahani, H., 1992b, 'Implementation of conjunctive and disjunctive fuzzy logic rules with neural networks', *Int. Journal for Approximate Reasoning*, vol. 6, pp. 221–240.

King, B., 1967, 'Step-wise clustering procedures', *Journal of the American Statistical Association*, vol. 69, pp. 86–101.

Klein, R.W. and Dubes, R.C., 1989, 'Experiments in projection and clustering by simulated annealing', *Pattern Recognition*, vol. 22, pp. 75–90.

Kosko, B., 1987, 'Adaptive inference in fuzzy knowledge networks', *Proc. IEEE First Int. Conf. on Neural Networks*, San Diego, CA, pp. 261–268.

Krisnapuram, R., Nasraoui, O. and Frigui, H., 1992, *The Fuccy C Spherical Shells Algorithm: A New Approach*, IEEE Trans. on Neural Networks, 3(5), pp. 663-671.

Kruskal, J.B., 1964, 'Multidimensional scaling by optimizing goodness of fit to a nonmetric hypothesis', *Psychometrika*, vol. 29, pp. 1–27.

Lance, G.N. and Williams, W.T., 1967, *A general theory of classificatory sorting strategies: II. Clustering systems*, Computer Journal 10, pp. 271-277.

Lee, R.C.T., Slagle, J.R. and Blum, H., 1977, 'A triangulation method for the sequential mapping points from N-space to two-space', *IEEE Transactions on Systems, Man, and Cybernetics*, vol. 3, pp. 288–292.

Lincklaen Westenberg, H.W., de Jong, S., van Meel, D.A., Quadt, J.F.A., Backer, E. and Duin, R.P.W., 1989, 'Fuzzy set theory applied to product classification by a sensory panel', *Journal of Sensory Studies*, vol. 4, pp. 55–72.

Mamdani, E.H. and Assilian, S., 1975, 'An experiment in linguistic synthesis with a fuzzy logic controller', *Int. Journal of Man-Machine Studies*, vol. 7, pp. 1–13.

Matthews, G. and Hearne, J., 1991, 'Clustering without a metric', *IEEE Transactions on Pattern Analysis and Machine Intelligence*, vol. 13, pp. 175–184.

McQueen, J.B., 1967, 'Some methods of classification and analysis of multivariate observations', *Proc. Fifth Berkely Symposium on Mathematical Statistics and Probability*, pp. 281–297.

Michalski, R.S., 1980, 'Knowledge acquisition through conceptual clustering: theoretical framework and an algorithm for partitioning data into conjunctive concepts', *Int. Journal of Policy Analysis and Information Systems*, vol. 4.

Michalski, R.S. and Stepp, R.E., 1983a, 'Learning from observations: Conceptual classification', in Michalski, R.S., Carbonell, T.M. and Mitchell, T.M., Eds., *Machine learning: an artificial intelligence approach*. R.S. Tioga, Palo Alto, CA.

Michalski, R.S. and Stepp, R.E., 1983b, 'Automated construction of classifications: conceptual clustering versus numerical taxonomy', *IEEE Transactions on Pattern Analysis and Machine Intelligence*, vol. 5, pp. 396–410.

Michalski, R.S., Mozetic, I., Hong, J. and Lavrae, N., 1986, 'The multipurpose incremental learning system AQ15 and its testing application to three medical domains',

Proc. AAAI86, pp. 1041–1045.

Milligan, G.W., 1981, 'A Monte Carlo study of thirty internal criterion measures for cluster analysis', *Psychometrika*, vol. 46, pp. 187–199.

Milligan, G.W. and Cooper, M.C., 1985, 'An examination of procedures for determining the number of clusters in a data set', *Psychometrika*, vol. 50, pp. 159–179.

Minsky, M., 1975, 'A framework for representing knowledge', in P. Winston, Ed., *The Psychology of computer vision*, McGraw-Hill, New York, NY, pp. 211–217.

Mizumoto, M. and Zimmermann, H.-J., 1982, 'Comparison of fuzzy reasoning methods', *Fuzzy Sets and Systems*, vol. 8, pp. 253–283.

Murtagh, F., 1985, *Multidimensional clustering algorithms*. Compstat lectures 4, Physica Verlag, Vienna.

Murthy and Krishna, 1981, *Contributions to the development of computational efficient clustering techniques*, PhD thesis, School of Autonation, Indian Institute of Science, Bangalore

Negoita, C.V., 1973, *On the application of the fuzzy sets separation theorem for automatic classification in information retrieval systems*, Inform. Sci., 5, pp. 279-286.

Negoita, C.V., 1984, *Expert systems and fuzzy systems*. Benjamin/Cummings, Menlo Park, CA.

Niemann, H., 1979, 'A fast converging algorithm for non-linear mapping of high-dimensional data to a plane', *IEEE Transactions on Computers*, vol. 28, pp. 142–147.

Panayirci, E. and Dubes, R.C., 1983, 'A test for multidimensional clustering tendency', *Pattern Recognition*, vol. 16, pp. 433–444.

Pao, H., 1989, *Adaptive pattern recognition and neural networks*. Addison-Wesley, Reading, MA.

Pitas, I., Mikos, E. and Venetsanopoulos, A.N., 1992, 'A minimum entropy approach to rule learning from examples', *IEEE Transactions on Systems, Man, and Cybernetics*, vol. 22, pp. 621–635.

Prim, R.M., 1957, 'Shortest connection networks and some generalizations', *Bell System Technical Journal*, vol. 36, pp. 1389–1401.

Quinlan, J.R., 1983, 'Learning efficient classification procedures and their application to chess and games', in R.S. Michalski, J.G. Carbonell, and T.M. Mitchell, Eds., *Machine learning: an artificial intelligence approach*. R.S. Tioga, Palo Alto, CA.

Quinlan, J.R., 1990, 'Decision trees and decision-making', *IEEE Transactions on Systems, Man, and Cybernetics*, vol. 20, pp. 339–346.

Rand, W.M., 1971, 'Objective criteria for the evaluation of clustering methods', *Journal of the American Statistical Association*, vol. 66, pp. 846–850.

Reggia, J., 1985, 'Abductive inference', *Proc. of Expert Systems in Government Symposium*, IEEE Press, New York, NY, pp. 484–489.

Ripley, B.D., 1979, 'Tests of randomness for spatial point patterns', *Journal of the Royal Statistical Society*, Series B, vol. 41, pp. 368–374.

Romaniuk, S.G. and Hall, L.O., 1993, 'SC-net: A hybrid connectionist, symbolic system', *Information Sciences*.

Romesburg, H.C., 1984, *Cluster Analysis for Researchers*. Lifetime Learning Publications, Belmont, CA.

Rosenfeld, A., 1975, *Fuzzy graphs*. In: Zadeh, L.A. et al., Fuzzy sets and their applications to cognitive and decision processes, Ac. Press, pp. 77–95.

Roubens, M., 1978, 'Pattern classification problems and fuzzy sets', *Fuzzy Sets and Systems*, vol. 1, pp. 239–253.

Roubens, M., 1982, 'Fuzzy clustering algorithms and their cluster validity', *Eur. Journal on Operational Research*, vol. 10, pp. 294–301.

Rumelhart, D.E., Hinton, G.E. and Williams, R.J., 1986, 'Learning representations by error propagation' in *Parallel Distributed Processing: Explorations in the microstructure of cognition*, Rumelhart, D.E. and McClelland, Eds., MIT Press, Cambridge, MA.

H. Ruspini, H., 1969, 'A new approach to clustering', *Information and Control*, vol. 15, pp. 22–32.

Sammon, J.W., 1969, 'A non-linear mapping for data structure analysis', *IEEE Transactions on Computers*, vol. 18, pp. 401–409.

Shafer, G., 1976, *A mathematical theory of evidence*. Princeton University Press, Princeton, NJ.

Shapiro, L.G. and Haralick, R.M., 1969, 'Decomposition of two-dimensional shapes by graph-theoretic clustering', *IEEE Transactions on Pattern Analysis and Machine Intelligence*, vol. 1, pp. 10–20.

Shekar, B., Murthy, N.N. and Krishna, G., 1987, 'A knowledge based clustering schema', *Pattern Recognition Letters*, vol. 5, pp. 253–259.

Sheppard, R.N. and Arabie, P., 1979, 'Additive clustering: representation of similarities as combinations of discrete overlapping properties', *Psychological Review*, vol. 86, pp. 87–123.

Smith, S.P. and Jain, A.K., 1984, 'Testing for uniformity in multidimensional data', *IEEE Transactions on Pattern Analysis and Machine Intelligence*, vol. 6, pp. 73–81.

Sneath, P.H.A. and Sokal, R.R., 1973, *Numerical taxonomy*. Freeman and Co., San Fransisco.

Srivastava, A. and Narasimhamurthy, M., 1990, *Knowledge-based clustering*, Techn. report, IISc-CSA-90-12, Dept. of Computer Science and Automation, Indian Institute of Science, Bangalore.

Srivastava, A. and Murthy, M.N., 1990, 'A comparison between conceptual clustering and conventional clustering', *Pattern Recognition*, vol. 23, pp. 975–981.

Staugaard, A.C., 1987, *Robotics in AI: an introduction to applied machine intelligence*. Prentice Hall, Englewood Cliffs, NJ.

Stillings, N.A., 1987, *Cognitive science: An introduction*. MIT Press, Cambridge, MA.

Szolovits, P. and Pauker, S.G., 1978, 'Categorical and probabilistic reasoning in medical diagnosis', *Artificial Intelligence*, vol. 11, pp. 115–144.

Tagaki, H. and Hayashi, I., 1992, 'NN-Driven Fuzzy Reasoning', *International J.ournal of Approximate Reasoning*, vol. 5, pp. 191–212.

Tamura, S., Higuchy, S. and Tanaka, K., 1971, 'Pattern classification based on fuzzy relations', *IEEE Transactions on Systems, Man, and Cybernetics*, vol. 1, pp. 61–66.

Tu, H.-K., Goldgof, D.B. and Backer, E., 1994, *Utilizing fuzzy c-shells for automatic approxiamte lv location for initialization of myocardial structure and function analysis alghorithm*, Medical Imaging 1994: Physiology and Function from Multidimensional Images.

Tukey, J.W., 1954, 'Unsolved problems in experimental statistics', *Journal of the American Statistical Society*, 49, pp. 706–731.

Tukey, J.W., 1977, *Exploratory data analysis*. Addison-Wesley, Reading, MA.

Van der Lubbe, J.C.A. and Backer, E., 1993, 'Human-like reasoning under uncertainty in expert systems', in R.J. Jorna, B. van Heusden, and R. Posner, Eds., *Signs, Search and Communication*, Walter de Gruyter, Berlin, pp. 113–133.

Wang, L.X. and Mendel, J.M., 1992, 'Generating fuzzy rules by learning from examples', *IEEE Transactions on Systems, Man, and Cybernetics*, vol. 22, pp. 1414–1427.

Watanabe, S., 1969, *Knowing and guessing*. John Wiley, New York.

Watanabe, S., 1985, *Pattern recognition*. Wiley Interscience, New York.

Weiss, S.M. and Kulikowski, C.A., 1991, *Computer Systems that learn*. Kaufmann, San Mateo, CA.

Wilson R., and Spann, M., 1990, 'A new approach to clustering', *Pattern Recognition*, vol. 23, pp. 1413–1425.

Wishart, D., 1986, *CLUSTAN User Manual*. Program Library Unit, Edinburgh University.

Yager, R.R., 1993, 'On a hierarchical structure for fuzzy modelling and control', *IEEE Transactions on Systems, Man, and Cybernetics*, vol. 23, pp. 1189–1197.

Yeh, R.T. and Bang, S.Y., 1975, 'Fuzzy relations, fuzzy graphs and their applications to clustering analysis', in Zadeh, L.A., Fu, K.S., Tanaka, K. and Shimura, M., Eds., *Fuzzy sets and their applications to cognitive and decision processes*, Academic Press, New York, pp. 125–149.

Zadeh, L.A., 1965, 'Fuzzy sets', *Information and Control*, vol. 8, pp. 338–353.

Zadeh, L.A., 1968, 'Probability measures of fuzzy events', *Journal Math. Anal. Appl.*, vol. 23, pp. 421–427.

Zadeh, L.A., 1973, 'The concept of a linguistic variable and its application to approximate reasoning', *Memorandum ERL-M 411*, University of California, Berkeley.

Zadeh, L.A., Fu, K.S., Tanaka, K. and Shimura, M., 1975, *Fuzzy sets and their applications to cognitive and decision processes*. Academic Press.

Zadeh, L.A., 1983, 'The role of fuzzy logic in the management of uncertainty in expert systems', *Fuzzy Sets and Systems*, vol. 11, pp. 199–227.

Zahn, C.T., 1964, *Approximating symmetric relation by equivalence relations*, SIAM Journ. of Applied Math., 12, pp. 840-847.

Zahn, C.T., 1971, 'Graph-theoretical methods for detecting and describing Gestalt clusters', *IEEE Transactions on Computers*, vol. 20, pp. 68–86.

Zeng, G. and Dubes, R.C., 1985, 'A comparison of tests for randomness', *Pattern Recognition*, vol. 18, pp. 191–198.

Appendix

Appendix

A | Incremental hierarchical knowledge organization: a proposal and an example

In Section 9.6, we discussed one of the most subtle problems in classification, identified as empirical learning (machine learning, concept learning) with the fundamental goal of inducing production rules from examples of data, which will be applicable to new samples of data.

This appendix reports a proposal for an incremental hierarchical knowledge organization. This includes inductive learning and incremental learning, as well as an inexact reasoning model while the fuzzy rules are still subject to modification (exception-tolerant reasoning).

The example of Section 9.4.2 has been used here to exemplify the proposed hierarchical knowledge organization and to show rule modification and inexact reasoning.

A1 Introduction

In fuzzy expert systems, knowledge is usually described as relations among fuzzy sets of objects (observations, actions, decisions, or other concepts), and organized as a list of relations (rule base), a network of relations (semantic network), or as distinct groupings of relations (frames).[1]

The organization of this knowledge is generally found to be a serious bottleneck because it usually involves in-depth analysis of interviews with an expert and the final formulation of relations encountered. In the absence of an explicit criterion for evaluating the relevant relations, their validity is very often not guaranteed. Instead of asking the expert to formalize his knowledge in terms of decision rules, a better approach – whenever possible – might be to ask the expert to provide examples of the decision-making process. In this case, the organization of knowledge is based on *inductive learning from examples*.[2]

Employing an inductive learning method for generating 'general' and 'representative' relations (rules) has the inherent advantage that the process of inducing rules is controlled by an explicit optimization criterion. The downside of such an advantage is

---

[1] See Section 8.3.
[2] See Section 9., (Ho et al., 1988). The learning scheme that we are going to employ originates from Ho et al. though it is modified in order to accomodate rule trees.

that generally a (very) large set of observations has to be supplied by the expert. Moreover, the learning set of observations will generally consist of a mixture of 'good' examples, 'rare' examples, and even 'exceptions'. Rare examples and exceptions are not due to inconsistency of the expert *per se*, but may originate from 'hidden causality'.

In recognition of the above, in this proposal we address the four major issues to overcome the difficulties mentioned:

– inductive learning from observations;
– incremental rule modification;
– hierarchical knowledge organization; and
– exception-tolerant (fuzzy) reasoning.

These issues have attracted the attention of many researchers in the past two decades and many interesting approaches and results can be found in the literature, Michalski (1980), Michalski and Stepp (1983a and 1983b), Zadeh (1983), Quinlan (1983 and 1990), Buchanan and Shortliffe (1984), Kosko (1987), Horikawa et al (1992), Yager (1993) and Van der Lubbe and Backer (1993). The main contribution of this appendix is to integrate and to mix some of the ideas that are brought forward, with the aim of proposing a conceptual framework in which incremental learning and rule modification are phrased in terms of (exception-tolerant) fuzzy rule-based reasoning. The performance of the resulting fuzzy expert system is made explicit through the choice of 'maximum coverage' as the optimality criterion.

Throughout this proposal, several choices are made, like the optimality criterion, the matching relation, the rule distance, the update thresholds, and so on. It should be understood that the domain of application may influence these choices. But even if some of the choices that have been made need justification from the application domain, the conceptual framework will not lose its generality.

The outline of this appendix is as follows. In Section A2, the problem is stated and the terminology is explained. Learning and incremental learning are discussed in Section A3. Then, in Section A4, we introduce the concept of rule trees as an appropriate representation of hierarchical knowledge organization. Based on the rule tree, we develop an inference mechanism and a model for inexact reasoning. In Section A5, we compare and contrast our formalism with the Dempster–Shafer scheme.

The appendix is concluded with an example (Section A6). The observations of Table 9.2 (Section 9.4.2) are used to derive a hierarchical knowledge organization and to exemplify rule modification and inexact reasoning.

A2 Statement of the problem

Let us assume that we have a set of training samples, W, from the entire instance space, I, supplied by the expert. The sample, W, is composed of 'good' examples and – to some extent – 'obscure' examples. The latter subset is refered to as the set of 'exceptions', E. We are not making any assumption as to how well our set of training samples, W, represents the instance space; in other words, W is simply considered to be just a collection of training samples. A member of the training set, W, is $\mathbf{w}$, and is represented as a n-dimensional attribute vector.

The attributes $\{a_1, a_2, ..., a_n\}$ may be quantitative or qualitative. For simplicity only, and without loss of generality, we assume that all qualitative attributes do posess the same domain of possible modalities $\{v_1, v_2, ..., v_p\}$.

An instantiation of attribute $\{a_i\}$ is refered to as $\{v_i^*\}$; in some cases, for reasons of clarity, we may further specify the instantiation as

$$\{..., a_i(\mathbf{w}), ..., a_j(\mathbf{w}), ...\} = \{..., v_i^* = v_{it}, ..., v_j^* = v_{js}, ...\}$$

where $(s,t) = \{1,2,...,p\}$, and $\mathbf{w}$ is the instance under consideration.

The domain of decisions (the concept space), D, is also finite, and has m elements, $\{d_1, d_2, ..., d_m\}$.

The problem is to find 'general' rules, $\{R\}$, in such a way that the rules do predict the appropriate decisions $\{d\}$ well if a set of attribute values $\{..., a_i = v_i^*, ..., a_j = v_j^*, ...\}$ is observed. How well the rule set, $\{R\}$, performs to predict the appropriate concepts will be refered to as *the degree of knowledgeability* and will certainly depend on the number of training samples presented (see Section A5).

Generally, the set of rules, $\{R\}$, is found by some sort of a *concept-driven* classification of the 'facts', as shown in Figure A1.[3] The facts are the observed attribute values, whereas the concepts (class labels, logical constraints) are supplied by the expert. Note that if the classification is in the form of a decision tree, the rule set, $\{R\}$, is almost made explicit. In all other cases, a rule induction mechanism is needed to obtain the rule set $\{R\}$. In what follows (more specifically, Section A4), the rule set, $\{R\}$, is organized – by definition – as a *rule tree*. The rule tree – even if incomplete – reflects *the knowledgeability* of the system and is the backbone of the representation of uncertainty and reasoning under uncertainty.

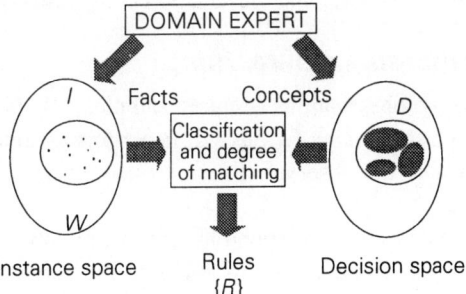

Figure A1: General scheme for rule induction.

We further assume that all modalities are represented by fuzzy sets. If so, the general format of a rule reads as follows:

$$\{D = d', f_{d'}(\mathbf{w}')\} \text{ IF } \{..., a_i = v_i^{*}, ..., a_j = v_j^{*}, ...\}$$

with $\{f_{d1}(\mathbf{w}'), f_{d2}(\mathbf{w}'), ..., f_{dm}(\mathbf{w}')\}$

represents the membership assignments to the domain of D.

If a_i is a qualitative attribute, its value is one of the domain modalities, say $v_i^{*} = v_{ik}$, whereas if a_j is a quantitative attribute, the membership assignments to all possible modalities are represented by

$$\{f_{v1}(v_{j'}), f_{v2}(v_{j'}), ..., f_{vp}(v_{j'})\}.$$

Then, when new instances become available, the rule set $\{R\}$ as well as the set of all membership functions $\{f\}$ are subject to modification. This process is called an 'incremental' rule learning from examples.

Not only do we need a rule induction mechanism, but we also must have a rule and membership modification algorithm that will be robust in the presence of 'exceptional' cases, $\{E\}$. The *maximum coverage principle*, which is discussed in the next section, turns out to be such a robust criterion which can be proven to yield the maximum performance achievable with the training set or with the augmented training set.

In summary, the questions that have to be answered are the following. Let $\mathbf{w}'$ be a new instance.

- Is this sample, $\mathbf{w}'$, covered by any of the 'general' rules in the rule set, $\{R\}$?
- If we have established a 'tentative' set of rules, $\{R\}$, does this sample, $\mathbf{w}'$, cause any modification of some of the existing rules, and to what extent?
- If this example, $\mathbf{w}'$, is found to be an 'exceptional' case, E', how would this affect future reasoning?

In what follows, we aim to provide some useful answers to these concerns.

A3 Learning and incremental learning

As stated before, the set of observations (training samples), W, is related to a set of m modalities of domain decisions, $D = \{d_1, d_2, ..., d_m\}$.[4] We assume – from the expert's point of view – that each of the observations is pertinent to one and only one of those domain modalities (concepts).

However, we allow the system to respond in terms of membership values to each of the possible modalities. This enables us to evaluate the 'crispness' of the ultimate

---

[4] As an example, we are going to use a domain decision on the presence of cvlustering tendency (in exploratory data analysis) of which it is known that the granularity is very well accomodated by the modalities {*very high, high, medium, low, very low*}. Note that $\{d_k\}$ may be a decision subspace like {*low* OR *very low*}.

decision-making. It seems logical to use a decision-oriented amount of fuzziness, I_d, for that purpose (see Backer, 1978).

From the viewpoint of an expert, however, the set of concepts $\{d_1, d_2, ..., d_m\}$ is then associated to a partition, P, of the training set, W, in m classes $\{P_1, P_2, ..., P_m\}$.[5]

Learning

Generally, conceptual clustering aims to construct these classes of observations in such a way that the relationships between the training samples in the classes are made explicit in the form of rules that relate the observations and the concepts involved, in a descriptive way. These methods are inherently of a supervised nature and are referred to as 'descriptive generalization' or *concept-driven learning*.

Clearly, as a result, a rule is said to be 'representative' of a concept or domain decision, d_m, if a sample, w, is recognized by that rule to be a member of d_m. Thus,

$$\{D(\mathbf{w}) = d_m\} \text{ IF } \{..., a_i(\mathbf{w}) = v_i^*, ..., a_j(\mathbf{w}) = v_j^*, ...\}.$$

For each domain decision, d_k, we may find a set of *representative rules*, $(k = 1, 2, ..., m)$. The set of all representative rules is denoted by $\{R_k\}$.

Assume that for the kth concept (i.e. the domain decision d_k) we have obtained a set of representative rules, $\{R_k\}$. Then, we define the 'coverage' of P_k as the portion of P_k that is recognized by rule R_k. Consequently, the rule R_{ki} is said to be 'more general' than the rule R_{kj} if the coverage of P_k due to rule R_{ki} is larger than the coverage of P_k when rule R_{kj} is applied. This is precisely the approach of Ho et al., who propose an algorithm that aims to find representative rules such that each of them covers the partition, P, best. Their (two-step) algorithm reads as follows:[6]

1. Determine the representative rule $R(\mathbf{w})$ from a training sample, $\mathbf{w}$.
Step 1. Select the attributes to make the rule discriminative.
Step 2. Choose the discriminant rule which covers P_k best.
2. Form the rule set $\{R\}$.
Step 1. Use 1. for all $\mathbf{w}$ in W.
Step 2. Find the maximum representative rule for each $\mathbf{w}$.

---

5 For the time being, we consider only one partition, P. At a later stage, we will consider a nested tree of partitions. The partition, P, may be $\{d_1, d_2\} = \{\{very\ low,low\}, \{medium,\ high,\ very\ high\}\}$, being very unspecific, or $\{d_1, d_2, d_3, d_4, d_5\} = \{\{very\ low\}, \{low\}, \{medium\}, \{high\}, \{very\ high\}\}$ as the most specific one.

6 The essential step is the following.
Let $P_k = \{w_1, w_2, ..., w_q\}$; $\{R_k\} = \varnothing$.
Having R_i generated by w_i, find w' in P_k for which
card$(P_k(R')) = $ max card$(P_k(R_i))$;
Add R' to $\{R_k\}$;
Repeat this step for $P_k = P_k(R')$ until $P_k = \varnothing$.
Then the rule set $\{R_k\}$ is found.

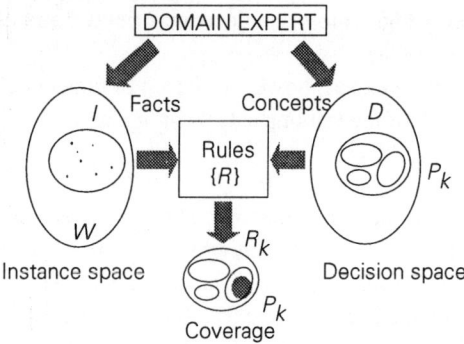

Figure A2: 'Coverage': the portion of P_k that is recognized by rule R_k.

The maximum coverage principle is illustrated in Figure A2.

At this stage, one should realize that incompleteness, 'rare' samples, and even 'exceptions' will obscure the net result. In order to rule out those effects, we also consider 'numerical' clustering. In these methods, similarity (or dissimilarity) between two training samples is expressed by the value of a numerical function applied to the attribute representation of these samples. The classes (clusters) generated are groups of samples such that the intra-cluster similarity is maximum and the inter-cluster similarity is minimum. Those methods are to be considered as almost completely unsupervised. They are refered to as 'data-driven' classification.

We adopt the pairwise similarity measure as proposed in Bezdek (1981). Let w_i and w_j be two training samples from W. Then, if $\{a_k\}$ is a qualitative attribute, the dissimilarity is defined by

$$\text{dis}(w_i, w_j | a_k) = 1 \text{ if } v_{ki} \text{ and } v_{kj} \text{ are different modalities;}$$

$$= 0 \text{ otherwise.}$$

If $\{a_k\}$ is a quantitative attribute, the dissimilarity is defined by

$$\text{dis}(w_i, w_j | a_k) = |\, v_{ki} - v_{kj}\,|.$$

As a result, the total pairwise dissimilarity is given by

$$\text{DIS}(w_i, w_j) = \sum_k \text{dis}(w_i, w_j | a_k).$$

Once we have established the full dissimilarity matrix representing the training set, W, we perform a fuzzy classification. The algorithm reads as follows:[7]

---

[7] Note that if we use the entire hierarchy, we obtain the tree of nested partitionings. If we apply the rule generation algorithm for each of the partitionings, we establish the rule tree. As said before, the rule tree forms the backbone of the reasoning part under uncertainty.

1. Consider all samples in W.
Step 1. Calculate all pairwise dissimilarities.
Step 2. Perform hierarchical clustering.
2. Determine at all cutting levels the amount of induced fuzziness.
Step 1. Calculate for each element in the 'hard' partition (from the hierarchy) its cluster membership values and determine the overall amount of induced fuzziness.
Step 2. Find the 'hard' partition that yields the smallest amount of induced fuzziness.

The net result of this algorithm is a 'hard' partition, P', with the fuzzy cluster membership structure, shown in Table A1. It is assumed that

$$\sum_k f_{ik} = 1$$

which is by definition the case in the above algorithm.

The membership values are assigned by means of the membership operator

$$f_k(x) = \frac{n_k - \sum \text{dis}(x,y) | y \text{ in } P_k}{n - \sum \text{dis}(x,y) | y \text{ in } P}$$

as described in Backer (1978).

The above procedure has to be followed by the following steps:

1. Find the correspondence between P and P', such that the smallest overall amount of induced fuzziness is achieved.

Table A1: Fuzzy partitioning of the instance space.

	P_1'	P_2'	P_3'	...	P_q'
w_1	f_{11}	f_{12}	f_{13}	...	f_{1q}
w_2	f_{21}	f_{22}	f_{23}	...	f_{2q}
w_3	f_{31}	f_{32}	f_{33}	...	f_{3q}
M				...	
w_t	f_{t1}	f_{t2}	f_{t3}	...	f_{tq}

2. Rule out those training samples that are not included in the intersection of P and P', and those samples that have no membership value to any of the classes above a certain threshold. These samples are either to be considered as 'exceptions' or as 'unreliable', respectively. They are considered in the exception set, $\{E\}$.

The above is shown in Figure A3. The reduced partition, P'', is then taken as the set of 'reliable' instances for generating the rule set, $\{R\}$, aiming at maximum coverage. In

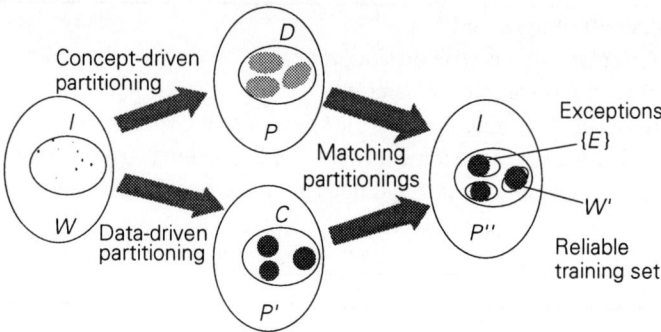

Figure A3: Matching the concept-driven partitioning and the the data-driven partitioning; ruling out the possible exceptions.

Figure A3, the process of matching the concept-driven partitioning and the data-driven partitioning of W is illustrated, resulting in the effective training set, W'. The set of exceptions, $\{E\}$, is kept aside.

Incremental learning

We now consider the case where a set of representative rules, $\{R\}$, has been inductively generated by the effective training set of observations, W_t, as described before. The ranking and selection of the representative rules have been based on the degree of coverage of the partition, P_t. As outlined above, P_t emerged from the matching between the concept-driven partitioning, P_c, and the data-driven partitioning, P_d, of W.

Consider one of the representative rules R_i. The rule is given by

$$R_i: \quad \{D = d_i\} \text{ IF } \{a_1, a_2,..., a_n\}_i.$$

An instantiation of R_i then reads as follows:

$$R_i: \quad \{D = d_i, f_i^*\} \text{ IF } \{a_1 = v_1^*, a_2 = v_2^*,..., a_n = v_n^*\}_i,$$

where the fuzzy decision membership assignment f_i^* is meant to represent the degree of coverage of R_i with respect to P_t. As such, f_i^* is the value which indicates the degree of generality of the rule (or the extent to which the rule is applicable or not).

An exception, E, is a rule for which the set of observations may have attribute values 'close' to those of a representative rule but may differ significantly in the decision assignment $\{D = d_i, f_i^*\}$.

The distance between a representative rule R_i and an exception E_i is given by $d(R_i, E_i)$ as follows:

$$d(R_i, E_i) = [(1/p) \sum_p |v_i^* - v_i| + \sum_m |f_i^* - f_i|]/2$$

satisfying the properties

1) $R_i = E_i \Leftrightarrow d(R_i,E_i) = 0$;

2) $R_i \neq E_i \Leftrightarrow d(R_i,E_i) \leq 1$.

Now, assume a new observation, $\mathbf{w}'$, is made and R_i is applied. Then, the following inference is being made:

USUALLY $\{D = d_i, f_i(\mathbf{w})^*\}$ IF $\{a_1(\mathbf{w}) = v_1(\mathbf{w})^*,. \ldots \}_i$

we observe $\{a_1(\mathbf{w}') = v_1(\mathbf{w}'), \ldots \}$

$\overline{\text{USUALLY } \{D = d_i, f_i(\mathbf{w}')\} \text{ IF } \{a_1(\mathbf{w}') = v_1(\mathbf{w}'), \ldots \}}$

The above inference will be dealt with in the next section. At this stage, we only consider incremental updating of the set of 'representative' rules, $\{R\}$, as a new observation, $\mathbf{w}'$, has been made. We may even, for the moment, disregard the fuzzy rule quantifier USUALLY.[8]

Let $\{S_j\}$ represent the system inferences, $j = 1, 2,\ldots,m$. So,

S_j: $\{D = d_i, f_i(\mathbf{w}')\}$ IF $\{\ldots\}_j$ when rule R_j is used.

The expert's judgement, however, might be

J: $\{D = d_k\}$.

Then, the distance between the rule inference and the expert's judgement is given by

$$D(S_j,J) = \sum_i |f_{ij}(\mathbf{w}') - d_k|/2$$

$$= |f_{kj}(\mathbf{w}') - d_k|.$$

If $\min_j [D(S_j,J)] = D(S_k,J)$

and $D(S_k,J) = \min_j [|f_{kj}(\mathbf{w}') - d_k|] < 0.5$

then the system responds, S_k, lines up with the expert's judgement, J, and R_k is found to cover $\mathbf{w}'$ also.

Let $D(S_m,J) = \min_j [D(S_j,J)]$. Now, the following update decisions are open:

---

8 In Section A4, we will consider the very general class of inference rules, the dispositional rules. These are rules in which one or more premises may contain, explicitly or implicitly, some rule quantifier, like USUALLY.

Step 1. Consider rule modification OR identification of an exception, E'. The update decisions are listed in Table A2. If $D(S_m,\mathrm{J}) < 0.5$, no modification is necessary, while if $D(S_m,\mathrm{J}) \geq 0.5$, either R_k is modified, or $\mathbf{w}'$ is found to be an exception, E'.

Step 2. Consider the list of exceptions, $\{E\}$, and match $\mathbf{w}'$ with the closest group of exceptions to form a 'potential' coverage. Thus, calculate the pairwise dissimilarity matrix for $\{E_j,\mathbf{w}'\}$, $j = 1,2,\ldots$ (as many as there are exceptions in $\{E\}$), and group them together. Keep track of 'potential' coverage for future consultation.

Step 3. Consider the case that $S_m = S_k$ with

$$D(S_m,\mathrm{J}) \geq 0.5.$$

The action that is anticipated is to modify R_k, aiming at coverage $\{P_k,\mathbf{w}'\}$.

Table A2: Update decisions when a new instance is added to the training set.

	$D(S_m,J) < 0.5$		$D(S_m,J) \geq 0.5$	
	$S_m = S_k$	$S_m \neq S_k$	$S_m = S_k$	$S_m \neq S_k$
	R_k covers $\{P_k,\mathbf{w}'\}$ correctly.	NA	R_k is found to be the 'best' rule; however, neither R_k nor any other rule covers $\{P,\mathbf{w}'\}$	R_m covers $\{P_m,\mathbf{w}'\}$ incorrectly; OR no rule covers $\{P,\mathbf{w}'\}$
	no modifications	NA	Modify R_k, aiming at covering $\{P_k,\mathbf{w}'\}$	$\mathbf{w}'$ is defined to be an 'exception', E', and added to the list $\{E\}$

1. Determine the representative rule $R(\mathbf{w}')$ from the new training sample, $\mathbf{w}'$.
Step 1. Select the attributes to make $R(\mathbf{w}')$ discriminative.
Step 2. Choose the discriminant rule that covers $\{P_k,\mathbf{w}'\}$ best.
Step 3. Compare the coverage of $\{R_k\}$ and $R_k(\mathbf{w}')$:
If the coverage of $R_k(\mathbf{w}')$ is larger than or equal to the coverage of $\{R_k\}$, then adopt $R_k(\mathbf{w}')$ as a new 'representative' rule and update $\{R_k\}$.
2. If $\{R_k\}$ is updated, new fuzzy membership values have to be induced using Backer's algorithm as described in the foregoing.

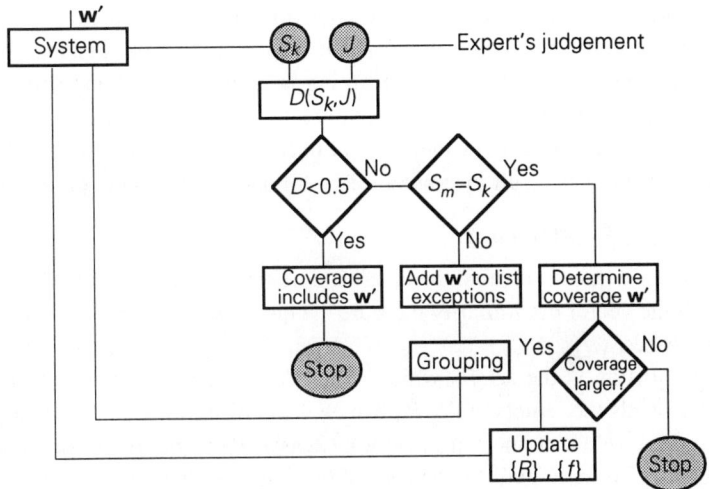

Figure A4: The process of rule updating and identifying exceptions.

The process of rule updating is shown in Figure A4.

Note that incremental inductive rule generation may suffer from the order in which new observations are added. However, the above method of incremental learning rules from observations always yields improvement of the 'overall' coverage of the representative rules.

A4 Hierarchical knowledge organization and a model for inexact reasoning

In this section, we introduce rule trees, derived from a tree of nested partitionings, as an appropriate representation of hierarchical knowledge organization. The way the tree of nested partitionings can be derived is described in the previous section. Note that the dissimilarity measure proposed by Bezdek is used for generating the hierarchical clustering (the nested set of partitions). The same measure is used to generate the decision membership values for each of the individual training samples and with respect to each of the decision subspaces occurring. This method of inducing membership values is discussed in depth in Backer and Jain (1981) and is found to be robust and very reliable.

Based on the rule tree, we develop an inference mechanism and a calculus for inexact reasoning. The calculus is motivated by the fact that knowledge (represented as rules) is thought to be organized and stored hierarchically, and that the rules are assumed to be generated from training samples (cases) supplied by an expert. The calculus is based on the range of uncertainty that comes with a decision, rather than some specific uncertainty value.

Furthermore, the calculus accomodates the principle that general rules (with low specificity) result in a large range of uncertainty, whereas specific instances have small ranges of uncertainty. As (incremental) learning is seen as embedding refinements and accomodating incompatibilities, learning aims to narrow the range of uncertainty. Refinements are defined to narrow the range of uncertainty while incompatibilities widen it. Likewise, certain facts result in smaller ranges of uncertainty than will uncertain facts.

The tree of nested partitionings

We assume that we have been able to construct a nesting of partitionings, P, based on the pairwise attribute vector dissimilarity measure,[9] which is constrained by non-overlapping decision subspaces, $\{d\}$.

The null partition is the set P_0 which contains all q training samples. To this null patition corresponds the 'empty' rule, R_0 (which is not discriminative at all, and therefore the 'most' general rule that one can imagine). Clearly, P_0 is the root of the tree of nested partitionings. Likewise, R_0 will be the root of the rule tree that has been generated from the tree of nested partitionings.

We denote the kth level of the tree of nested partitionings by

$$\{P_i^k\}_{i=1,m^k}$$

where k is the index for the tree level, and m^k the number of disjunct subsets in the partitioning. Clearly, we have that

$$P_i^k \subset P_j^k = \varnothing \text{ for all } i,j = \{1,m^k\}.$$

If we denote the number of training samples that is contained in P_i^k as q_i^k, then

$$\sum_{i=1}^{m^k} q_i^k = q.$$

As we consider a nesting of partitions, then for each specific pair

$$(i,j), i = \{1,m^{k+1}\}, j = \{1,m^k\},$$

we have

$$P_i^{k+1} \subset P_j^k.$$

It is important to notice that for the specific pair (i,j), we have

$$(1) \quad q_i^{k+1} \leq q_j^k$$

---

[9] Bezdek; quantitative attributes are measured by the absolute difference of the attribute values; qualitative attributes are measured by exact match of modality.

(2) P_j^k contains q_j^k training samples, including the q_i^{k+1} samples of P_i^{k+1}.

Or in other words, P_j^k and P_i^{k+1} share the q_i^{k+1} samples.[10]

The rule tree and specificity

For each partitioning, $\{P_i^k\}_{i=\{1,mk\}}$, it is found to be feasible to generate corresponding rule sets, $\{\{R_{ih}^k\}_{h=\{1,hk\}}\}_{i=\{1,mk\}}$, controlled by maximum coverage of $\{P_i^k\}_{i=\{1,mk\}}$, and showing maximum discriminative power by some conjunction of tied attributes. The above rule sets are said to contain representative rules. Furthermore, the partitioning subset, P_i^k, is said to be fully covered by the rule set $\{R_{ih}^k\}_{h=\{1,hk\}}$, where R_{i1}^k is found to cover the largest portion of P_i^k, R_{i2}^k covers the second largest portion, and so on. All hk rules are covering P_i^k completely.

Consequently, the rule R_{i1}^k covers q_{i1}^k samples, R_{i2}^k covers q_{i2}^k samples, and so on, leaving the rule R_{ihk} to cover the remaining q_{ihk}^k training samples.[11] Thus

$$\sum_{j=1}^{h^k} q_{ij}^k = q_i^k .$$

Note that the 'empty' rule has no attributes tied (non-discriminative) whereas the rule representing a singleton has all attributes tied (most discriminative).

The nesting $P_i^{k+1} \subset P_j^k$ for some pair (i,j), $i = \{1,m^{k+1}\}$, $j = \{1,m^k\}$, results in corresponding rule sets

$$\{R_{ih}^{k+1}\}_{h=\{1,hk+1\}} \text{ and } \{R_{ih}^k\}_{h=\{1,hk\}}.$$

For all rules $R_{ij'}$ in the rule set $\{R_{ih}^{k+1}\}_{h=\{1,hk+1\}}$ there exists a rule R_{ij} in the rule set $\{R_{ih}^k\}_{h=\{1,hk\}}$ for which it is said that R_{ij}' is a refinement of R_{ij}. In other words, $R_{ij'}$ is *more specific* than R_{ij}. This can be shown as follows.

Let R_{ij} be IF $\{a\}_j$ THEN $\mathbf{w} \Rightarrow \{d_i^k\}$

and

let $R_{ij'}$ be IF $\{a\}_{j'}$ THEN $\mathbf{w} \Rightarrow \{d_i^{k+1}\}$,

where $\{a\}_j$ and $\{a\}_{j'}$ are tied attribute subsets for the rules R_{ij} and R_{if} respectively.

Such a rule reads as follows:

The unknown sample, $\mathbf{w}$, is assigned to a decision subset, $\{d_i^k\}$ IF the tied

---

[10] The nesting property is fundamental to the definition of refinement.

[11] These properties are forced by minimizing the criterion function of the learning scheme. The criterion function is formulated as the ratio between the cardinality of erroneous coverage and the cardinality of desired coverage.

attribute subset, $\{a\}_j$, is satisfied.

As a consequence of the nesting of partitions we know

$$\{d_i^{k+1}\} \subset \{d_i^k\}$$

(specificity with respect to decision-making).

Also, as a consequence of the rule generating process, we know that the number of tied attributes at level $k+1$ is greater than or equal to the number of tied attributes at level k, thus

$$|\{a\}_j| \leq |\{a\}_{j'}|.$$

Then, in case of $\{a\}_j \subset \{a\}_{j'}$, the refinement is obvious. If this is not the case, we just have to realize that $R_{ij'}$ performs equally well as .

$$R_{ij'}\colon \text{ IF } \{(\{a\}_j),\{a\}_{j'}\} \text{ THEN } \mathbf{w} \Rightarrow \{d_i^{k+1}\}$$

considering $(\{a\}_j)$ as free attributes. As a result, again $\{a\}_j \subset \{(\{a\}_j),\{a\}_{j'}\}$, and therefore, the refinement is shown.

Note that refinement with respect to the attributes corresponds to refinement with respect to decision-making. In view of that, refinement implies increasing specificity.

Note also that R_{ij} is being generated by a smaller number of training samples (or equal at the most) than R_{ij}. This is because of

$$q_{ij'}^{k+1} \leq q_{ij}^k \quad {}^{12}$$

Rule of inference; upper and lower bounds

In Sections A2 and A3, it has been assumed that the system's response should be in the form of a set of decisions $\{d = d_k, f_k(\mathbf{w})\}_{k=\{1,m\}}$. Thus, for all partitionings, $\{P_i^k\}_{i=\{1,m^k\}}$, it is possible to generate the above membership values for each of the q training samples available.

In effect, the rule

$$R_{ih}^k\colon \mathbf{w} \Rightarrow \{d_i^k\} \text{ IF } \{a\}_{ik}$$

(the unknown sample, $\mathbf{w}$, is assigned to the decision subset $\{d_i^k\}$ IF the tied attribute subset $\{a\}_{ik}$ is satisfied) represents the evidence obtained from a subset of training samples

$$w_j, j = \{1, q_i^k\}$$

---

[12] It is therefore that the range of uncertainty due to the upper and lower bounds to be defined, is related to the training samples from which the rules originated.

Therefore, $R_{ih}{}^k$ is interpreted as

$$R_{ih}{}^k : \{\{f_i{}^k(w_j)\}_{j=\{1,q_i{}^k\}}\}_{i=\{1,m^k\}} \text{ IF } \{a\}_{ik}\text{ [13]}$$

For any sample, $\mathbf{w}'$, we use the rule of inference given by

$$f_i{}^k(\mathbf{w}') = \max_j = \{1,q_i{}^k\}[\min[r(\{v',v_j{}^*\})_{ih}, f_i{}^k(w_j)]]$$

where $r(\{v',v_j{}^*\})_{ih}$ constitutes the matching between the observed attribute values and the tied attribute values of the rule $R_{ih}{}^k$.

$r(v',v_j{}^*) = [0,1]$ is called the premise matching coefficient[14]

Without loss of generality, we may choose here

$$r(v',v_j{}^*) = 1 \text{ if } v' = v_j{}^*$$

$$= 0 \text{ otherwise}$$

This choice enhances the clarity of the bounds to be generated. Any more nuanced matching will – of course – affect the actual bounds, though it will not alter the ordering between them.[15]

Assume that the new training sample, $\mathbf{w}'$, comes with a decision

$$d = d_{i'} , i' = \{1,m\}$$

and that this decision $d_{i'}$ is an element of the decision subset $\{d_i{}^k\}$. Then, if $\{a\}_{ih}$ is satisfied, rule $R_{ih}{}^k$ implies

$$f_{\text{OUB}}{}^k(\mathbf{w}') = \max_{j=\{1,q_i{}^k\}}[f_i{}^k(w_j)]$$

which is called the *optimistic upper bound* (OUB).

Let this maximum be found for w_u, $u = \{1,q_i{}^k\}$. Then

$$\{f(i)^k(\mathbf{w}')\}_i = \{1,m_k\} = \{f_i{}^k(w_u)\}_i = \{1,m_k\}$$

which includes the OUB(i').

Likewise, we determine

$$\max_j = \{1,q_i{}^k\}[\{f_i{}^k(w_j)\}_i = \{1,m_k\}; i = i']$$

---

[13]   This is equivalent to classification inference or the interpolation rule of inference.

[14]   See Bezdek; in essence it is the same measure that is constituting the pairwise attribute vector dissimilarity measure encountered before.

[15]   At a later stage, we will show that through the matching coefficient we may input observed uncertain attribute facts.

Let this maximum be found for w_l, $l = \{1, q_i^k\}$. Then

$$f_{\text{PLB}(i')}{}^k(\mathbf{w}') = 1 - \sum_{\substack{i=1 \\ i \neq i'}}^{m^k} f_i^k(w_l)$$

which is called the *pessimistic lower bound* (PLB).

As a consequence, we obtain

$$\{f_i^k(\mathbf{w}')\}_i = \{1, m_k\} = \{f_i^k(w_l)\}_i = \{1, m\}$$

including the PLB(i').

We always obtain:
$$\mathbf{w}' \Rightarrow \{f_{\text{OUB}(i')}{}^k(\mathbf{w}'), f_i^k(\mathbf{w}'), f_{\text{PLB}(i')}{}^k(\mathbf{w}')\}$$
constrained by $f_{\text{OUB}}(\mathbf{w}') < f(\mathbf{w}') < f_{\text{PLB}}(\mathbf{w}')$.

Note that $f_{i'}(\mathbf{w}') = [\sum_{j=1}^{q_i^k} f_{i'}^k(w_j)]/q_i^k$

Refinement and specificity

Now, let us reconsider the refinement as discussed before.

Suppose, we consider the partitionings $\{P_i^k\}_{i=\{1,mk\}}$ and $\{P_i^{k+1}\}_{i=\{1,mk+1\}}$. The pair (P_i^k, P_i^{k+1}), for some (i,i'), $i = \{1, m^k\}$, $i' = \{1, m^{k+1}\}$, are nested subsets of the partitionings considered. Further, let R_{ij}^k and R_{ij}^{k+1} be two representative rules with respect to that pair of nested partitioning subsets.

Clearly, R_{ij}^k is related to the decision subspace $\{d_i^k\}$, and R_{ij}^{k+1} is related to the decision subspace $\{d_i^{k+1}\}$ for which we have that

$$\{d_i^{k+1}\} \subset \{d_i^k\}.$$

Then, for a given $d_{i'} \in \{d_i^k\}$, and thus also $d_{i'} \in \{d_i^{k+1}\}$, we obtain the following inferences:

$$R_{ij}^k \text{ generates } \{f_{\text{OUB}(i')}{}^k; f_{\text{PLB}(i')}{}^k\}, \text{ and}$$

$$R_{ij}^{k+1} \text{ generates } f_{\text{OUB}(i')}{}^{k+1}; f_{\text{PLB}(i')}{}^{k+1}\}.$$

It then follows that

$$1 \geq (f_{\text{OUB}(i')}{}^k - f_{\text{PLB}(i')}{}^k) \geq (f_{\text{OUB}(i')}{}^{k+1} - f_{\text{PLB}(i')}{}^{k+1}) \geq 0$$

from the fact that $q_i^{k+1} \leq q_i^k$, on the basis of which the bounds were derived.

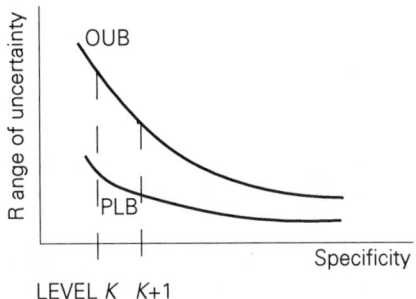

Figure A5: Uncertainty bounds versus specificity.

The quantity $f_{OUB} - F_{PLB}$ is cvalled the *range of uncertainty*. Evidently, this range will be equal to 1 for the 'empty' rule (no discriminative power, 'most' general, 'least' specific), and will be equal to 0 for a singleton (maximum discriminative power, 'most' specific but not general at all). By going through the rule tree, from the root to the layer of singletons, we observe a narrowing of the range of uncertainty. In other words, increasing specifity goes with a decrease in the range of uncertainty. Conversely, increasing generality implies an increase in the range of uncertainty. This is pictured in Figure A5, showing the bounds narrowing the range of uncertainty when specificity increases.

Combining rules

Let us next consider how combining rules affects the bounds. We denote $R_{jj'} = R_{ij}^k \circ R_{ij'}^{k+1}$ as the combination of the rules R_{ij}^k and $R_{ij'}^{k+1}$.

Suppose, first, that $R_{ij'}^{k+1}$ is a *refinement* of R_{ij}^k. If so, we know that the related decision subspaces match in the sense that

$$\{d_i^{k+1}\} \subset \{d_i^k\}$$

Assuming that the conditions of both rules are satisfied, we obtain $R_{jj'}$ generating

$$\{\min[f_{OUB(i')}^k, f_{OUB(i')}^{k+1}], \max[f_{PLB(i')}^k, f_{PLB(i')}^{k+1}]\}$$

which is in essence copying the upper bound and lower bound of $R_{ij'}^{k+1}$, being the refined rule.

In contrast to the above, we next consider the case where $R_{ij'}^k$ and $R_{ij'}^k$ are two rules which are related to different decision subspaces, $\{d_i^k\}$ and $\{d_{i'}^k\}$, respectively, and for which we therefore have that

$$\{d_i^k\} \cap \{d_{i'}^k\} = \varnothing$$

Then, $R_{ij'}^k$ and $R_{ij'}^k$ are said to be *incompatible*. They can only be combined by evaluating their common predecessor in the tree, say at level $k - 1$.

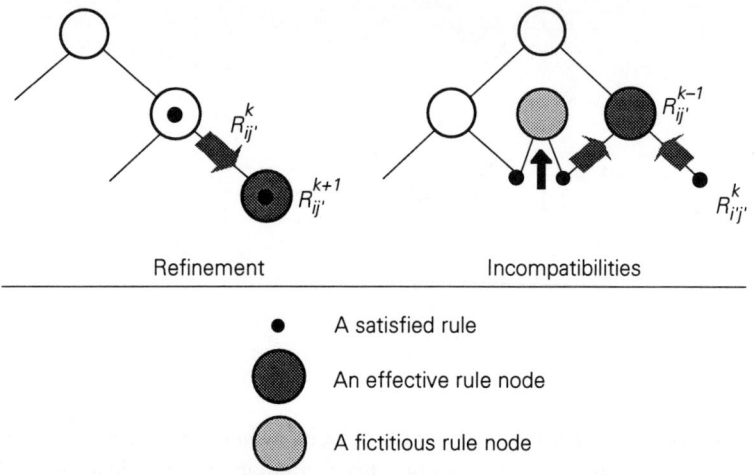

Refinement   Incompatibilities

● A satisfied rule

● An effective rule node

● A fictitious rule node

Figure A6: The effects of refinement and incompatibilities.

Thus $R_{jj'} = R_{ij}^k \circ R_{ij'}^k$ generates

$$\{\max[f_{\mathrm{OUB}(i')}{}^k, f_{\mathrm{OUB}(i')}{}^{k-1}], \min[f_{\mathrm{PLB}(i')}{}^k, f_{\mathrm{PLB}(i')}{}^{k-1}]\}$$

knowing that $d_{i'} \in \{d_{i'}^{k-1}\} = \{d_i^k\} \cup \{d_{i'}^k\}$.

This is, in essence, copying the upper and lower bound of the common predecessor in the tree, on level $k - 1$, $R_{jj'}$. Both the refinement and incompatibility are pictured in Figure A6, showing the effect on the bounds for a refinement and for an occurring incompatibility.

In conclusion, a refinement $(k+1 \Leftarrow k)$ results in

$$\{d_i^{k+1}\} \text{ AND } \{d_i^k\} \text{ yielding}$$

$$\{\min[f_{\mathrm{OUB}}{}^k, f_{\mathrm{OUB}}{}^{k+1}], \max[f_{\mathrm{PLB}}{}^k, f_{\mathrm{PLB}}{}^{k+1}]\}$$

whereas an incompatibility at level k results in

$$\{d_i^k\} \text{ OR } \{d_{i'}^k\} \text{ which is a decision subspace } \{d_{i'}^{k-1}\}, \text{ yielding}$$

$$\{\max[f_{\mathrm{OUB}}{}^{k-1}, f_{\mathrm{OUB}}{}^k], \min[f_{\mathrm{PLB}}{}^{k-1}, f_{\mathrm{PLB}}{}^k]\}$$

The case where $\{d_{i'}^{k-1}\}$ is not an existing decision subspace as brought forward by generating the tree of nested partitionings, $R_{jj'}$, constitutes a fictitious rule for the decision subspace $\{d_{i'}^{k-1}\}$. Logically, that rule should read

$$R_{jj'} : \mathbf{w} \Rightarrow \{d_{i'}^{k-1}\} \text{ IF } \{a\}_i \text{ OR } \{a\}_{i'}$$

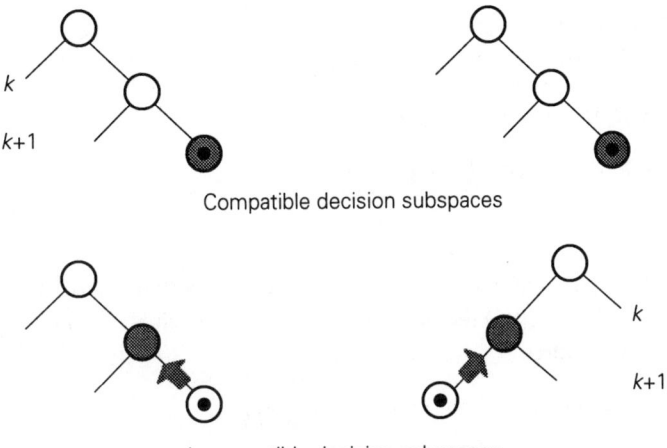

Compatible decision subspaces

Incompatible decision subspaces

Figure A7: (a) Compatible and (b) incompatible decision subspaces.

It is needless to say that rules from different (independent) rule trees can be combined as long as the trees share (partly) the same decision subsets. This is of particular interest if one has a blackboard architecture in mind where independent knowledge sources contribute to the decision-making or diagnosis (see Section A6 for an example).

Figure A7 shows two cases where compatible and incompatible decision subspaces result in

$$\{\max[f_{\text{OUB}(1)}^{k+1(k)}, f_{\text{OUB}(2)}^{k+1(k)}], \min[f_{\text{PLB}(1)}^{k+1(k)}, f_{\text{PLB}(2)}^{k+1(k)}]\}$$

though at different levels in the trees.

Propagation of uncertainty

Next, we consider the case where uncertain attribute values are observed, either from the environment or generated by the system itself. The question arises how that would affect the range of uncertainty.

Assume that one of the decisions $d = d_k$, $k \in \{1,m\}$ is supposed to be one of the attributes $\{b\}$ in rules from another rule tree. For simplicity, we assume the rule under consideration has only one tied attribute, b, for which

$$b^* = d_k \quad [16]$$

As discussed before, the bounds were infered assuming that the premise matching coefficient $r(v', v^*)$ was either one or zero.

---

[16] Note that rules are derived under the assumption of exact attribute value matches.
If $b^* = d_k$ is supposed to be such an attribute value, it is required that $r(b', b^*) = 1$; thus $f_k(w) = 1$.

However, uncertain attribute values would never justify the above.

If the system's response were $\{f_k(\mathbf{w}')\}_{k=\{1,m\}}$, we could set

$$b' = d_k \text{ if } f_k(\mathbf{w}') = \max\{f_i(\mathbf{w}')\}_{i=\{1,m\}}$$

so that $r(b',b^*) = 1 \text{ if } b' = b^* = d_k$

$$= 0 \text{ otherwise}$$

We thereby completely ignore the uncertainty that comes with the attribute value, $b = d_k$.

By setting $r(b',b^*) = f_k(\mathbf{w}')$ the non-perfect attribute match is clearly exhibited. Then, the inference rule becomes

$$f_i(\mathbf{w}') = \max_{j=\{1,q\}}\{\min[f_k(\mathbf{w}'),f_i(w_j)]\}_{j=\{1,q\}}$$

Since $f_k(\mathbf{w}') \leq 1$, we find

$$f_{\text{OUB}(i)}(w_u) - f_{\text{OUB}(i)}(\mathbf{w}') \leq 1 - f_k(\mathbf{w}')$$

whereas

$$f_{\text{PLB}(i)}(w_l) - f_{\text{PLB}(i)}(\mathbf{w}') \geq 1 - f_k(\mathbf{w}')$$

and therefore, the effect of uncertain attributes can only be a widening of the range of uncertainty.

Updating of the uncertainty

Up to now, we have considered a static system in which rule sets have been generated and membership assignments to all nested decision subspaces have been established. When the conditions of a rule are satisfied, we know that an instance under consideration, $\mathbf{w}'$, results in a triple of membership values regarding a specific decision subspace, say $\{d_i\}$.

The triple includes the membership value, $f_i(\mathbf{w}')$, which is to be interpreted as the degree of certainty of $\mathbf{w}'$ being assigned to the decision subspace $\{d_i\}$, as well as the range of uncertainty that comes with that decision, expressed by $(f_{\text{OUB}(i)}(\mathbf{w}') - f_{\text{PLB}(i)}(\mathbf{w}'))$.

If more training instances become available, one at a time, the coverage of the tree of nested partitionings will change gradually. Even the tree itself may change, and consequently, the rule tree.

Modification of the rules (tying more or other attributes, adding new rules, and so on) has been discussed in Section A3 and will not be discussed here. Here, it is sufficient to restate the fact that rule modification is controlled by the degree of matching of the expert's judgement, J, and the system's response, S, defined as

$$D(S,J) = \sum_{i=1}^{m} |f_i(\mathbf{w}') - d_i|.$$

Suppose we have a rule R_{ij} related to decision subspace $\{d_i\}$, for which the conditions are satisfied by the instance under consideration, $\mathbf{w}'$. Then $R_{ij'} : \mathbf{w} \Rightarrow \{d_i\}$ IF $\{a\}_i$ generates

$$\{f_i(\mathbf{w}')\}_{i=\{1,m\}} \text{ IF } \{a\}_i.$$

If the expert's judgement is $d = d_i$, then if

$$D(S,J) = |f_i(\mathbf{w}') - d_i| < 0.5$$

the system's response, S, is said to line up with the expert's judgement, J. R_{ij} is said to cover $\mathbf{w}'$ correctly and it should be the case that

$$f_i(\mathbf{w}') = \max[\{f_j(\mathbf{w}')\}_{j=\{1,m\}}]$$

Then, no modification is to be undertaken.[17]

But even in the case of no modification, the coverage of the correct decision subspace is altered by adding the instance $\mathbf{w}'$. As a consequence of the properties of the chosen operator to induce membership assignments to existing decision subspaces, we always obtain

$$f_i^{t+1}(\mathbf{w}') \geq f_i^t(\mathbf{w}')$$

where t is the time index before and $t + 1$ is the time index after an instance $\mathbf{w}'$ is supplied by the expert. As a result, the more cases are correctly covered, the higher the corresponding membership values with respect to the decision subspace under consideration.

A5 The upper and lower bound formalism compared and contrasted with the Dempster–Shafer scheme

In order to justify the formalism based on upper and lower bounds as described in the previous section, we first compare and contrast the formalism with the Dempster–Shafer scheme for reasoning with uncertainty. This discussion will be followed by a closer look at the desiderata to be satisfied by the formalism for representing uncertainty and making inferences with uncertain information in a desired way.

---

[17] As described in Section A3, rule modification is controlled by

    (1) $D \geq 0.5$

and (2) $f_i(w') \leq \max[\{f_j(w')\}_{j=\{1,m\}}]$

If a modification is foreseen, coverage and nested trees are partly re-examined, and consequently, existing rule sets are also partly re-examined in order to obtain improved coverage. After each modification, new membership assignments are performed reflecting the improved coverage.

For ease of notation, we denote the calculus as introduced in the foregoing as BA, and the Dempster–Shafer formalism as DS.

Formalisms are either adopted from well-established fields, such as probability theory, by often accepting unrealistic global assumptions, or proposed as *ad hoc* solutions without formal justification. DS is related to probability theory, though it posesses *ad hoc* updating and propagation. The BA formalism is to be considered as a fully ad hoc formalism and relies entirely on the list of desiderata[18] that the formalism should try to satisfy.

The BA calculus is based on the range of uncertainty that comes with a decision, rather than with a specific uncertainty value. The range is established by taking the difference of the optimistic upper bound (OUB) and the pessimistic lower bound (PLB).

If f denotes the degree of certainty associated with the hypothesis that the hard evidence $\{E\}$ is covered by some concept $\{X\}$, then the range of uncertainty is given by $f_{OUB} - f_{PLB}$. Therefore, f is assigned to concept $\{X\}$, given the evidence $\{E\}$ which is assumed to be hard evidence. The more specific the concept $\{X\}$, the smaller f generally turns out to be. That is to say, if we hypothesize the evidence to be covered by a very specific concept, then we become less and less certain that the hypothesis is really true. This implies that the more specific the concept, the more the system needs to know to cover the instance with high certainty. Thus, given a certain level of knowledgeability, there is a limit to what the system can infer to be certain.

It is therefore important to understand that concepts with increasing specificity are nested. The formalism is based on the property that

$$f_{\{X\}} = \sum_{\{Y\}<\{X\}} f_{\{Y\}},$$

where $\{Y\}$ are nested, more specific concepts than the concept $\{X\}$; thus $\{Y\}$ is implied by $\{X\}$.

Back in 1967, Dempster attempted to model uncertainty by a range of probabilities rather than as a single probabilistic number (Dempster, 1967). He established this range on a measure of belief associated with a concept $\{X\}$, denoted by $\mathrm{BEL}(\{-X\})$, being the degree of belief in evidence $\{E\}$ associated with $\{X\}$, where $\mathrm{Bel}(\{X\})$ is assigned to the evidence $\{E\}$.

Dempster's formalism is based on the property that

$$\mathrm{BEL}(\{X\}) = \sum_{\{Y\}<\{X\}} m(\{Y\})$$

---

[18] The proposed formalism
- accepts linguistic term sets with fuzzy-valued semantics;
- suits a hierarchical knowledge representation that allows incremental updating;
- deals with refinement (specificity), conflicts (incompatibilities) and ignorance;
- produces (even if the expert is required to provide precise and consistent assessment data) imprecise output very much dependent on the level of knowledgeability of the system itself.

where $m(\{Y\})$ is the mass assignment (the basic probability assignment) of the concept $\{Y\}$.

The common denominator of the BA calculus and the DS theory can be found in the fact that both assume that there is a fixed set of mutually exclusive and exhaustive elements called the environment, T. Each subset of T can be interpreted as a possible concept (a possible answer to a question) where there can be only one correct answer to a question. Of course, not all possible concepts may be meaningful; however, the important point to realize is that the subsets of the environment are all possible valid concepts in the universe of discourse.

Each subset can be considered to be an implied proposition such as

• the correct answer is $\{X,Y,Z\}$
• the correct answer is $\{X,Y\}$

and so on.

In the DS theory, it follows that given evidence $\{E\}$, we have

$$\text{BEL}(\{X,Y,Z\}) > \text{BEL}(\{X,Y\})$$

whereas in the BA calculus given evidence $\{E\}$, we have

$$f\{X,Y,Z\} > f\{X,Y\}.$$

This means that the possibility of the answer being correct is ordered in the above sense. It should be noted that the DS theory relates to a hierarchical lattice with T at the top and the null set at the bottom. Thus, for $\{X,Y,Z\}$, we have the lattice shown in Figure A8. Rather than exploring the full lattice, the BA calculus tends to explore the tree that has emerged from the training set and allow for incremental revision (see Figure A9). However, this tree can still be captured in the DS theory by assigning zero mass to the remaining subsets in the lattice.

In the DS theory, the mass assignment (the basic probability assignment), m, to any subset yields the belief in that subset being correct.

If one attempts to relate the mass assignment, $m(\{X\})$, to the possibility of being correct, $f_{\{X\}}$, then one should realize that we are not using the potential freedom of the masses in the sense that the membership assignments (possibilities of being correct) f sum to unity, and, as a consequence, one always obtains

$$f_{\{X,Y,Z\}} = f_{\{X\}} + f_{\{Y,Z\}} = f_{\{X\}} + f_{\{Y\}} + f_{\{Z\}} < 1$$

whereas in the DS theory, one may obtain

$$m(\{X,Y,Z\}) + m(\{Y\}) + m(\{Y,Z\}) = 1$$

The difference in the interpretation should be noted. In DS:

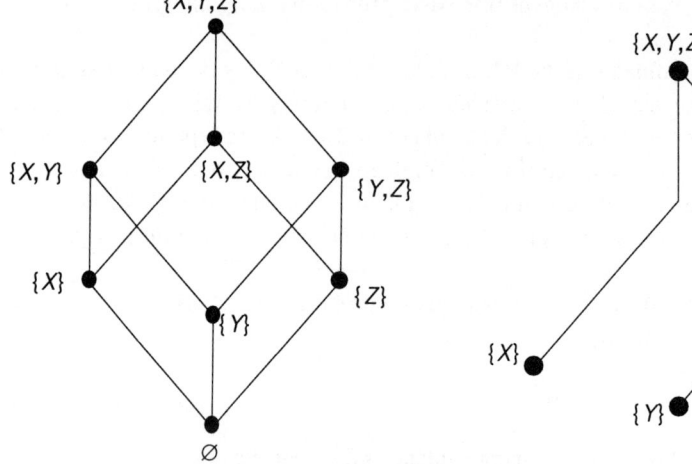

Figure A8: Hierarchical lattice in the   Figure A9: The generated tree in the BA
                DS theorie.   calculus.

$m(\{Y\}) = 0.9$ expresses the belief that the target (**w**) is concept $\{Y\}$,
and *only* concept $\{Y\}$, given $\{E\}$;

however,

$m(\{Y,Z\}) = 0.07$ (expressing the belief that the target is concept $\{Y\}$ OR $\{Z\}$)

and

$m(\{X,Y,Z\}) = 0.03$ (expressing the belief that the target is concept
$\{X\}$ OR $\{Y\}$ OR $\{Z\}$)

convey additional information and may contribute to a belief in $\{Y\}$. As a result, we
obtain a range of belief [known belief in $\{Y\}$, plausible belief in $\{Y\}$] = [0.9, 1]. The true
belief is assumed somewhere in the range 0.9 to 1.

In BA:

$f_{\{X,Y,Z\}} = 0.84$ expresses the possibility that the target (**w**) is covered by concept
$\{X\}$ OR $\{Y\}$ OR $\{Z\}$, given $\{E\}$;

$f_{\{Y,Z\}} = 0.66$ expresses the possibility that the target is covered by concept $\{Y\}$ OR
$\{Z\}$, and $f_{\{Y\}} = 0.48$ expresses the possibility that the target is covered by concept
$\{Y\}$ *only*, given $\{E\}$.

It is important to realize that the number of training instances contributing to the
concept space $\{X,Y,Z\}$ is large, as is the variability. It is also important to realize that the
number of training instances is related to the knowledgeability, α, where

$$\alpha = 1 - \exp(-q/Q - q),$$

and where q is the number of different training instances that have been covered by the concept space, and Q is the cardinality of the meaningful possible instance space. It is clear that the uncertainty assessment is related to α, and the bounds are related to the variability of the training instances which have been encountered so far. Consequently, the range of possibility ($f_{OUB} - f_{PLB}$) or the range of uncertainty will be large.

On the other hand, the number of training instances contributing to the concept $\{Y\}$ only is small, as is the variability; consequently, the range of uncertainty will also be small.

Both formalisms will result in smaller ranges of uncertainty if more evidence becomes available. In DS, the rule of combination is

$$m_1 + m_2(\{Z\}) = \Sigma_{X \& Y = Z}\, m_1(\{X\}) \cdot m_2(\{Y\})\ [19]$$

as illustrated in Table A3. However, in BA, the rule of combination is

$$f_1 + f_{2\{Z\}} = MIN_{X \& Y = Z}\, \min(f_{1\{X\}}, f_{2\{Y\}}),$$

as illustrated in Table A4.

Table A3: Combining under DS.

	$m_2(\{Y\}) = 0.9$	$m_2(\{X,Y,Z\}) = 0.1$	$\Sigma = 1$
$m_1(\{Y,Z\}) = 0.7$	$m\{Y\} = 0.63$	$m\{Y,Z\} = 0.07$	
$m_1(\{X,Y,Z\}) = 0.3$	$m\{Y\} = 0.27$	$m\{X,Y,Z\} = 0.03$	
$\Sigma = 1$	0.9	0.1	

Table A4: Combining under BA.

	$f_{\{Y\}} = 0.6$	$f^{\{X,Z\}} = 0.4$	$\Sigma = 1$
$f_{\{Y,Z\}} = 0.7$	$f_{\{Z\}} = 0.1$	$f_{\{X,Y\}} = 0.9$	
$f_{\{X,Y,Z\}} = 1$	$f_{\{X\}} = 0.3$	$f_{\{Y\}} = 0.6$	
Σ	0.4		

---

[19] To establish this relation, the DS approach requires a normalization process, which has been extensively criticized in the literature, [.].

From these tables, we observe that $m(\{Y,Z\})$ and $m(\{X,Y,Z\})$ convey additional information with respect to $\{Y\}$ and may contribute to the belief in $\{Y\}$, whereas $f_{\{X,Y,Z\}}$ is fixed, $f_{\{X,Y,Z\}} = f_{\{Y,Z\}} + f_{\{X\}}$ and $f_{\{Y,Z\}} = f_{\{Y\}} + f_{\{Z\}}$.

In conclusion, even if the DS and BA formalisms rely on the identical assumption of the fixed environment, T. They greatly differ in the assignment of the mass, m, and the concept membership, f, respectively. It is therefore concluded that the formalisms (assignment and combining)

BA: $f_{\{X\}} = \Sigma_{\{Y\}<\{X\}}\, f_{\{Y\}}$

$$f_1 + f_{2\{Z\}} = \Sigma_{X\&Y=Z}\, \min(f_{1\{X\}}, f_{2\{Y\}})$$

and DS:

$$\text{BEL}(\{X\}) = \Sigma_{\{Y\}<\{X\}}\, m(\{Y\})$$

$$m_1 + m_2(\{Z\}) = \Sigma_{X\&Y=Z}\, m_1(\{X\}).m_2(\{Y\})$$

result in quite different interpretations and views of the problem.

Desiderata

We next consider the desiderata to be satisfied by the formalism for representing uncertainty and making inferences with uncertain information.

The following requirements have motivated the origin of our formalism.

1. There should be an explicit representation of uncertainty reflecting
    (i) the amount of evidence at any level of reasoning and/or decision-making;
    (ii) the knowledgeability of the system.
2. There should be an explicit representation of uncertainty reflecting
    (i) specificity (rule refinement);
    (ii) incompatibility of concepts;
    (iii) ignorance.
3. The representation must appear natural to the user and allow for natural interpretation.
4. There should be a function that has clear semantics, which is used to propagate and summarize uncertainty.
5. There must be a hierarchical multi-layer structure allowing a variety of intermediate decision-making.
6. It should be possible to make pairwise comparisons of uncertainty, as the induced ranking is needed for performing any kind of decision-making activities and/or modification.
7. There must be a function, α, that accounts for the knowledgeability of the system, such that $\alpha = 0$ means that no training instances of a particular concept have been encountered, and $\alpha = 1$ means that we have encountered so many training instances that the entire instance subspace related to some particular concept is fully covered.

Granularity

Szolovits and Pauker (1978) note that 'while people seem quite prepared to give qualitative estimates of likelihood, they are often notoriously unwilling to give precise numerical estimates to outcomes'. This seems to indicate that any scheme that relies on the user to provide consistent and precise numerical quantifications of the confidence level of his or her conditional or unconditional statements is bound to fail. It is reasonable to expect that the user provides linguistic estimates of the likelihood of given statements.

A recent study on panel attribute assessment for food quality control (Lincklaen Westerberg et al., 1989) has shown the feasibility of a verbal scale for rating the degree of certainty in a given rule or piece of evidence. In that study, the verbal scale {*very low, low, medium, high, very high*} seems to be the utmost precision of assessment. One might conclude that verbal scales may vary from three up to nine meaningful terms.

Our formalism is therefore assumed to possess a set of attribute modalities[20] corresponding to the uncertainty/precision granularity, appropriate to the problem at hand.

Defining the calculus

The calculus is expected to provide answers to the following questions.

(1) *When the premise of a rule is composed by conjunction of different attributes, how can we aggregate the degree of certainty of the facts matching the premise?*

Given the granularity of the input, the evidence, $\{E\}$, is expected to match the attribute modalities as defined, exactly. Thus, if the premise includes attribute a_i, and $a_i = v_i^*$ ('high') is the condition, then, the observed fact v_i' has to be v_i^*. Thus,

$$r(v_i^*, v_i') = 1 \text{ if } v_i' = v_i^* \text{ ('high')} \text{ [21]}$$

$$= 0 \text{ otherwise,}$$

where $r(.,.)$ is defined to be the premise matching coefficient.

(2) *When a rule represents some form of plausible implication, how can we aggregate the strength of the rule implication?*

The strength of a rule implication is brought forward by its conclusion part

$$\text{IF } \dots \text{ THEN } \{f_i(\mathbf{w'})\}_{i=1,m}$$

which constitutes a possibility assignment, f_i, to each possible conclusion, d_i. The fact that these assignments are generated by a limited training set is made explicit through an

---

[20] Input granularity: the attributes $\{a_1, a_2, \dots, a_n\}$ are specified by the modalities $\{v_{i1}, v_{i2}, \dots, v_{ip}\}$; the rule premise reads then: IF "$\{\dots, a_i, \dots\} = \{\dots, v_i^*, \dots\}$" THEN ...

[21] Treating $r(.,.) = [0,1]$ is theoretically justified, though it appears to be counter-intuitive once we have defined the input granularity.

'optimistic' upper bound and a 'pessimistic' lower bound, the range of uncertainty.

Both $f_i(\mathbf{w}')$ and $f_{OUB(i)} - f_{PLB(i)}$, are very much dependent of the number of training samples involved (or in other words depend on the knowledgeability (alpha) of the system trained so far).

(3) *When the same conclusion is established by multiple rules with various degrees of implication strength, how can we aggregate these contributions into a final degree of certainty/precision?*

When the same conclusion, $\{d_i\}$, (or in short (i)), is provided by more than one rule (say k), we have

$$\{f_{PLB(i)}, f_{OUB(i)}\}^1,$$

$$\{f_{PLB(i)}, f_{OUB(i)}\}^2,$$

$$\vdots$$

$$\{f_{PLB(i)}, f_{OUB(i)}\}^k,$$

for which it should be the case that

$$\{f_{PLB(i)}, f_{OUB(i)}\}\text{total} = \{\max_k f_{PLB(i)}, \min_k f_{OUB(i)}\}.$$

Upper and lower bounds

In the foregoing, it has been established that the calculus accommodates the principle that general rules (with low specificity) result in a large range of uncertainty, whereas specific instances have small ranges of uncertainty. As (incremental) learning is seen as embedding refinements and accommodating incompatibilities, learning is aiming at narrowing the range of uncertainty. Refinements are defined to narrow the range of uncertainty, while incompatibilities widen the uncertainty range. Likewise, certain facts result in smaller uncertainty ranges than uncertain facts will do.

This range of uncertainty is defined as the difference between the optimistic upper bound (OUB) and the pessimistic lower bound (PLB).[22]

$$0 < f_{OUB} - f_{PLB} < 1.$$

A few properties of the bounds are of immediate importance here.

a. $f_{PLB(i)} = 1 - f_{OUB(\neg i)},$

where (i) denotes the decision subspace $\{d_i\}$ and $(\neg i)$ denotes the complement in the environment, T.

---

[22] These bounds are found to be essential in reflecting the nature of incremental learning: uncertainty management under the construction of 'knowledgeability'.

b. The pair

$$\{f_{PLB(i)}, f_{OUB(i)}\} = \{1, 1\}$$

indicates that the decision subspace $\{d_i\}$ is certain.

c. $\{f_{PLB(i)}, f_{OUB(i)}\} = \{0, 0\}$ indicates that $\{d_i\}$ is impossible. Through a), this implies

$$\{f_{PLB(\neg i)}, f_{OUB(\neg i)}\} = \{1, 1\}, \text{ saying that } (\neg i) \text{ is certain.}$$

d. For

$$\{f_{PLB(i)}, f_{OUB(i)}\} = \{a, A\}, \text{ and}$$

$$\{f_{PLB(j)}, f_{OUB(j)}\} = \{b, B\}, \text{ then}$$

decision (i) is preferred to decision (j) if $a > b$, and (optionally), $A > B$.

Next, looking back at the desiderata, we find:

(1) *Explicit representation of the amount of support for any given hypothesis, H.*
By establishing the triplet $\{f_{PLB\{H\}}, f_{\{H\}}, f_{OUB\{H\}}\}$, we believe that such a triplet is an explicit representation of the support as well as the range of support for a given hypothesis, $\{H\}$.

(2) *Explicit representation and measure of incompatibility.*
As we expected to find

$$f_{OUB(i)} > f_{PLB(i)}, \text{ (refinement)}$$

an incompatibility results in a violation of this relation, thus

$$f_{PLB(i)} - f_{OUB(i)} > 0,$$

which is at the same time a measure of incompatibility. Undoubtedly, $f_{OUB(i)} - f_{PLB(i)}$ explicitly represents and measures ignorance.

> It is appropriate to clarify the semantic impact of
> - the degree and the range of support for a given hypothesis, the range of uncertainty;
> - the specificity and generality of induced rules;
> - ignorance;
> - the system's 'knowledgeability', α, in the context of incremental knowledge construction in the form of trees of decision subspaces within the lattice of Depster–Shafer's environment, Θ.

Assume we have encountered only a limited number of training samples (limited 'knowledgeability') then the rule tree is still weakly developed. As the 'knowledgeability' is said to reflect the number of training instances with respect to a specific hypothesis (decision subspace), then the system's 'knowledgeability' represented by rules at the top of the tree will be (much) higher than the 'knowledgeability' of the rules that are at the bottom of the tree.

Thus, general rules at the top of the tree are expected to be supported by a reasonable number of training instances, and are associated with very broad, not very discriminative conceptual hypotheses (broad, non-specific decision subspaces). The degree of support for such a broad hypothesis is high, as is the degree of certainty that the conclusion about the hypothesis is correct. At the same time, the more general (or the less specific) the rule appears to be, the larger the range of support. So, we conclude that, though the certainty is likely to be high, the range of uncertainty (range of support) is also high, measuring the rule's lack of discriminative power or ignorance. General rules are conditioned by a small fraction of all available attributes. That is why they fail to possess discriminative power with respect to the decision subspaces. In other words, ignorance is proportional to the range of uncertainty, defined by the difference between the upper and lower bound.

Specific rules at the bottom of the tree, on the other hand, are conditioned by a substantial fraction of available attributes (if not all), however may originate from just a few training instances. Therefore, the 'knowlegeability' is very limited, though they possess high discriminative power. As a consequence, the range of uncertainty (the range of support) will be small, as will the ignorance. As the 'knowledgeability' is small, the degree of support (the certainty of a very specific hypothesis being correct) will be small too.

If we assume that the number of training instances encountered is very high, then the 'knowlegeability' is said to be very high at all levels. The lack of discriminative power or ignorance still is represented and measured by the range of support or range of uncertainty; however, the certainty assignment to each of the decision subspaces is no longer a function of the 'knowledge-ability' but depends entirely on the intrinsic structure of the decision problem at hand.

In essence, incremental learning aims to enlarge the 'knowledgeability' of the system and will result in a stable rule tree in which general rules always possess less discriminative power than more specific rules. So, the range of uncertainty (ignorance) will always be larger for general rules than for specific rules at the bottom of the tree. This is exactly what the measure $(f_{OUB(i)} - f_{PLB(i)})$ represents.

(3) *Natural interpretation of the representation.*
We believe that the semantics of the lower bound (pessimistic, risk-avoiding) and the upper bound (optimistic, gambling) allow certainly natural interpretation. In terms of training samples encountered, we identify the representation of the 'worst case' and the 'best case', respectively.

(4) *Clear semantics for combining and propagation of uncertainty.*
We also believe that the intersection of the ranges of support (uncertainty) is still an operation that is of a humanistic nature.

(6) *Pairwise comparisons*
Pairwise comparisons are at least considered to rely on the lower bounds but more elaborate rankings are of course possible, including the upper bounds as well.

(5) and (7) *The origin of the formalism.*
We believe that hierarchical knowledge structures have at least the intuitive resemblance with human knowledge structures for thinking and reasoning. The extent to which the hierarchical structure has been developed is referred to as 'knowledgeability'.

Conclusions
We have proposed a hierarchical knowledge structure which is the backbone of the uncertainty formalism. In this structure, rule trees are just specific instances in the Dempster–Shafer's environment, Θ.

Although we have found some resemblance between the DS and the BA formalism, the latter is still a fuzzy possibilistic approach and does not relate to the probabilistic concept as DS does. Skipping any *ad hoc* normalization, the BA formalism is more straight forward.

The optimistic upper bound and the pessimistic lower bound were found to satisfy the most significant desiderata of uncertainty formalisms. Unlike the proposal of Bonisone (1987), here the representation, inference and control are intertwined at each level of the tree of decision subspaces. However, the structure of the knowledge organization (rule trees) lends itself to reasoning which is easy to control. Refinements (top-down forward reasoning) and resolving incompatibilities (bottom-up backward reasoning) are the major issues that we have addressed.

Even if there is no formal proof and justification for the formalism as outlined, experiments show that the formalism leads to satisfactory results.

A6 An example

We conclude this discussion with an example which is directly related to the subject matter of this book. The example exhibits the functionality of the hierarchical knowledge organization (the rule tree) combined with the calculus as discussed in the foregoing. The training samples are derived as cases from exploratory data analysis as given in Section 9.4.2., Table 9.2.

The system under consideration, shown in Figure A10, is designed as described in Section 9.4 (initiation level; support level; diagnostic level; Figure 9.2). The rule trees for each of the processing levels are derived in essentially the same way, though with different attribute and decision (sub)spaces. In this example, we consider the initial processing only.

We have five attributes, $\{a_1, a_2, ..., a_5\}$ and the modalities are
- *very high* (VH)
- *high* (H)
- *medium* (M)
- *low* (L)
- *very low* (VL)

For this purpose, an expert was asked to convert numerical values into the modalities given. The decision space is refered to clustering tendency (CT). Also the CT, ($\{d\}$), was decided to be *very high* (VH), *high* (H), *medium* (M), *low* (L), *very low* (VL). As a result, Table A5 shows the training set used.

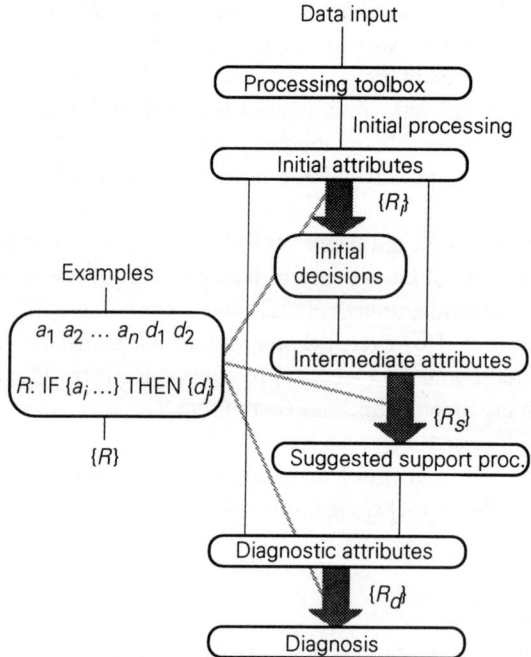

Figure A10: System under consideration.

Table A5: The initial training set.

w	a_1	a_2	a_3	a_4	a_5	d
1	H	H	L	VL	H	VL
2	H	H	L	VL	H	VL
3	H	H	L	VL	H	VL
4	H	H	L	VL	M	VL
5	M	M	L	VL	M	L
6	H	H	L	VL	M	L
7	M	H	H	M	M	M
8	L	H	H	M	L	M
9	L	H	H	H	L	M
10	L	H	H	L	M	H
11	L	H	H	M	H	H
12	L	H	H	M	L	H
13	L	H	H	H	H	H
14	L	L	H	H	H	H
15	L	H	H	VH	H	VH
16	L	H	H	VH	L	VH
17	L	L	H	VH	H	VH
18	L	H	H	H	H	VH
19	L	H	H	VH	H	VH

The nesting of partitions, P, (decision subspaces ordered with increasing discriminative power or increasing specificity) is shown in Figure A11a. Each existing partition in the nesting generates one particular rule set accordingly. Consequently, P, generates a rule tree, R, shown in Figure A11b. P is the result of a hierarchical clustering of the training data represented by the pairwise dissimilarity matrix (see Section A3). Then, the rule tree is generated using the maximum coverage algorithm.

The rules that have been generated are the following.

$\{R\}_0$: the empty rule;

$\{R\}_1$: the rule set that covers the partition subset $\{VL,L\}$

R_1:   IF a_3 = L

THEN $\mathbf{w} \Rightarrow \{VL,L\}$; [6/6]

R_2:   IF a_4 = VL

THEN $\mathbf{w} \Rightarrow \{VL,L\}$; [6/6]

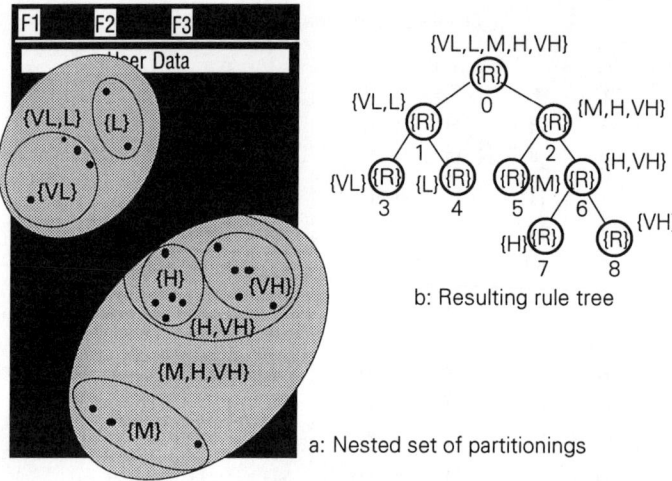

b: Resulting rule tree

a: Nested set of partitionings

Figure A11: The nesting of partitions (a) and the generated rule tree (b).

$\{R\}_2$: the rule set that covers the partition subset $\{M,H,VH\}$

R_8:   IF $a_1 = L$

THEN $\mathbf{w} \Rightarrow \{M,H,VH\}$; [13/13] =

R_9:   IF $a_3 = H$

THEN $\mathbf{w} \Rightarrow \{M,H,VH\}$; [13/13] =

$\{R\}_3$: the rule set that covers the partition subset $\{VL\}$

R_3:   IF $a_1 = H$ AND $a_5 = H$

THEN $\mathbf{w} \Rightarrow \{VL\}$; [3/4]

R_4:   IF $a_1 = H$ AND $a_2 = H$ AND $a_3 = L$ AND $a_4 = VL$ AND $a_5 = M$

THEN $\mathbf{w} \Rightarrow \{VL\}$; [1/4]

!   IF $a_1 = H$ AND $a_2 = H$ AND $a_3 = L$ AND $a_4 = VL$ AND $a_5 = M$

THEN $\mathbf{w}\{L\} \Rightarrow \{VL\}$

$\{R\}_4$: the rule set that covers the partition subset $\{L\}$

R_5:   IF $a_1 = H$ AND $a_2 = H$ AND $a_3 = L$ AND $a_4 = VL$ AND $a_5 = M$

THEN $\mathbf{w} \Rightarrow \{L\}$; [1/2]

!   IF $a_1 = H$ AND $a_2 = H$ AND $a_3 = L$ AND $a_4 = VL$ AND $a_5 = M$

THEN $\mathbf{w}\{VL\} \Rightarrow \{L\}$

R_6:　IF $a_1 = M$ AND $a_5 = M$

　　　THEN $\mathbf{w} \Rightarrow \{L\}$; $[1/2] =$

R_7:　IF $a_2 = M$ AND $a_5 = M$

　　　THEN $\mathbf{w} \Rightarrow \{L\}$; $[1/2] =$

$\{R\}_5$: the rule set that covers the partition subset $\{M\}$

R_{23}:　IF $a_1 = M$ AND $a_4 = M$ AND $a_5 = M$

　　　THEN $\mathbf{w} \Rightarrow \{M\}$; $[1/3]$

R_{24}:　IF $a_1 = L$ AND $a_2 = H$ AND $a_3 = H$ AND $a_4 = M$ AND $a_5 = L$

　　　THEN $\mathbf{w} \Rightarrow \{M\}$; $[1/3]$

!　　　IF $a_1 = L$ AND $a_2 = H$ AND $a_3 = H$ AND $a_4 = M$ AND $a_5 = L$

　　　THEN $\mathbf{w}\{H\} \Rightarrow \{M\}$

R_{25}:　IF $a_1 = L$ AND $a_4 = H$ AND $a_5 = L$

　　　THEN $\mathbf{w} \Rightarrow \{M\}$; $[1/3]$

$\{R\}_6$: the rule set that covers the partition subset $\{H,VH\}$

R_{10}:　IF $a_3 = H$ AND $a_5 = H$

　　　THEN $\mathbf{w} \Rightarrow \{H,VH\}$; $[7/10]$

R_{11}:　IF $a_4 = VH$ AND $a_5 = L$

　　　THEN $\mathbf{w} \Rightarrow \{H,VH\}$; $[1/10]$

R_{12}:　IF $a_1 = L$ AND $a_5 = M$

　　　THEN $\mathbf{w} \Rightarrow \{H,VH\}$; $[1/10] =$

R_{13}:　IF $a_4 = L$ AND $a_5 = M$

　　　THEN $\mathbf{w} \Rightarrow \{H,VH\}$; $[1/10] =$

R_{14}:　IF $a_1 = L$ AND $a_2 = H$ AND $a_3 = H$ AND $a_4 = M$ AND $a_5 = L$

　　　THEN $\mathbf{w} \Rightarrow \{H,VH\}$; $[1/10]$

!　　　IF $a_1 = L$ AND $a_2 = H$ AND $a_3 = H$ AND $a_4 = M$ AND $a_5 = L$

　　　THEN $\mathbf{w}\{M\} \Rightarrow \{H,VH\}$

$\{R\}_7$: the rule set that covers the partition subset {H}

R_{15}:   IF $a_4 = L$ AND $a_5 = M$

THEN $\mathbf{w} \Rightarrow \{H\}$; [1/1] =

R_{16}:   IF $a_1 = L$ AND $a_5 = M$

THEN $\mathbf{w} \Rightarrow \{H\}$; [1/1] =

R_{17}:   IF $a_4 = M$ AND $a_5 = H$

THEN $\mathbf{w} \Rightarrow \{H\}$; [1/1]

R_{18}:   IF $a_1 = L$ AND $a_2 = H$ AND $a_3 = H$ AND $a_4 = M$ AND $a_5 = L$

THEN $\mathbf{w} \Rightarrow \{H\}$; [1/5]

!   IF $a_1 = L$ AND $a_2 = H$ AND $a_3 = H$ AND $a_4 = M$ AND $a_5 = L$

THEN $\mathbf{w}\{M\} \Rightarrow \{H\}$

R_{19}:   IF $a_1 = L$ AND $a_2 = H$ AND $a_3 = H$ AND $a_4 = M$ AND $a_5 = L$

THEN $\mathbf{w} \Rightarrow \{H\}$; [1/5]

!   IF $a_1 = L$ AND $a_2 = H$ AND $a_3 = H$ AND $a_4 = M$ AND $a_5 = L$

THEN $\mathbf{w}\{VH\} \Rightarrow \{H\}$

R_{20}:   IF $a_2 = L$ AND $a_4 = H$

THEN $\mathbf{w} \Rightarrow \{H\}$; [1/5]

$\{R\}_8$: the rule set that covers the partition subset {VH}

R_{21}:   IF $a_4 = VH$

THEN $\mathbf{w} \Rightarrow \{VH\}$; [4/5]

R_{22}:   IF $a_1 = L$ AND $a_2 = H$ AND $a_3 = H$ AND $a_4 = H$ AND $a_5 = H$

THEN $\mathbf{w} \Rightarrow \{VH\}$; [1/5]

!   IF $a_1 = L$ AND $a_2 = H$ AND $a_3 = H$ AND $a_4 = H$ AND $a_5 = H$

THEN $\mathbf{w}\{H\} \Rightarrow \{VH\}$

Note that [m/n] denotes that m out of n training samples are covered by the rule. Rules that have been indicated by '=' are to be considered as identical. Finally, rules that are preceded by an '!' are covering samples erroneously.

In order to evaluate new samples, we generate the fuzzy membership structure for each partition encountered in the rule tree, as described in Section A3. As an example we illustrate the membership structure $\{f_{\{VL,L\}}, f_{\{M,H,VH\}}\}$, and $\{f_{\{VL,L\}}, f_{\{M\}}, f_{\{H,VH\}}\}$, shown in Table A6.

Let us recall the essence of the calculus based on upper and lower bounds. Assume we consider R_1 from $\{R\}_1$, the rule set that covers the partition subset $\{VL,L\}$.

$$R_1: \text{IF } a_3 = L \text{ THEN } \mathbf{w} \Rightarrow \{VL,L\}$$

is to read as a qualified rule like

$$\text{USUALLY IF } a_3 = L \text{ THEN } \mathbf{w} \Rightarrow \{VL,L\}$$

This qualified rule represents the set of encountered training samples:

$$\{\text{IF } a_3 = L \text{ THEN } [f_{\{VL,L\}}(w_i), f_{\{M,H,VH\}}(w_i)]\}_i$$

Note that the rule is satisfied for $i = 1,2,3,4,5,6$.

If we observe a new sample $\mathbf{w}'$ which satisfies $a_3(\mathbf{w}') = v_3'$, we infer

$$[f_{\{VL,L\}}(\mathbf{w}'), f_{\{M,H,VH\}}(\mathbf{w}')], \text{ where}$$

Table A6: Induced membership values for some of the partitions encountered in the rule tree.

w	$f_{\{VL,L\}}$	$f_{\{M,H,VH\}}$	$f_{\{M\}}$	$f_{\{H,VH\}}$
1	0.59	0.41	0.07	0.34
2	0.59	0.41	0.07	0.34
3	0.59	0.41	0.07	0.34
4	0.65	0.35	0.11	0.24
5	0.69	0.31	0.04	0.27
6	0.65	0.35	0.11	0.24
7	0.21	0.79	0.26	0.53
8	0.10	0.90	0.24	0.66
9	0.10	0.90	0.22	0.68
10	0.15	0.85	0.20	0.65
11	0.16	0.84	0.18	0.66
12	0.10	0.90	0.24	0.66
13	0.16	0.84	0.16	0.68
14	0.09	0.91	0.14	0.77
15	0.16	0.84	0.14	0.70
16	0.10	0.90	0.20	0.70
17	0.09	0.91	0.12	0.79
18	0.16	0.84	0.16	0.68
19	0.16	0.84	0.14	0.70

$$f_{\{VL,L\}}(\mathbf{w}') = \max_i[\min[r(v',v_i^*), f_{\{VL,L\}}(w_i)]] \text{ , and}$$

$$f_{\{M,H,VH\}}(\mathbf{w}') = \max_i[\min[(r(v',v_i^*), f_{\{M,H,VH\}}(\mathbf{w}')]].$$

Here, $r(v',v_i^*)$ is the 'premise matching coefficient'.
  In our example, $r(v',v_i^*) = 1$ if $v' = L$

$$= 0 \text{ otherwise.}$$

Then, from Table A5, we observe that

$$f_{OUB}(\mathbf{w}') = \max_i[f_{\{VL,L\}}(w_i)] = 0.69, \text{ and}$$

$$f_{PLB}(\mathbf{w}') = \max_i[f_{\{M,H,VH\}}(w_i)] = 0.41$$

$$(i = 1, 2, 3, 4, 5, 6)$$

Schematically:

$R_1(\mathbf{w}') \Rightarrow f_{\{VL,L\}}(\mathbf{w}') = 0.69$ AND $f_{\{M,H,VH\}}(\mathbf{w}') = 0.31$ as the optimistic upper bound (OUB),

and $\quad \Rightarrow f_{\{VL,L\}}(\mathbf{w}') = 0.59$ AND $f_{\{M,H,VH\}}(\mathbf{w}') = 0.41$ as the pessimistic lower bound (PLB).

Note that the OUB is 'less fuzzy' ('crisper') than the PLB.
  Moreover, the OUB reflects the fact that the rule is USUALLY true. If the rule is ALWAYS true, than the OUB and the PLB approach [1, 0]. The PLB accounts for the fact that the rule is NOT ALWAYS true and that cases are encountered that show less distinct behaviour.
  Clearly, one 'counter-example' or 'exception' in the training set may cause a PLB near to 0.5. However, at the same time, such an exception also lowers the OUB. So, the presence of exceptions is noticable when the OUB approaches the PLB from above.
  Now let us assume a new training sample $\mathbf{w}'$, given below

$\mathbf{w}$	a_1	a_2	a_3	a_4	a_5	d
$\mathbf{w}'$	L	H	H	M	L	H

The system inference (for the given training set and the instance $\mathbf{w}'$ above) yields the following bounds.
  Rule R_8 is fired and we find the bounds valued as follows.

$R_8(\mathbf{w}') \Rightarrow f_{\{VL,L\}}(\mathbf{w}') = 0.09, f_{\{M,H,VH\}}(\mathbf{w}') = 0.91$ as the optimistic upper bound;

$\quad \Rightarrow f_{\{VL,L\}}(\mathbf{w}') = 0.21, f_{\{M,H,VH\}}(\mathbf{w}') = 0.79$ as the pessimistic lower bound.

If we denote the system's OUB response as S_{OUB}, and the system's PLB response as S_{PLB}, and the expert judgement as J (here $d = H$), then

$$D_{OUB}(S_{OUB},J) = |0.91 - 1| = 0.09 \; (\ll 0.5) \text{ and}$$

$$D_{PLB}(S_{PLB},J) = |0.79 - 1| = 0.21 \; (\ll 0.5).$$

So, the rule (R_8) lines up very well with the expert's judgement, J, and no modification of this rule is needed.

We now add $\mathbf{w}'$ to the training set and establish new membership assignments on the basis of the augmented training set.

On a lower level in the rule tree, the effects of $\mathbf{w}'$ may be more noticeable than on the highest level. On the highest level, the rules are more robust as the discriminative impact is much lower. On a lower level, however, we have a smaller number of training samples and the discriminative impact is much higher. Also, at lower levels the chance of 'misjudgement' is much higher.

As an example, we consider those rules which discriminate between subsets of {M,H,VH}: for instance, rules that account for {M} ($R_{23} - R_{25}$), and rules that account for {H,VH} ($R_{10} - R_{14}$). Only the premises for rules R_{14} and R_{24} are satisfied. So these rules are in effect.

Note that if *no* rule is satisfied, we simply have to add the full $R(\mathbf{w}')$ to the rule set of the appropriate concept which aims to cover that concept completely.

Note also that the premises of rule R_{14} and rule R_{24} are the same; they differ in the concept consequent. In other words they are each other's exception!

Now, R_{14} reads as follows:

R_{14}:   IF $a_1 = L$ AND $a_2 = H$ AND $a_3 = H$ AND $a_4 = M$ AND $a_5 = L$

THEN $\mathbf{w} \Rightarrow$ {H,VH}.

The system inference (for the given training set and the instance $\mathbf{w}'$ above), yields

$$R_{14}(\mathbf{w}') \Rightarrow f_{\{VL,L\}}(\mathbf{w}') = 0.10 \text{ AND } f_{\{M\}}(\mathbf{w}') = 0.24 \text{ AND}$$

$$f_{\{H,VH\}}(\mathbf{w}') = 0.66$$

Since only one case is in the training set, we find the OUB to be the same as the PLB. If the OUB and the PLB are the same then either we have only one case in the training set (like here) or as many for which all tied attributes hold. Note that this number of 'identical' cases do have impact on the membership assignment to the corresponding concept.

Again, $D(S,J) = |0.66 - 1| = 0.34 \; (< 0.5)$, and no modification is needed.

As before, we add $\mathbf{w}'$ to the training set and establish new membership assignments on the basis of the augmented training set.

R_{24} results in the same way in

$$R_{24}(\mathbf{w}') \Rightarrow f_{\{VL,L\}}(\mathbf{w}') = 0.10 \text{ AND } f_{\{M\}}(\mathbf{w}') = 0.24 \text{ AND}$$

$$f_{\{H,VH\}}(\mathbf{w}') = 0.66$$

However, now the expert's judgement no longer lines up with the system decision, as follows from

$$D(S,J) = |0.24 - 1| = 0.76 \ (> 0.5).$$

So, with respect to concept {M}, this rule does not respond in a favourable way. Either, the case from which R_{24} originated is to be considered as an exception, or future instances (with concept consequent {M}) have to show the relevance of rule R_{24}. By now, the case encountered remains alone in the training set.

    Note that in neither of the above cases did new representative rules have to be found in order to cover the decision space completely.

    We next consider $\mathbf{w}'$ being given as

$\mathbf{w}$	a_1	a_2	a_3	a_4	a_5	d
$\mathbf{w}'$	M	H	M	M	M	M

We observe two facts:

(i) $\mathbf{w}'$ is neither covered by $R_1(R_2)$ nor by $R_8(R_9)$, and

(ii)   the only rule that is satisfied is R_{23}.

The consequence of these facts is that the set of representative rules for {M,H,VH} needs to be reconsidered since $\mathbf{w}'$ will not be covered by the existing rules. Also, the result of R_{23} needs to be analyzed.

    We obtain

$$R_{23}(\mathbf{w}'): \{f_{\{VL,L\}}(\mathbf{w}') = 0.21, f_{\{M\}}(\mathbf{w}') = 0.26, f_{\{H,VH\}}(\mathbf{w}') = 0.53\}.$$

From this it follows that

$$D(S,J) = |0.26 - 1| = 0.74 \ (> 0.50),$$

meaning that the rule is found to cover $\mathbf{w}'$ incorrectly and thus needs to be reconsidered.

    With respect to {M}, we reconsider finding the most representative rules to cover the concept {M}. Thus, we have:

$\mathbf{w}$	a_1	a_2	a_3	a_4	a_5	d
7	M	H	H	M	M	M
8	L	H	H	M	L	M
9	L	H	H	H	L	M
$\mathbf{w}'$	M	H	M	M	M	M

Then, the following representative rules are found:

R_{23}':   IF $a_1 = M$ AND $a_4 = M$

THEN $\mathbf{w} \Rightarrow \{M\}$; [2/4]

R_{25}:   IF $a_1 = L$ AND $a_4 = H$ AND $a_5 = L$

THEN $\mathbf{w} \Rightarrow \{M\}$; [1/4]

R_{24}:   IF $a_1 = L$ AND $a_2 = H$ AND $a_3 = H$ AND $a_4 = M$ AND $a_5 = L$

THEN $\mathbf{w} \Rightarrow \{M\}$; [1/4]

R_{23}' is the modified rule R_{23}, becoming less specific (or more general) and tying another attribute.

For $\{M,H,VH\}$ a new representative rule R' is needed for complete coverage of all training samples in this set:

R':   IF $a_3 = M$

THEN $\mathbf{w} \Rightarrow \{M,H,VH\}$; [1/14],

leaving $R_{8[13/14]}$ or $R_{9[13/14]}$ unchanged.

Note that the dissimilarity matrix is now 20×20 and the values of the $\mathbf{w}'$ entries will affect all the membership assignments.

As a result, given the properties of the membership operator, we observe that

$$\{f_{\{M\}}(wj)\}_{j\,=\,\{1,q\{M\}\}}$$

will slightly improve.

B | EDAPLUS functionality

This appendix summarizes the commands and functions for performing useful actions in EDAPLUS. Additionally, examples will be given in order to demonstrate the main functionality.

EDAPLUS is designed to be an educational tool to clarify, exemplify and illustrate nearly all the methods discussed in this book. Moreover, the package comes with a rule-based decision network to allow the reader to make explicit his understanding of issues like clustering tendency and clustering validity in the form of rules. An easy rule language has been offered in Chapter 9. A public domain editor has been integrated with EDAPLUS and offers continuously the possibility of expanding and modifying the rule base in operation.

EDAPLUS runs on a standard IBM PC or compatible with at least 1024 kB of memory. The system runs from version DOS 2.0 and up. EDAPLUS is distributed on a single DS,HD diskette.

For proper functioning, the following files must be available on disk:
- EDAPLUS.EXE
- EDITOR.EXE
- HELP.EDA
- PARAMS.EDA
- RULEBASE.EDA
- *.DAT

There should be sufficient disk space for the *.DIS (the *dissimilarity matrix*) which is generated automatically after loading the *.DAT (the *input data matrix*).

B1 Commands and functions

We now describe how EDAPLUS comes into operation.

After the system prompt, just type EDAPLUS ‹CR›. After typing this command, the program (EDAPLUS) and the EDITOR will be loaded. Additionally, the help file, the parameter file (PARAMS) and the rule base (RULEBASE) are loaded into the program. Then, the main menu appears:

– PARTITIONAL ⟨P⟩

– HIERARCHICAL ⟨H⟩

– LOAD DATA ⟨L⟩

– TRANSFORM ⟨T⟩

– STATISTICS ⟨S⟩

– QUIT ⟨Q⟩

After selecting ⟨L⟩, the directory mask appears:

for example A:*.DAT

All data files will be displayed and left/right-up/down arrow manipulation leads to the selection of the *name*.DAT file as desired. These data will be refered to as *user data*. As mentioned before, *name*.DIS will automatically be generated (using the standard Euclidean distance coefficient). If another coefficient is desired, the stand-alone program DATA can be used to generate a different *name*.DIS file. Once such a dissimilarity file is detected on the disk (after loading the *name*.DAT file), generating a standard *name*.DIS will be skipped.

   If we select from the main menu the partitional clustering approach, using the commands, ⟨P⟩, we directly enter the options menu, ⟨F6⟩. See Figure B1.

   From the options menu we may select

– upper window / lower window

– user data / random data / Gaussian data

– *K*-means algorithm / Minimal Spanning Tree Algorithm.

The selection ⟨upper window⟩/⟨user data⟩/⟨*K*-means algorithm⟩ yields the result as shown in Figure B2.

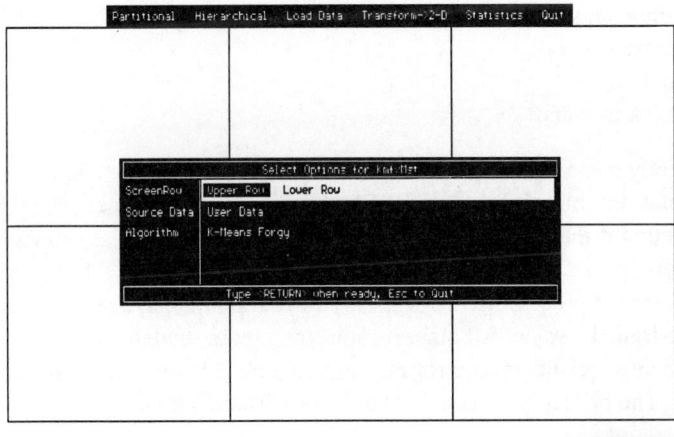

Figure B1

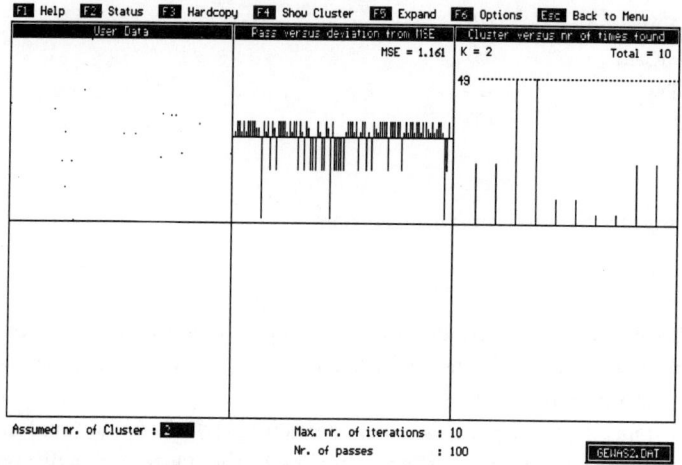

Figure B2

In the upper window, we see the user data (mapped on a 2D, a priori defined, projection space), the deviation from the MSE over all passes, and the frequencies of cluster appearances (cluster dominance), respectively. At the bottom of the screen, the user has specified the number of clusters (*K*) to look for, the maximum number of iterations for each pass, and the number of passes; 2, 10 and 100, respectively.

Also in Figure B2, we see the function key menu for partitional clustering.
- F1   help
- F2   status
- F3   hardcopy
- F4   show cluster
- F5   expand
- F6   options
- esc   back to menu

F1 offers briefly some help information and refers to specific page numbers if applicable. Due to limitations of memory space, F5 has not been implemented in this version. Generally, if that is the case, a message will appear on the screen, as shown in Figure B3.

⟨F2⟩(status) gives the status information so far; see Figure B4. As a result of the foregoing *K*-means processing, we will only have *alpha(2)*, *m user data*, and *perc user data* being assigned a value. All other parameters remain ⟨undefined⟩.

⟨F3⟩(hardcopy) results in an error message (Figure B5) when the printer is missing or is turned off. The printer is expected to be Epson 80 compatible.

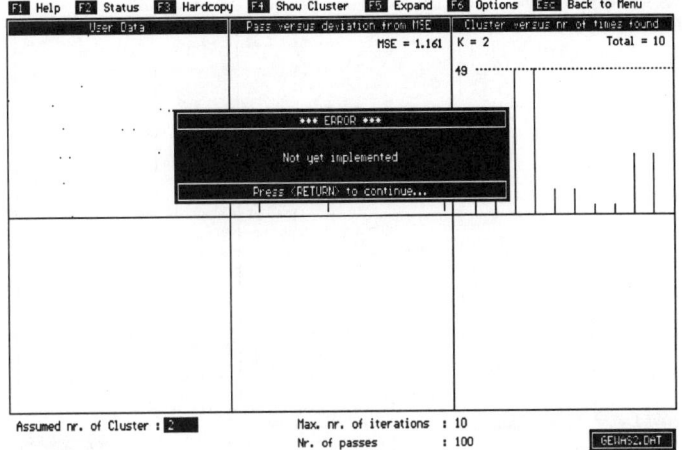

Figure B3

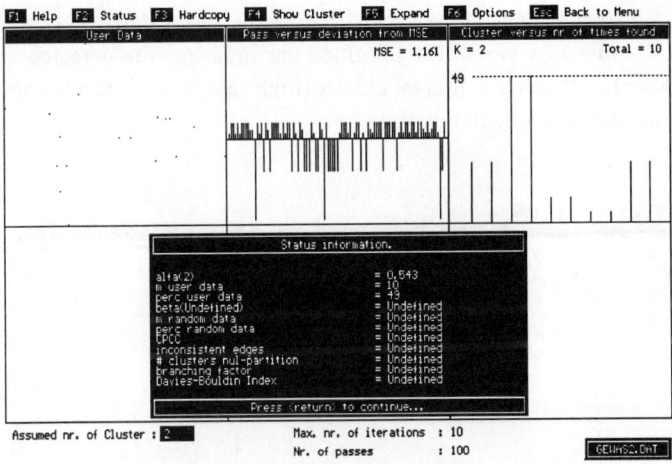

Figure B4

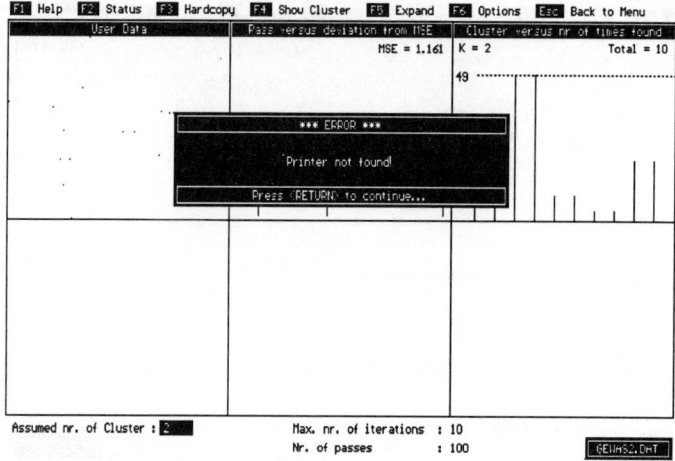

Figure B5

⟨F4⟩(show cluster) enables the user to visualize any of the detected clusters by using left/right arrow control. A vertical highlighted bar indicates the detected cluster to be displayed. Figure B6 shows a dominant cluster (high rate of occurrences) and Figure B7 shows an odd cluster (low rate of occurrences).

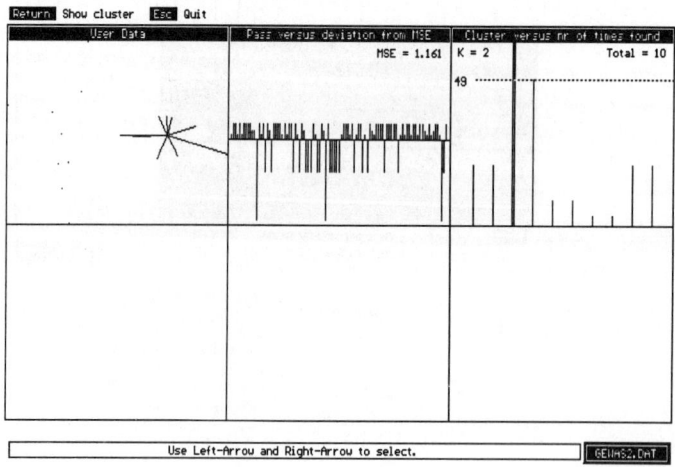

Figure B6

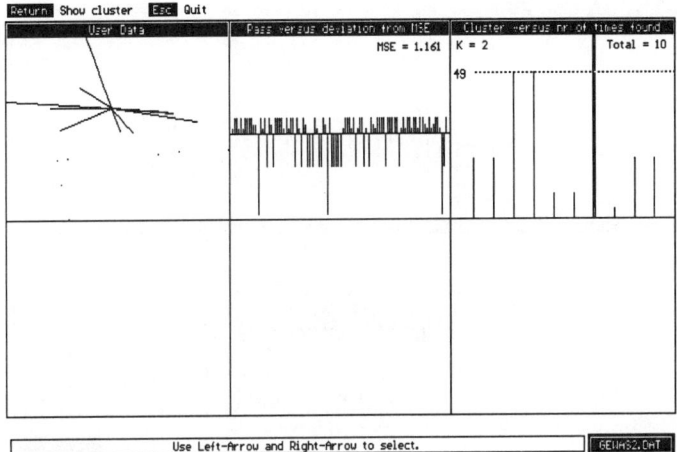

Figure B7

Finally, ‹F6›(options) enables us to select further processing steps. As Figure B8 shows, we wish to compare our previous findings with the results derived from *random data*. Random data will be generated with the same sample size, same dimensionality, and within a cubic sampling window derived from the user data. The randomly generated data will be displayed on the same (user data defined) 2D projection. The result is shown in Figure B9 (lower window), again for $K = 2$.

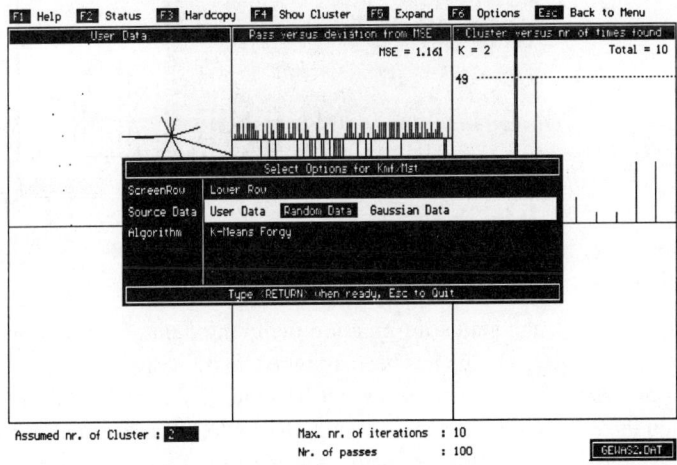

Figure B8

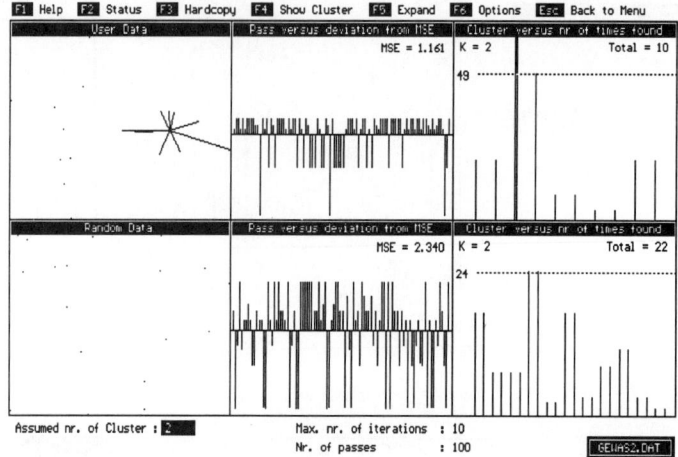

Figure B9

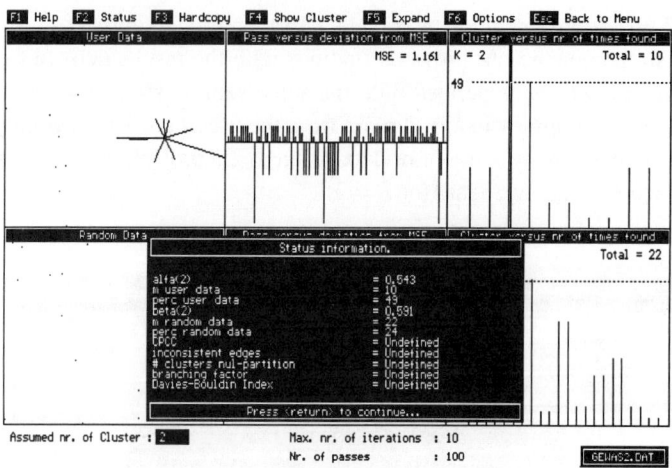

Figure B10

‹F2›(status) will show the status information including *beta(2)*, *m random data*, and *perc random data* to which a value has been assigned. See Figure B10.

Through ‹F6›, we now may wish to inspect some MST results; see Figure B11. We firstly generated the *edge-length distribution* and the *CPCC distribution* for random data (upper window) and repeated this operation for the user data. At the bottom of the screen, the MST results can be evaluated on a 1–5 scale (1 = *very low*, 2 = *low*, 3 = *medium*, 4 = *high*, 5 = *very high*). These results will also appear in the status information

(⟨F2⟩), as shown in Figure B12.

After ⟨esc⟩, we may select ⟨S⟩, in order to compute the Davies–Bouldin index. Not surprisingly, a message will appear saying that some dendrogram level has to be specified in advance (Figure B13).

At this stage it might also be tempting to consult the decision network.

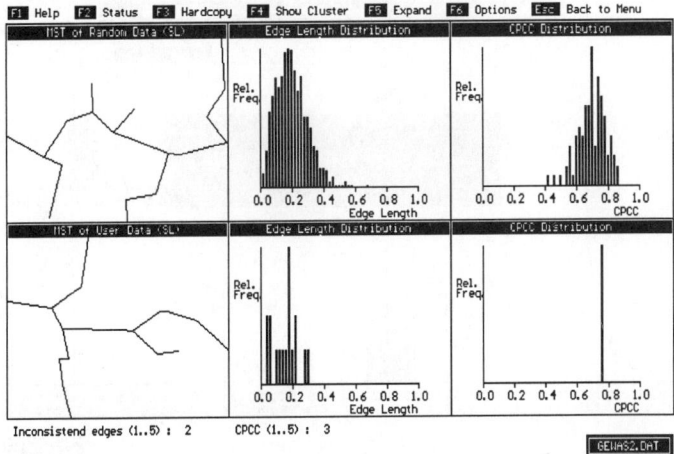

Figure B11

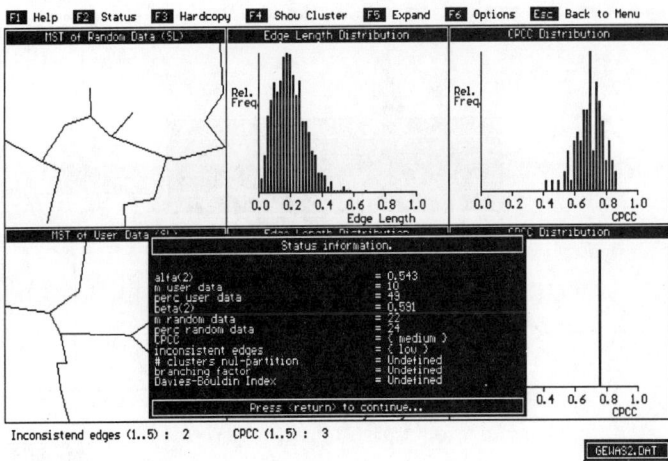

Figure B12

‹F2›(status)‹classify› starts the rule-based reasoning; (Figure B14). The rule base must include messages describing how to proceed when no conclusions can be drawn yet. Such a message is shown in Figure B15. After that, the reasoning process/decision making is aborted, as shown in Figure B16. Obviously, so far, no conclusions have been reached. This is displayed in Figure B17.

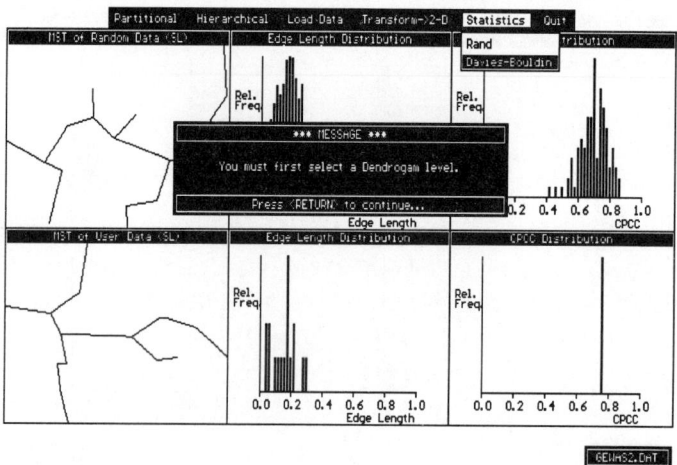

Figure B13

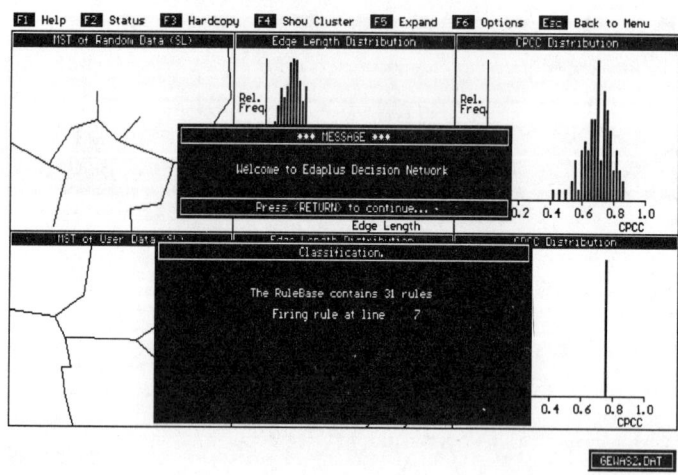

Figure B14

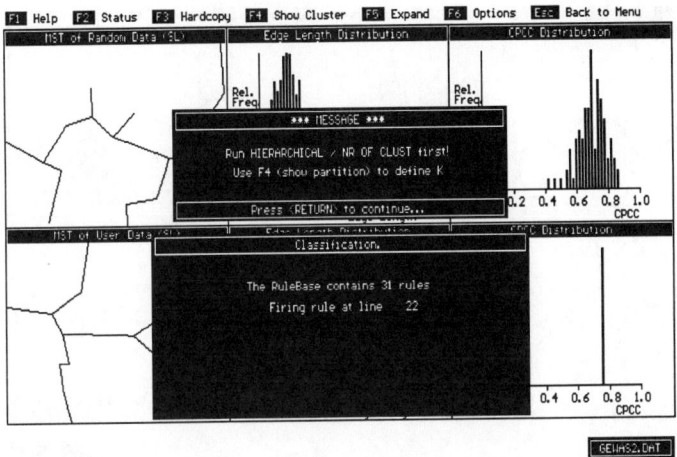

Figure B15

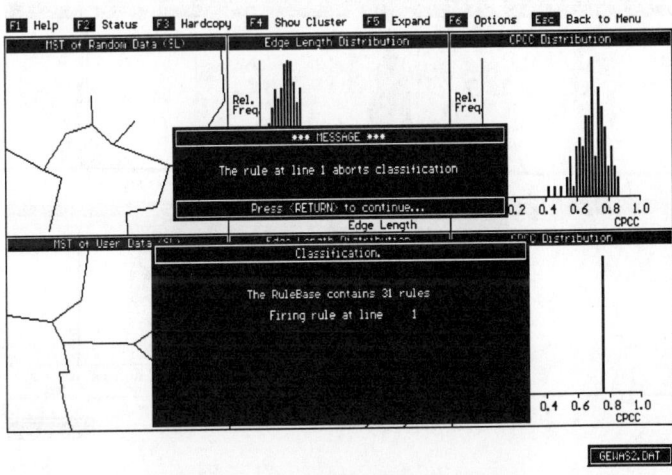

Figure B16

Through ‹esc› and ‹H›, we enter the hierarchical clustering approaches. Based upon complete linkage and the Davies–Bouldin index, we may estimate a best guess for the *number of clusters*. This is achieved by using the menu option ‹# of clusters›. The result will be displayed on the screen, as shown in Figure B18.

344   *Computer-assisted reasoning in cluster analysis*

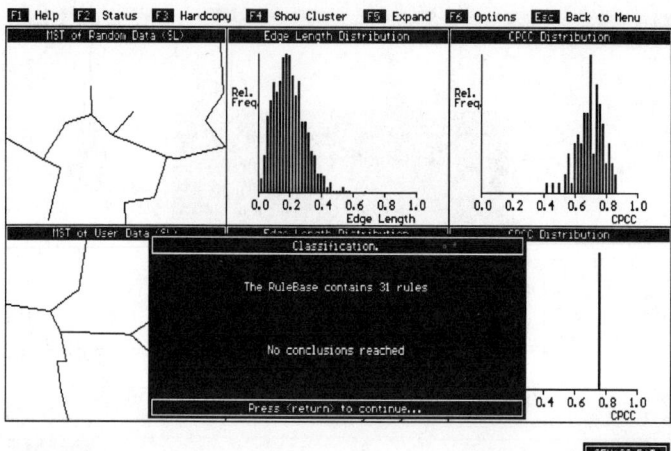

Figure B17

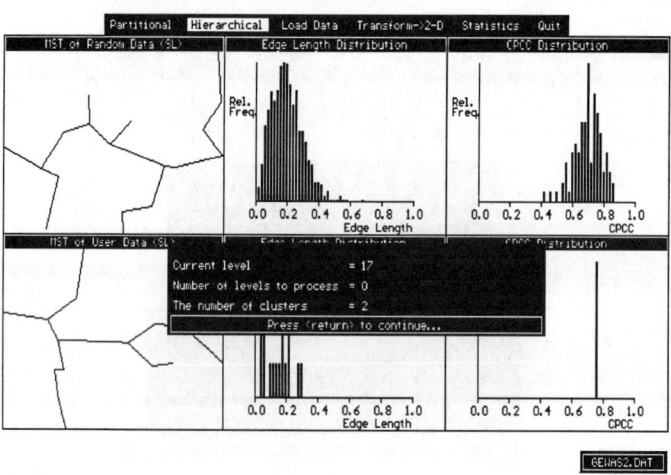

Figure B18

When using the menu option ‹dendro›, we may produce single-linkage, complete-linkage, and average-linkage clusterings. As an example, Figure B19 shows the Complete Linkage dendrogram. Here we have the following function keys:

  – F1   help   – F2   status
  – F3   hardcopy   – F4   show partition
  – esc back to menu

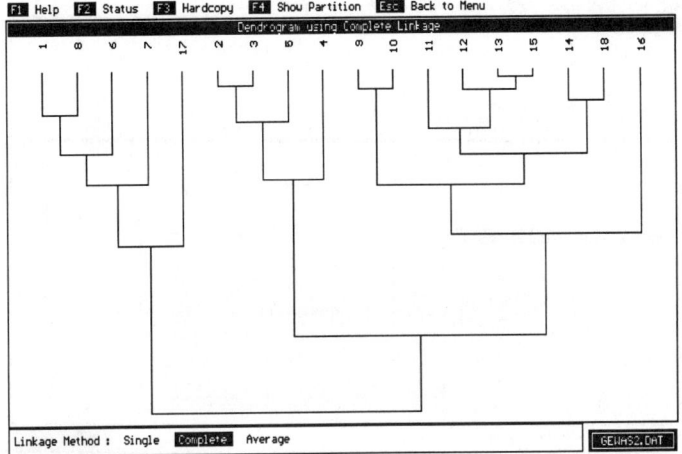

Figure B19

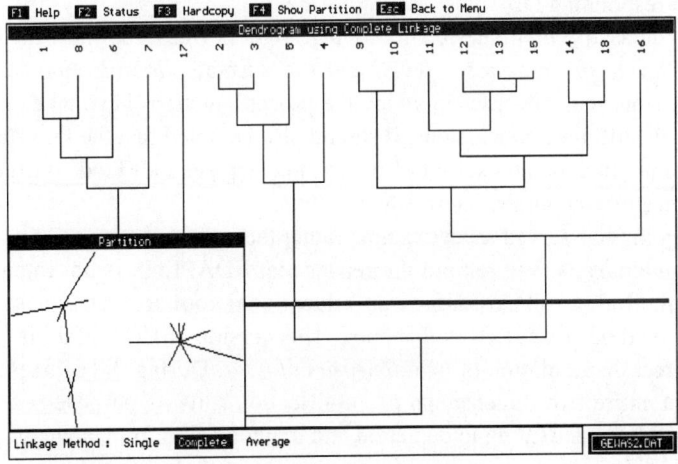

Figure B20

‹F4›(show partition) enables us to select a level on the basis of which the corresponding partition can be displayed on the projection plane derived from the user data. A high-lighted horizontal line represents the level and can be displaced by the up/down arrow keys. Figures B20 and B21 show two partitioning levels. Once a level has been selected, the command sequence

‹esc›‹STATISTICS›‹Davies Bouldin›

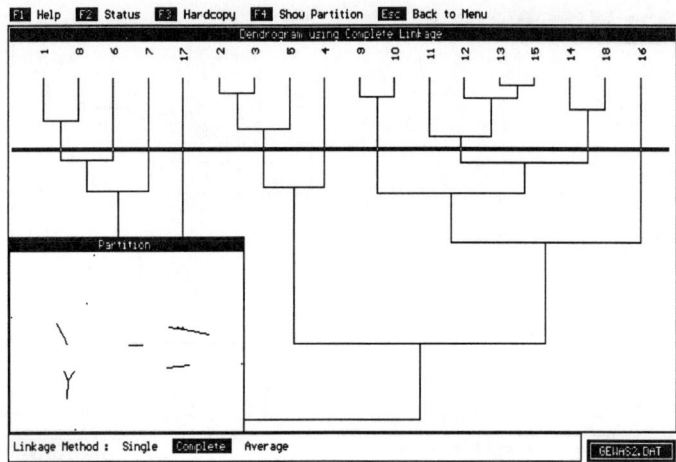

Figure B21

yields the corresponding DB value (see Figure B22).

Next, ‹F2›‹classify› continues the reasoning. Figures B23–B26 show the *initial level, the support level, the diagnosis level,* and the ultimate *classification* (conclusions), respectively. Note that the order in which the processing steps have taken place has no influence on the ultimate conclusions. It should also be noted that the knowledge base to be designed may include any subset of processing steps and all kinds of interpreting or evaluating questions about the intermediate results.

The foregoing has served as an example rather than prescribing a general strategy for performing an analysis. We remind the reader that EDAPLUS is not intended to be a research tool. Rather, EDAPLUS is an educational tool to clarify, exemplify and illustrate the methods discussed in this book. This version of EDAPLUS is intentionally *limited* to direct manipulation of *two-dimensional data*. During the courses, it has been observed that entire two-dimensional exemplification suits its purpose best. Moreover, this choice has been activating to depict on and to understand from, some of the inherent problems of higher dimensional data, in a two-dimensional space. Some examples will be discussed in the next section.

An earlier version, EDA, is also supplied on the accompanying diskette. This version includes the *projections* (eigenvector projection, discriminant projection, Sammon projection, and multi-dimensional scaling) and allows data to be manipulated in up to *ten* dimensions.

EDA has the same command structure as EDAPLUS, except that rule-based reasoning is not supported.

Finally, in support of EDA and EDAPLUS, five *stand-alone programs* are available. They are

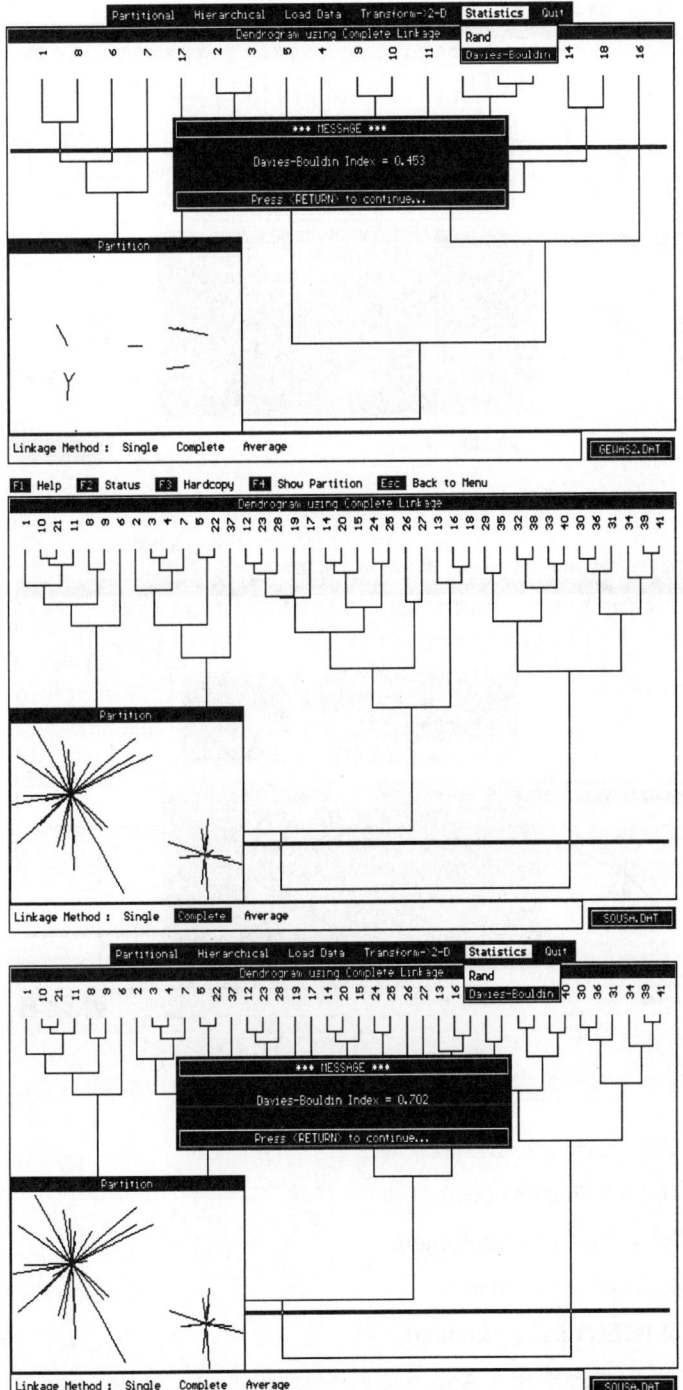

Figure B22

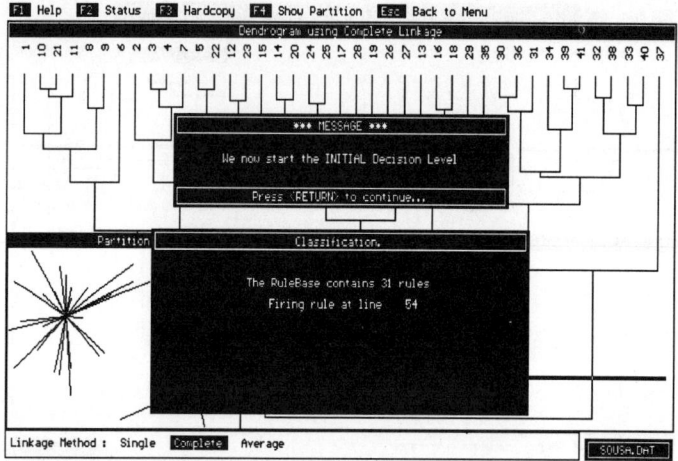

Figure B23

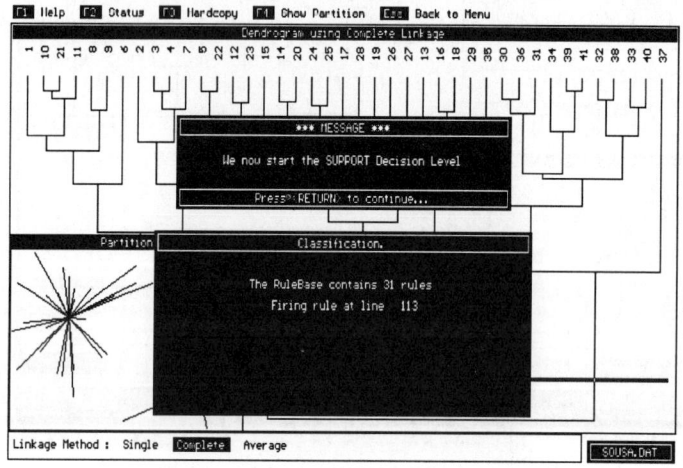

Figure B24

DATA (data conversion)

MISDAT (missing data)

FEATSEL (feature selection)

PROJECT (projections)

DATGEN (data generation)

and will be discussed in Section A4.

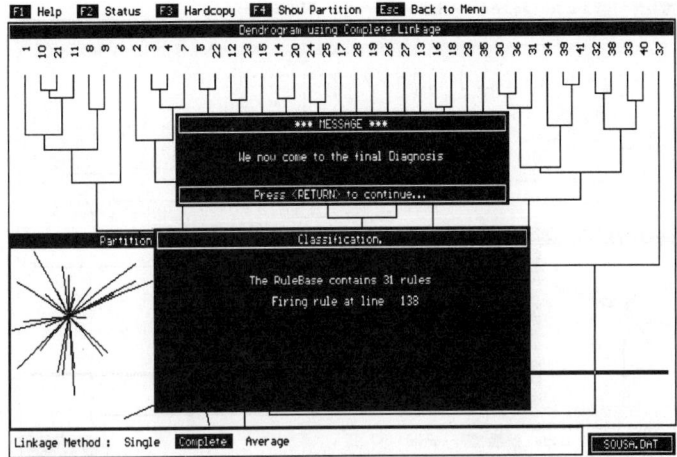

Figure B25

B2 Useful actions and logging

In the previous section, the series of actions were straightforward. After a session, the following information is retained:

input
A:\>type gewas2.dat ‹CR›

```
18
2
-133        34        1
-164        31        1
-154        28        1
-183        24        1
-173        23        1
-123        38        1
-172        52        1
-133        48        1
-121        25        1
-118        21        1
-94         14        2
-94         25        2
-71         32        2
-94          8        2
-89         21        2
-72         14        2
-94         47        2
-94         19        2
```

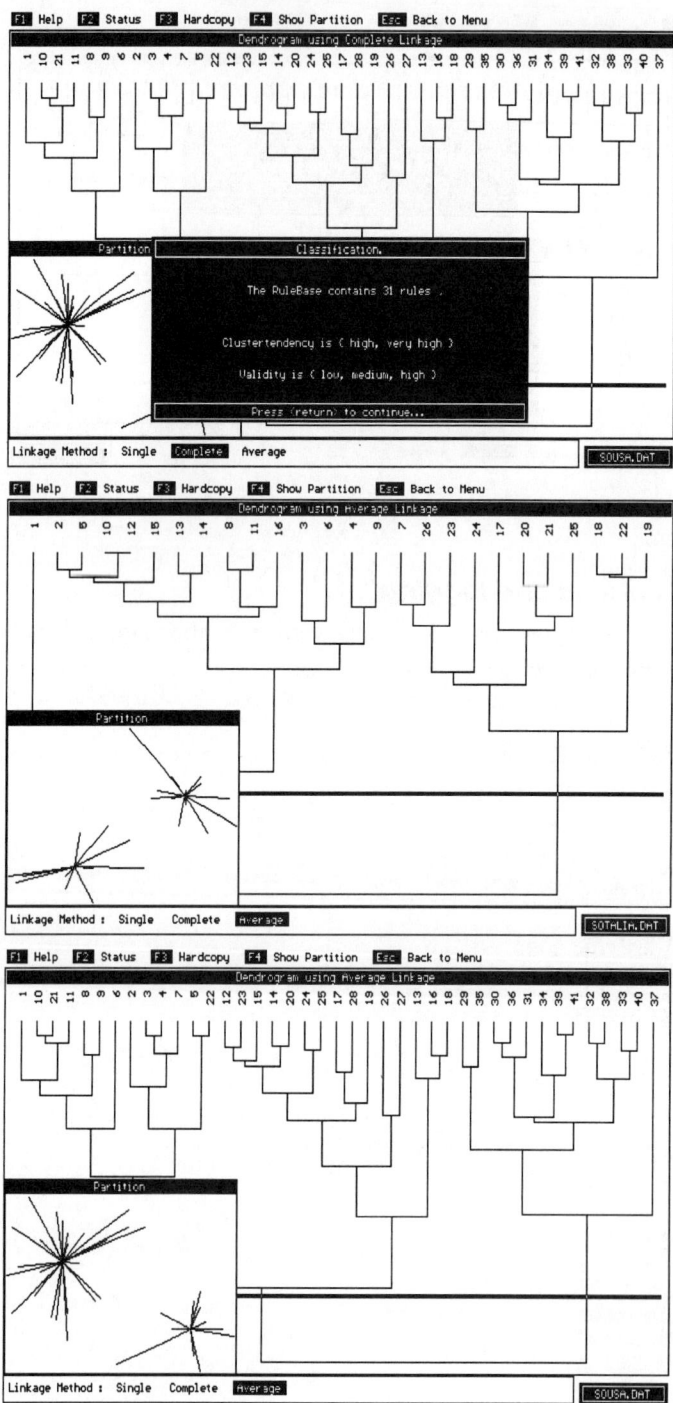

Figure B26

status information
The status information as obtained in the session is stored in the file *name*.STA which is generated automatically.

A:\>type gewas2.sta ‹CR›

GEWAS2.DAT
alpha(2)	0.547
m user data	12
perc user data	45
beta(2)	0.647
m random data	28
perc random data	18
CPCC	{medium}
inconsistent edges	{low}
# clusters nul-partition	3
branching factor	0.500
Davies-Bouldin Index	undefined

conclusions
The session yielded conclusions like

CT is {*low,medium*}
VAL is {*very low,low,medium*}

As mentioned before, *name*.DIS (the dissimilarity matrix) is generated automatically and can be retrieved by

A:\>type gewas2.dis ‹CR›

Now we describe some actions which account for the true data (GEWAS.DAT) where GEWAS2.DAT originated from, the label information, and the true dissimilarity matrix (GEWAS.DIS).

I. Using label information
The true data is four-dimensional as follows:

GEWAS.DAT
18
4
−133	−83	76	34	1
−164	−84	67	31	1
−154	−85	68	28	1
−183	−103	49	24	1
−173	−95	64	23	1

−123	−50	81	38	1
−172	−69	81	52	1
−133	−71	73	48	1
−121	−57	67	25	1
−118	−53	67	21	1
−94	−24	60	14	2
−94	−38	64	25	2
−71	−45	57	32	2
−94	−56	61	8	2
−89	−35	64	21	2
−72	−36	43	14	2
−94	−30	90	47	2
−94	−67	56	19	2

GEWAS2.DAT originated from the above data by selecting the best two features (1 and 4), given the labelling (labels 1 and 2). Note that this was done using FEATSEL (see Section B4).

As labels are assigned to the data, we may opt for a *discriminant projection*. This is performd by EDA; its output is depicted in Figure B27. Here we have the option to switch from object numbers to labels. The screen of Figure B27 also shows a dominant clustering (lower window). Clearly, object 17 (label 2), and objects 9 and 10 (label 1) are misclassified by the clustering. By moving the highlighted bar horizontally, we can inspect all detected clusterings.

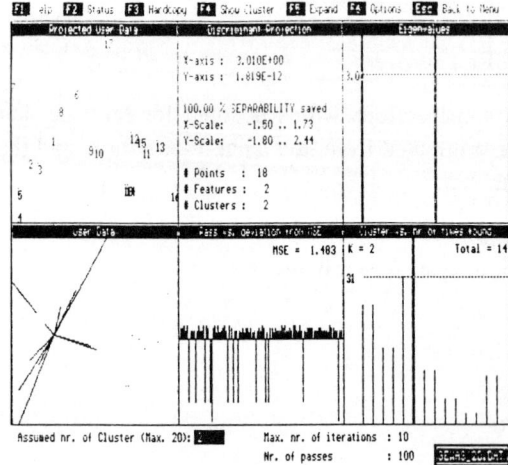

Figure B27

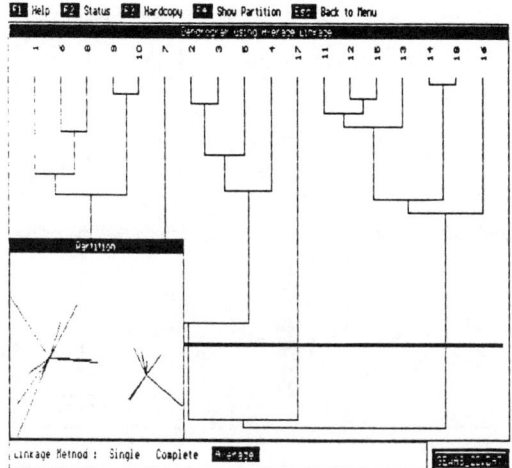

Figure B28

The dendrogram using average linkage is shown in Figure B28. Leaving object 17 as an outlier, no misclassifications are observed while cutting the dendrogram at the appropriate level (highlighted cross bar).

If we delete GEWAS_2D.DIS and replace it by GEWAS.DIS we impose the dissimilarity matrix from the original (four-dimensional) data on the projected data. Thus after

A:\>copy gewas.dis gewas_2d.dis ‹CR›

the program (EDA or EDAPLUS) will detect, after loading GEWAS_2D.DAT, that the file GEWAS_2D.DIS already exists and will use that information for further processing. This is demonstrated in Figure B29. The true (four-dimensional) MST is projected onto the two-dimensional data representation.

If we next type

A:\>delete gewas_2d.dis ‹CR›

and ‹LOAD DATA›GEWAS_2D.DAT the corresponding GEWAS_2D.DIS (from the two-dimensional data) will be generated. The result (Figure B30) shows the true MST of the two-dimensional data.

Comparing Figure B29 and B30 shows that 'local' structures have been preserved to a certain extent. One might even count the number of preserved edges (9 out of 17). However, most likely only a few edges are obscured dramatically by performing the discriminant projection. Note that this projection does not aim to preserve MST properties.

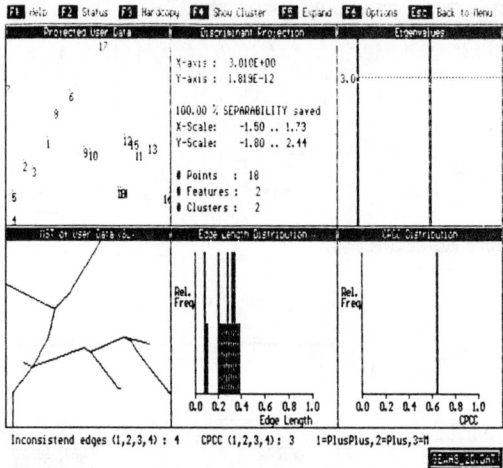

Figure B29

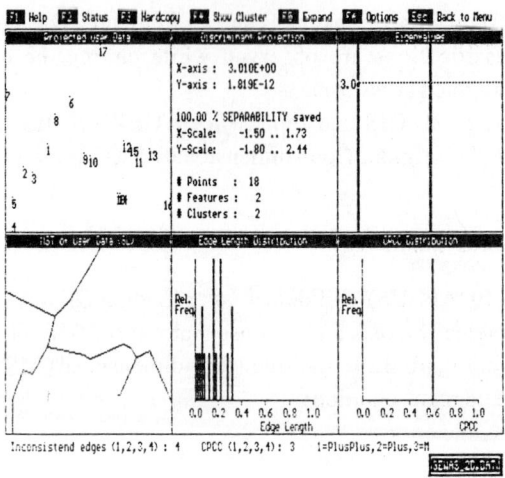

Figure B30

II. Analyzing the attributes (features)

For some applications, both an analysis of objects (*Q-analysis*) and an analysis of attributes (*R-analysis*; Sneath and Sokal (1973)) of the same data matrix can be informative.

Assume that we have a pattern matrix of 28 objects, measured on 16 attributes. The format will be

28 (*number of rows*)
16 (*number of columns*)
$x(1,1) x(1,2) x(1,3) \ldots x(1,16)$
$x(2,1) x(2,2) x(2,3) \ldots x(2,16)$

.

.

.

$x(28,1) x(28,2) x(28,3) \ldots x(28,16)$

Q-analysis leads to the interpretation of 28 objects in a 16-dimensional space.
    If we reverse the rows and the columns, we obtain

16
28
$y(1,1) y(1,2) \ldots y(1,28)$
$y(2,1) y(2,2) \ldots y(2,28)$

.

.

.

$y(16,1) y(16,2) \ldots y(16,28)$

which is equal to

16
28
$x(1,1) x(2,1) x(3,1) \ldots x(28,1)$
$x(1,2) x(2,2) x(3,2) \ldots x(28,2)$

.

.

.

$x(1,16) x(2,16) x(3,16) \ldots x(28,16)$

Then, R-analysis will show which attributes (features) gave similar values, and may be useful for locating redundant attributes.
    In order to use EDAPLUS (or EDA), we first compute a dissimilarity matrix (*name*.DIS) using the program DATA. Next, we generate a *dummy*.DAT file using DATGEN.
    Then type

A:\rename *name*.DIS FEATURE.DIS ‹CR›

A:\rename *dummy*.DAT FEATURE.DAT ‹CR›

and use the EDA(PLUS) command

‹LOAD DATA›FEATURE.DAT .

We are now able to generate any dendrogram to inspect feature redundancy. Note that cutting the dendrogram and showing the partition (F4) is meaningless as we have *dummy*.DAT as the projection base.

If desired, we can obtain an appropriate projection base through multi-dimensional scaling (EDA) which takes a dissimilarity matrix as input. Note that in this way any relation matrix on objects or features can be inputted to the system.

B3 The rule base editor

To begin with, EDAPLUS accepts an *argument*. Suppose we have two rule bases on disk, named RULES1.EDA and RULES2.EDA. Suppose further that we want to use RULES2.EDA. Then, we use the DOS command

A:\›EDAPLUS RULES2.EDA ‹CR›

Note that PARAMS.EDA should contain all *parameters* to be used in RULES2.EDA. If not, an error message will occur, like

A:\RULES2.EDA(*line number*) *parameter name* undefined

Once we are in EDAPLUS, we may choose to edit the loaded rule base at any time. We only have to select ‹F2›‹edit rules› (see Figure B31).

Then the full screen is used to display the active rule base:

```
RULES2.EDA.....t...F1-Help
if a then
begin
     a := not a; (*reset a*)
     abort;
end;
if undefined(COUNT) then
begin
     Message('Welcome to Edaplus Decision Network');
     COUNT := 1;
end;
if undefined(b) and undefined(K) then
     b := Query_yesno(
     'input # of clusters (K) yourself?',
```

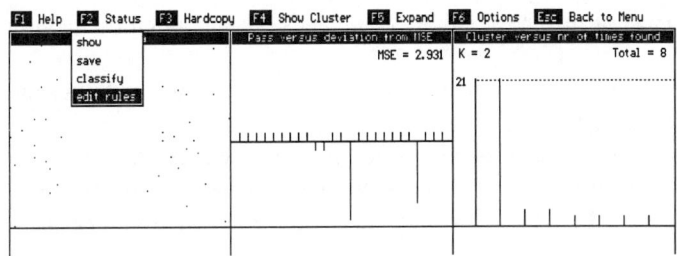

Figure B31

which we can edit by using available editing commands. Use ‹F1› Help.

A	set mark, append file;
B	format, right margin;
C	shell, execute line;
D	cursor right, line end;
E	cursor up, file top;
F	forward search, backward search;
G	goto line, goto column;
H	delete before, delete under;
I	tab, tab insert;
J	play macro, record;
K	block copy, block cut;
L	refresh screen, status;
M	return;
N	toggle insert mode;
O	toggle mode;
P	command prefix;
Q	command prefix;
R	replace, replace all;
S	cursor left, line begin;
T	delete word, delete eol;
U	block paste, block write;
V	page down, page up;
W	window, window cancel;
X	cursor down, file bottom;
Y	block delete line;
Z	exit, edit other file;

Each letter represents a control key. Each key corresponds to one or two editing commands (separated by a comma). Press the control key for the first editing function, or control_P and the control key for the second editing function.

‹Press any key to continue›

Notice that most of the above editing commands are supported by traditional keyboard functions.

We now are at the stage where a rule base editing example will illustrate most of the occuring difficulties in practice.

Assume that we plan to develop and write a new rule base to test how rules are processed in EDAPLUS. We shall name this new rule base RULETEST.EDA.

Before entering EDAPLUS, we copy the existing rule base and rename it, as follows:

A:\›copy RULEBASE.EDA RULETEST.EDA ‹CR›

Now, start EDAPLUS by typing

A:\›EDAPLUS RULETEST.EDA ‹CR›

In EDAPLUS, after ‹LOAD DATA›data and ‹H› or ‹P›, we select ‹F2›‹edit rules›. Next, we see

RULETEST.EDA .....t..F1 Help
{...
    old rulebase
                 ...}
Delete {........} and start editing the new rulebase.

IF undefined(count) then
begin
    message('This is a rulebase test');
    count := 1;
end;

IF undefined(alqha) then
begin
    alpha := 0.385;
    beta := 0.579;
    k := 2;
end;

[1.00]IF alpha < beta then
begin
    V1 := {low,medium,high,very high};
    message('rule 3 fired');
    count ;= 2;
end;

[0.8]IF alpha < 1/k then

```
begin
      V1 := {medium,high,very high};
      message('rule 4 fired');
      count := 3;
end;

[0.6]IF alpha < 1/k and beta > 1/k then
begin
      V1 := {high,very high};
      message('rule 5 fired');
      count := 4;
end;

IF undefined(V2) then
      message('program interrupt');

IF count = 4 then
begin
      CT := V1;
end; (*end of the program*)
---
```

When this editing is done we exit by typing

control_Z and
Save file (yes/no/cancel): Y

and we are back in EDAPLUS. Through ‹F2›‹classify›, we are immediately able to test the rule base.

As a result of the above editing, we obtain three (main) error messages:

A:\RULETEST.EDA(7): alqha undefined;

A:\RULETEST.EDA(18): syntax error;

A:\FATAL error: No value assigned to CT and VAL in the rule base.

We can correct RULETEST.EDA by typing

A:\›EDITOR ‹CR›

Line 7 should read:

      IF undefined(alpha) then

Line 18 should read:

```
count := 2;
```

The value assignement to CT and VAL should read:

```
begin
     CT := V1;
     VAL:= {high}; (*just a dummy*)
end; (*end of the program*)
```

After correcting these errors, the program runs properly.

When executed (‹F2›‹classify›), the following messages will successively appear on the screen:

This is a rulebase test

rule 3 fired

(certainty factor [1.00])

program interrupt

rule 4 fired

(certainty factor [0.8])

rule 5 fired

(certainty factor [0.6])

CT = {high,very high}

VAL = {high}

If one desires to change any value or rule, or to add new rules, the only command is ‹F2›‹edit rules›. After completing the edit action, just type ‹control_Z›,‹Yes›, and a new session (‹F2›‹classify›) will automatically include the updated rule base.

B4 Stand-alone programs

Five stand-alone programs are available on the accompanying diskette. They are

        DATA.EXE (data conversion)
        MISDAT.EXE (missing data)
        FEATSEL.EXE (feature selection)
        PROJECT.EXE (projections)
        DATGEN.EXE (data generation)

The module DATA is a stand-alone program which converts all kinds of raw data into EDAPLUS format. All distance measures and matching coefficients documented in Section 2.1 are available to produce the desired pattern matrix and dissimilarity matrix. The program is self-explanatory. Start up with

A:\>DATA ‹CR›

The module MISDAT is a stand-alone program which computes a dissimilarity matrix from pattern matrices containing missing values. Following Section 2.1, the methods used are based on Dixon method, 3 and 4. The program is self-explanatory. Startup with

A:\>MISDAT ‹CR›

The module FEATSEL is a stand-alone program for the selection of feature subsets. Following Section 2.2, the program includes sequential forward, sequential backward, and branch and bound feature evaluation. The program is self-explanatory.

The program expects a two-class problem where the pattern matrix should be ordered such that the patterns of class 1 come first, say n_1 in number. This number, n_1, is asked for; the remaining patterns are considered to belong to class 2. Start up with

A:\>FEATSEL ‹CR›

The module PROJECT is a stand-alone program for linear projections of multi-dimensional data. The program includes eigenvector projection and discriminant projection. The program expects to find the labels in the EDAPLUS data format. Start up with

A:\>PROJECT ‹CR›

Finally, the module DATGEN is a stand-alone program which generates both, uniformly distributed random data and Gaussian distributed random data. The program is self-explanatory. Start up with

A:\>DATGEN ‹CR›

B5 Data

Four different data sets are available on the accompanying diskette. The data may serve as predictable cases to exemplify the methodology as discussed throughout the book. They are

URL01.DAT
SOUSA.DAT

GEWAS.DAT

URL01.DAT contains 36 two-dimensional pattern vectors; SOUSA.DAT contains 41 two-dimensional pattern vectors; GEWAS.DAT contains 18 four-dimensional pattern vectors. All data originate from real measurements.

Index

a posteriori knowledge 167
a posteriori probability 180
a priori knowledge 167
abduction 175
acquisition 229
action frame 172
affinity 90
affinity decomposition 91
agenda 165
agglomerative approach 39
ambiguity 18, 179
analogical reasoning 177
antecedent 166
anti-reflexive 97
approximate reasoning 6, 78, 254
approximation 105
artificial intelligence 124, 152
artificial neural networks 250
assignment 202
associative nets 170
attributes 170

backpropagation algorithm 251
backward chaining 166, 177
basic probability assignment 186
Bayes' theorem 180
belief 180, 184
belief function 187
belief revision 146, 148
best' approximation 105
between-group scatter 32
biconditional 173
bottom-up reasoning 177
branch and bound algorithm 33
branches 173
bridge set 104

category labels 30
certain belief 183
certain evidence 183
certainty factor 147, 179, 184
chaining 67
classification 73, 74, 154, 346
classification inference 178
classification intention 198

cluster analysis consultant 152
cluster dominance 154, 223
cluster indices 54
cluster membership function 88
cluster significance 79
cluster validation 154
cluster validity 77
cluster-oriented conclusions 161
cluster-oriented reasoning 156
clustering criterion 43
clustering expertise 151
clustering strategy 7
clustering technique 74
clustering tendency 7, 18, 54, 77, 79, 131, 154
clusters 154
combining evidence 182
common-sense knowledge 195
common-sense reasoning 175
compactness 54
compiled knowledge 168
complement 82
complete linkage 124
complete subgraphs 40
completeness 231
complex network 161
computational cost 39
concentration 191
concept definition 198
concept-driven classification 295
concept-driven learning 297
concepts 216
conceptual association 214
conceptual clustering 8, 198
conclusion 18, 172, 351
conditional 173
conditional inference 116
conditional part 166
conditional probability 180
conjunction 173
conjunction of evidence 182
connectedness 47, 103
consistency 231
constraining relation 195
context-sensitivity 198

contextual knowledge 168
continuity property 132
contradiction 173
cooling schedule 38
cophenetic correlation coefficient (CPCC) 51
cophenetic matrix 51
cophenetic proximity measure 58
correctness 231
cost criterion 105
coupling coefficient 94
covering 243
CPCC distribution 340
criterion function 86
cross-partition 110

data collection 7
data points 154
data processing systems 156
data-driven reasoning 177
Davies–Bouldin index, 131, 208
Davies–Bouldin statistic 13
De Morgan's law 176
decision tree 173, 242
declarative programming 168
deduction 175
deductive propositional logic 175
deep knowledge 168
degree of belief 180, 186
degree of knowledgeability 295
Dempster–Shafer theory 185
dendrogram 5, 39
density of points 4
descriptive generalization 243
deterministic 179
diagnosis 148, 157, 177
diagnostic level 213, 214, 346
dilation 192
disbelief 184
discriminant projection 352
discriminant rules 242
disjunction 173
disjunction of evidence 182
dispositional rules 195
dissemblance 98
dissimilarity 22, 98
dissimilarity matrix 334
domain 167
domain knowledge 5

ε-complete 104
edge-length distribution 340
embedded simulation 78

empirical learning 242
equivalence relation 58
errors of reasoning 179
Euclidean distance 23
evaluation 218, 236
events 179
exact reasoning 172, 178
exclusive clustering 76
experience 162
experimental probability 180
expert inconsistency 180
expert system 151, 152, 156
expert system development life cycle 229
expert's judgement 236
explanation facility 165
explicit conceptual association 216
explicit functional association 154
exploratory data analysis 5
external index 154, 264

fact base 167, 203
facts 157
false clusters 124
features 354
formalized intuition 162
forward chaining 166, 177
frame 171
frame of discernment 186
framing 18, 71
framing hypothesis 258
free attributes 243
functional association 154
functional knowledge 168
fuzzy classification 78, 115
fuzzy cluster 104
fuzzy clustering 8, 76, 132
fuzzy constants 201
fuzzy expert system 254
fuzzy graph 101
fuzzy inference 115
fuzzy K-means algorithm 89
fuzzy K-shells algorithm 89
fuzzy logic 19, 117, 133, 175, 182, 193
fuzzy membership function 81
fuzzy neural networks 252
fuzzy partition 78, 89, 90
fuzzy proposition 189
fuzzy qualifier 189
fuzzy region 139
fuzzy relation 78, 82
fuzzy relaxation 133
fuzzy rules 246

fuzzy sets 19, 78
fuzzy value 204

generalization 198, 215
generalizing rule 145
goal-driven reasoning 177
grade of membership 78
graph 101

hard *K*-means clustering 86
hard partition 86
heuristic 6, 123, 162, 175
heuristic knowledge 164
heuristic rule 145
heuristic search 242
hierarchical 4
hierarchical lattice 186
hierarchical methods 18, 74
hybrid approaches 213
hybrid system 242
hypothesis of completeness 243
hypothetico-deductive approach 73, 152

identification 229
IF ... THEN rules 164
ill-defined classes 79
image analysis 8
implication 117, 173
implicit conceptual association 216
implicit functional association 154
impossible belief 183
imprecise description 79
inclusion 82
incompatibility 309
incompleteness 179
inconsistency 48, 146
inconsistent edges 48
incorrectness 179
increased belief 184
increased disbelief 184
indices of confidence 74
induced fuzziness 93
induced fuzzy sets 90
induction 175
inductive learning 245, 293
inexact reasoning 178, 183
inference 124
inference by abduction 178
inference engine 164, 165
informal conclusions 6
information 149, 216
information function 266

information-based systems 156
inheritance 171
initial level 346
initial partition 44
initial screening 7
initiation level 213, 214
input data matrix 334
instance space 245
instantiation 167
integrated intelligent processing 255
intelligent computer programs 19
intelligent hybrid system 254
intensification 192
intention of classification 162, 168
internal indices 154, 264
interpretation 5, 8
intersection 82, 83
intuition 6, 175
inverse relations 98
isolation 54
iterative optimization 94

K-means algorithm 124
knowledge 149, 167, 216
knowledge acquisition facility 165
knowledge base 152, 164
knowledge program 230
knowledgeability 295
label information 351
law of contrapositive 176
law of the syllogism 176
law of inference 176
learning 8, 197, 214, 254
level set 82, 83
likelihood 180
littoral animals 257
local minimum 5, 45
logic 172
logic programming 172
logical constants 201

Mahalanobis distance 23
mapping 39
Markov chain 38
matching coefficients 24
max function 182
max min composition 193
max min rule 193
maximum coverage principle 296
maximum method 195
measure of fuzziness 82
measurement errors 179

measurements 73
measures of association 22
medical consultant 152
minimum distortion 101
minimum fuzziness 78
minimum path length 40
minimum spanning tree 48
minimum variance 76
missing values 25
model-driven reasoning 6
modelling 157
modelling paradigm 150, 157
modus ponens 173, 175
modus tollens 175
monotonic reasoning 175
multi-dimensional space 4
multi-layer perceptron 250
multi-valued logic 193
multiple assignment 236
multiple testing 5
multivariate data analysis 6
mutual neighbourhood value 83

natural grouping 124, 208
natural language 188
negation 173
neighbourhood 154
neural network 242
nodes 173
non-deterministic 179
non-exclusive clustering 76
non-monotonic reasoning 175
non-uniqueness 80
normalization 10, 192
null set 186
number of clusters 13, 239, 343
numerical clustering 198
numerical constants 201

object partition 90
object representation 168, 198
objective knowledge 162, 168
object 73
object–attribute–value triple 171
observations 214
operational knowledge 168, 198
optimal partition 132
optimistic upper bound 307
optimization problem 38
ordered sets 204
outlier 46

parameter file 202
parameter 356
partitional 4
partitional methods 18, 74
partition 154
pattern 166
pattern matrix 4, 5, 10, 22
pattern space 22
pelagic dolphins 257
performance 213
performance measure 93
pessimistic lower bound 308
planning 71
plausible belief 183
population 179
possibility 116
possible belief 183
power set 186
predefined concept 198
predicate logic 173
premise 172
probability 115, 179
probable belief 183
problem knowledge 5
procedural programming 168
procedure 202
processing paradigm 150, 156
production rules 165
projection 10, 346
property 170
proposition 173
proximity matrix 4, 5, 10

Q-analysis 354
quality 244
quantifier 173

R-analysis 354
Rand coefficient 132, 264
random 51, 179
random data 11, 339
randomness 131
range of uncertainty 309
ranking 113
reasoning 152
reclassification function 94
recognition 8
reflexivity 58, 97
region 154
reinforcement 146, 225
repeated experimentation 143
representation 7, 154

representative rule 245, 297
research hypothesis 258
research planning 259
research question 258
resemblance 117
resemblance relation 97
resemblance coefficient 74
retroduction 73
rule induction 242
rule of combination 187
rule space 245
rule-based reasoning 12
rule-based system 199
rules 8, 18, 198
rules of thumb 162

S-function 190
sample size 73, 208, 238
sampling window 53, 131
scaling 10
scatter 154
schema 171
semantic network 170
separability 208
shallow knowledge 164
significance 114, 138
similarity 4, 22, 177, 198
similarity relation 97
simulated annealing 38
simulation 19
singleton 186
situational frame 172
slot 171
software package 12
spanning tree 48
spatial clustering 8
square error 43
square-error criterion 5
standardization 10, 74
statistical methods 115
statistics 179
status information 351
status parameters 202
stochastic relaxation 38
strategy 18
stress 35
string constants 201
strong clustering 51
strong(est) patterns 110
structural variables 154, 214
sub-structure 244

subjective decision making 260
subjective knowledge 162, 168
subjective probability 180
subjectivity 6
sufficiency 243
sufficient condition 244
support 157
support level 213, 214, 346
surface knowledge 168
symmetry 58, 97
syntax 200

tautology 173
tendency 6
tied attributes 243
token 200
toolbox 150
top-down reasoning 177
training set 115
transitive closure 99
transitivity 58, 97
tree 47
trial 179
trial and error 175
triangle inequality 97
ultrametric inequality 59
uncertain evidence 181
uncertainty 18, 115, 167, 178
uncorrelated features 30
union 82, 83
unreliability 179
unsupervised learning 4
user data 335
user interface 165
utility 113

validation 5, 8, 11, 18, 131, 236
validity 79
visual examination 29
visualization 10

weak clustering 51
weak(est) patterns 110
weakening 146, 225
Wilks lambda statistic 32, 132
within-cluster variation 43
within-group scatter 32
working memory 165

Π-function 191